BEEKEEPING BETWEEN TWO QUEENS

Nottinghamshire 1901 – 1952

John Stuart Ching

BEEKEEPING BETWEEN TWO QUEENS
Nottinghamshire 1901 – 1952

© John Stuart Ching

All rights reserved. No part of this publication may be reproduced, stored in a retrieval system, transmitted in any form or by any means electronic, mechanical, including photocopying, recording or otherwise without prior consent of the copyright holders.

By the same author

Bee and Bee Interest Postage Stamps
(with Dorothy Shaw)

Beekeeping on Two Fronts
1914-1919

Beekeeping in Victorian Nottinghamshire
1837-1901

ISBN 978-1-912271-33-7

Published by Northern Bee Books, 2018
Scout Bottom Farm
Mytholmroyd
Hebden Bridge HX7 5JS (UK)

Design and artwork by DM Design and Print

Printed by Lightning Source UK

BEEKEEPING BETWEEN TWO QUEENS

Nottinghamshire 1901 – 1952

John Stuart Ching
BA, BA (Hons), PGCE, MEd
Archivist, Nottinghamshire BKA

Following on from the work "Beekeeping in Victorian Nottinghamshire" which covered the reign of Queen Victoria from her accession in 1837 until her death in 1901, this work covers the half-century from then to the death in 1952 of King George VI.

During the period of this work Britain was ruled by ruled by three kings plus one who relinquished the crown – hence the title.

	Born	Crowned	Died
Edward VII	1841	1901	1910
George V	1865	1910	1936
Edward VIII	1894	--	1972
George VI	1895	1936	1952

FOREWORD

This latest book by Stuart Ching has the inspired title of "Beekeeping between Two Queens" for, sandwiched between the long reigns of Queen Victoria and Queen Elizabeth, four Kings occupied the throne, and collectively their time in office was less than each of the queens, both preceding and succeeding them.

Undoubtedly, the half century between 1901-1952 was the most horrific in British history. In 1901, the tail end of the Boer War was being fought - a war which proved to be '*the longest, costliest, bloodiest and most humiliating*' since the Battle of Waterloo in 1815, and which Kipling remarked gave the British '*no end of a lesson*'. No one would have believed that further horrors were to come with a decade being taken up with the two World Wars, which were far worse than anything that had happened previously.

Against this background Stuart Ching delves into a variety of sources to give snapshots of Nottinghamshire beekeeping through these difficult times, though to learn more about the happenings during WWI, I would refer readers to his earlier work '**Beekeeping on Two Fronts 1914-1919**'.

So what occupied beekeepers during these times? Certainly, during the reigns of Edward VII and George V, a great change in beekeeping was taking place with more and more beekeepers giving up their skeps in favour of wooden moveable frame hives. The practice of driving was commonplace - not only amongst skeppists, but by those who used this method to transfer bees to the modern hives. So many designs came on the market, notably the WBC hive - and others - the Meadow's, Wells and Harrison, made by local manufacturers and suppliers of equipment, and which were often given as first prizes in honey shows. The philanthropist, Baroness Burdett-Coutts, also set up a Prize Hive Competition for deserving cottagers. However, difficulty arose as to the meaning of the words 'cottage member'. The Competition Committee decided that for the purposes of this competition the words mean '*any member of an association whose house-rent does not exceed 6s. per week.*'

Regarding disease, Foul Brood was the bane of many beekeepers' lives and various methods of control were used including carbolic, Dioxogen, and even chlorine, though one interesting method, the validity of which I am certain, was the placing of diseased combs outside the hive 'where after some hours' exposure to direct sunlight bacilli and spores are destroyed, this beneficial effect being due to the action of the ultra-violet rays on the oxygen of the air, resulting in the formation of peroxide of hydrogen - a powerful disinfectant - and not to the heat of the sun.' Meetings regarding Foul Brood legislation began in 1910 and eventually an an Act was passed which allowed inspectors to check colonies, much to the disproval of many beekeepers, the result being that infected colonies together with their honey had to be destroyed. Many beekeepers pointed to colonies in skeps as the main source of infection.

However, it was the Isle of Wight Disease which pre-occupied beekeepers around the time of WWI, and such was the loss of colonies that stocks had to be imported, mostly from Holland, an operation that went on until Spring 1922. The importation of foreign bees into the UK, especially queens, had been very popular since late Victorian times, often for fanciful reasons, with the beautifully coloured Italians being favoured, but others, including queens from Tunisia, were in demand. Some of the passages in this book deal with the management of these bees and describe their various benefits or failings.

In this first half century we see the rise of one of the most prominent beekeepers William Herrod-Hempsall, author of many books, and who was always in demand as a speaker and honey judge. Another beekeeper who appears on the scene in later years is ROB Manley, and mention is given to Joseph Tinsley, West of Scotland College of Agriculture, who did tremendous work there by rehabilitating disabled soldiers via beekeeping.

Both wars made the provision of sugar for winter feed difficult for beekeepers. The inauguration of the sugar beet industry in 1919 *'with 5,000 acres of land at Kelham (Notts) being purchased for the purpose at a cost of about £500,000 towards which the Treasury contributed £125,000'* was an important step with the first sugar factory being opened in 1921. Despite this, however, beekeepers still needed help to feed their stocks during WWII and, whilst the government helped them out, we can read for March 1940: *'Bees in Demand. Sugar rationing is largely responsibly the unusual demand for bees in country areas. Realising that honey can help the sugar deficiency, and that beekeeping is a profitable way to supplement the larder, many people are now becoming amateur beekeepers. Beehive manufacturers are having difficulty in obtaining enough suitable wood to meet the demand.'*

Much of the content of this volume covers the everyday business of the Nottingham and nearby associations, their meetings, honey shows, and the questions raised and answered by beekeepers derived from the sources diligently searched by Stuart Ching. It is interesting to find that a topic, still sometimes discussed, preoccupied the NBKA for some time - i.e. was it worthwhile for the Association to be a member of the national BBKA - especially since their Association benefited with support from the Nottinghamshire County Councils and Nottingham City. No doubt some of this money was used for educational purposes and we see that beekeeping in schools began to take place, though not always to the liking of non-beekeeping schoolmasters.

BEEKEEPING BETWEEN TWO QUEENS

There is a very long piece included in Stuart's book on school beekeeping and complete essays on various beekeeping matters. Other inclusions, though only a sentence or two in length, add enormously to give us a reliable and informative picture of how beekeeping developed during that half century within the county.

Odd snippets tell us something of the times which are thankfully over - the birching of a boy who had the effrontery to deck himself in a beekeeper's protective clothing and deprive his hive of honey - whenever did birching stop? Also we can read about a stolen chicken and four eggs, the woman responsible for the thefts being allowed to choose between a fine of 10s or ten days hard labour. What was hard labour, I wonder in 1902?

One item though from 1909 I find to be extremely relevant for today: the use of paper pulp honey pots! Their use today may well help to stop the enormous waste and pollution caused by single use plastics. What a good idea from 110 years ago!

This lengthy book is full of fascinating information, with material which has never before been gathered together in one publication. Perfect for dipping into, or reading in full the longer extracts from the archives.

I am greatly looking forward to the next volume which will bring us up to date with all that has happened in beekeeping during the longest reign of any monarch in history: yes, another Queen!

John Phipps
September 2018

INTRODUCTION

The Nottinghamshire Beekeepers' Association (NBKA) was founded in 1879 but soon fell into disarray. The second attempt at forming an Association in 1884 was more successful and it is that organisation which continues to this day.

NBKA has a poor reputation with regard to its records. Few original documents exist from its early start. Indeed, there is a story that one of its first Minute Books was later found for sale in a second-hand bookshop in Stamford, Lincolnshire! Where I have been able to obtain information from the Minute Books it is noted in the text.

However, as this information was hand-written in ink (and at one time in pencil!) sometimes this was quite difficult to read. Unfortunately the correspondence noted in the Minute Books has not survived. There are hints in the surviving records that the Regional Branches of NBKA also had Minute Books but these have not been located. It is of note that, in this volume, there is no information about NBKA activities for 1933 and 1938.

The documentary history of NBKA did not really start until the production of the first newsletter in 1958 (after the final date of this work) but, apart from the first issue of BEE-MASTER as it was called, no complete issues survive until 1968.

SOURCES

I have access to the British Bee Journal (BBJ) from 1873–1922. This publication by the British Beekeepers' Association (BBKA) covers the time from the original ill-founded attempt to form an Association but did not completely cover the gap up to the date of the first newsletter. I have also been able to search, through the British Newspaper Archives, the local newspapers of the relevant times. The Nottinghamshire Evening Post (NEP) and the Nottinghamshire Guardian (NG) are good sources of beekeeping activities in the county as they often gave differing reports of the same event. The NG ran an anonymous "Beekeepers' Corner" for some time.

However, it should be remembered that, due to the two World Wars and several national industrial strikes which occurred during the period covered by this book, there was severe rationing of newsprint and this restricted the reporting of many events. Indeed, the strikes involving railway workers affected the attendance at some meetings of the BBKA!

RESEARCH

This history is written in chronological order as extracts will be taken from sources which are in themselves arranged in such an order (dates of publication). As people are named in the extracts, a diversion will be made to include biographies of them. This will not be possible in all cases but those who made a contribution to the well-being of NBKA are worthy of some acknowledgement. However, the timeline for this book may allow people to contribute to the history of NBKA from their own family memories.

Access to the various census records via the Internet has enabled me to find details of some of the people involved with NBKA. Unfortunately this has not been achieved with all of them which I regret. This task is not helped by the capricious nature of the spelling of names in many cases.

One of the decisions I have had to make is about the deaths of members whose beekeeping activities occurred mainly during the time covered by this book but who died after it. Out of respect I have included as many of them as I can find even though it may disrupt the smoothness of the story being told. However, there are other beekeepers whose deaths are recorded in the newsletters and whose periods of activity within NBKA are not specified but must have occurred within the timeframe. There are some others whose names only appeared in the later newsletters but not in the records. This is another example of the poor recording by the NBKA. Some of these former members will be commemorated in an Appendix.

ACKNOWLEDGEMENTS

Once again, Penny Forsyth, who took over the editorship of **BEEMASTER** when I had to relinquish the post after twenty plus years, continues to give my literary efforts the "thumbs up!"

I am grateful for the financial support given by the Council of the NBKA in the production of this book.

One is always grateful to a publisher like Jeremy Burbidge. His professional advice and encouragement made me determined to carry on with this work which will bring a history of beekeeping in Nottinghamshire into the modern era. No other Association has such a detailed recorded history and this would not have happened for NBKA were it not for Jeremy.

Average weekly cash wages in 1901 for ordinary agricultural workers for a 10-hour day excluding extra payments for piecework, hay and corn harvests, overtime and the value of allowances in kind was 14s 11d.

KING EDWARD VII

Albert Edward was born in 1841, the first son of Queen Victoria. He chose the regnal name of Edward. His coronation, scheduled for 6th June 1902, was postponed until the 9th August due to his serious medical problems which included an emergency operation (on the billiard table in the Music Room at Buckingham Palace!). By the time of his re-arranged coronation most of the foreign guests had returned home so the ceremony became a more domestic one.

East Bridgford Parish magazine, February 1901
Our new king, Edward VII, was proclaimed on January 23rd in London and on the 25th in Nottingham.

BBJ 21 Mar 1901
Prior to the commencement of the ordinary business of the BBKA Council the Hon and Rev Henry Bligh moved:
"That on the first occasion of meeting since the death of our beloved Queen, the members of the BBKA beg to be allowed to offer their respectful and sincere condolence to their gracious Majesties the King and Queen and other members of the Royal Family on the sad loss which they have sustained, and with grateful remembrance of the kind interest which her gracious Majesty Queen Victoria took in the Association and it's work of establishing the improved and humane system of bee-keeping, would venture to express the hope that his Majesty the King will continue to take the same interest as he has already shown in the BBKA, and extend to it his Royal patronage."

No mention of any such sentiments was found in the records of the NBKA so this statement is placed here.

NEP 26 Jan 1901
The AGM of the Notts. Agricultural Society was held this afternoon at the Exchange Hall, Nottingham. The NBKA held an Annual Exhibition and gave demonstrations in bee driving.

BEEKEEPING BETWEEN TWO QUEENS

The Exchange Building was demolished in 1920 to make way for the current Council House which was built in 1929. The original Exchange Building was built between 1724 and 1726 replacing a shambles of buildings on the same site. It cost £2,400 at the time and comprised a four-storey, eleven bay frontage 123 feet (37 m) long.

NG 27 Jan 1901
I was asked if I could tell a beekeeper where he could purchase candy without being at the trouble of making it. The other day I purchased 45 lb at 3d per pound from Mr Roger Buxton, confectioner, Chesterfield-street, Nottingham and it was the best I have ever purchased.

Buxton's, wholesale confectioners, Lister-gate, Nottingham also Chesterfield-street

Minute Book - Annual Meeting held in the Peoples' Hall on February 9th, 1901. This meeting was arranged to take place on February 2nd but, owing to the death of Queen Victoria, it was postponed till February 9th.

Mr William S Ellis having taken the chair called upon the meeting to commence business.

Mr J McKinnon objected to Mr Ellis occupying the chair and to the committee for selecting a member of this association for that duty; he was, however, ruled out of order and the business proceeded.

The minutes of the previous meeting were read when Mr Scattergood proposed and Mr Deacon seconded the same be confirmed by the signature of the chairman. Mr Ellis proposed that the Balance Sheet as submitted to each member be passed as satisfactory. Mr William Herrod seconded and it was carried.

The secretary then read the Annual Report after which Mr Joseph Herrod proposed and Mr Trimmings seconded, Mr Scattergood supporting, that the same be printed with the Annual Report, etc.

The election of officers then followed. Mr Arthur G Pugh proposed and Mr Robert Mackender seconded that the very best thanks of this association be accorded Viscount St. Vincent for his services as President during the past year, for his generous help in the prize list and for the two hives sent for the prize drawing and that he be elected President for the ensuing year. Carried with acclamation.

Mr T Marshall proposed and Mr Scattergood seconded that Mr George Hayes be re-elected secretary and treasurer. Mr McKinnon proposed and Mr Alex Harrison seconded that the thanks of the association are due and hereby accorded to Mr

Scattergood and that he be re-elected hon. Auditor. Mr Scattergood proposed and Mr Turner seconded that Messrs. Hayes and Pugh be re-elected delegates to the BBKA meetings.

It was then resolved that the following gentlemen form the committee: Messrs Faulconbridge, Harrison, Hesslewood, Marriott, Pugh, Puttergill, Skelhorne, Swann, Scattergood and Wadsworth.

A discussion took place as to awarding medals but no alteration was made.

The meeting then adjourned for tea at which about 60 members and friends sat down.

After the adjourned meeting the first business was to award medals, certificates and prizes which were as follows:
> BBKA Silver and Bronze Medals and certificates all to Mr AG Pugh, Beeston
> NBKA silver pendants to Mr GE Puttergill of Beeston and Mr W Swann of Eastwood

and in the one bottle competition at that meeting:
> 1st, 2/6, AG Pugh: 2nd, 100 labels, Mr GM Bolton

About 20 lb of honey was sent to the Children's' Hospital

A discussion arose as to the price of labels and after well ventilating the subject it was decided to make no alteration.

Mr W Herrod, apiarist of Swanley College, Kent, next gave a very interesting and instructive paper on "The production, preparation and disposal of honey". After which Mr Puttergill proposed and Mr Turner seconded that a hearty vote of thanks be accorded Mr Herrod for his very interesting paper and his kindness in giving it.

BEEKEEPING BETWEEN TWO QUEENS

The meeting was brought to a close by the usual prize drawing which was as follows:

		Given by	**Won by**
1st	Hive	President	Mr Loughton, Southwell
2nd	Hive	President	Mr G Smith, Bradmore
3rd	Useful articles	Mr Trimmings	Mr G Halstead, Newark
4th	Crumb tray and brush	Mr Harrison	Mr W Lee, Southwell
5th	Feeder	Mr Pugh	Mr C Berry, Newthorpe
6th	"Weed" foundation	Mr Hayes	Mr J Rawson, Snr, Selston
7th	200 labels	NBKA	Mr Goodchild, Nottingham

J McKinnon (also known as Sergeant McKinnon)
The wonderful memory of Lord Roberts was further illustrated at the recent inspection of the North Midland Brigade, now under training at Farnborough Common, Aldershot. His lordship was accompanied General Sir Buller, General Sir Evelyn Wood, Major General Ian Hamilton, Major-General Kelly Kenn, and a brilliant staff. Upon reaching the VBSF the Commander-in-Chief reined up to ask Sergeant-Instructor J. McKinnon, E Company, the name of his previous regiment. The sergeant instructor was previously in "The 95th," and served under Lord Roberts in India. Sergeant-Instructor McKinnon enjoys the distinction being the only man in the brigade to whom the Commander-in-Chief spoke that morning.

NEP 9 Aug 1901
The 95th/45th Regiment of Foot was the official title of "The Notts and Derbys Regiment (the Sherwood Foresters)

William S Ellis was born in 1860 in Basford. He owned a lace making business.

The second President of NBKA was Carnegie Robert Parker Jervis, 5th Viscount St. Vincent who held office from 1890 to 1907. He was born in 1855 in Cheshire and lived at Norton Disney which although in Lincolnshire, was administered from Newark and he died (in Nottingham Place, Marylebone) in 1908.

He appears to have been popular with members, being heartily thanked for his services at every AGM. Doubtless one of the reasons why he was asked to be President was that he was a practising beekeeper. This is evidenced by the many times his name appeared in the prize list for numerous shows. Around the turn of the century he was paying an annual subscription of £21 as a member of Newark district.

Arthur G Pugh was a Railway Contractor and was born in 1856 in Tuttley, Gloucestershire.

George E Puttergill was born in Caunton in 1865. He was a teacher of woodwork and technical studies.

NEP 9 Feb 1901
The Annual Meeting of NBKA was held on February 9th in the Peoples' Hall. Mr. WS Ellis, vice-President, in the chair. Among those present were Mrs. Hemsley, Mrs. Faulconbridge, Mrs. Turner, Miss Hunt, Mrs. T. Herrod, Messrs. G. Hayes, P. Scattergood, GE Skelhorne, W Smeeton, T Marshall, AG Pugh, TN Harrison, E Mackender, J Herrod, W Herrod, G Puttergill, AR Hesslewood, J Gray, C Forbes, W Marriott, JC Wadsworth, Hemsley, J Mackinnon and AE Trimmings.

The Balance Sheet for 1900 showed total receipts during the year of £62 1s 4d comprising £24 11s 6d by subscription, the remainder in donations, grants and sundries. The expenditure, including £18 17s in prizes at shows which, after adding value of assets, reached £65 1s 4d left a deficit of £1 on the year's working.

The Secretary read his report which showed the membership to be 172. He thought that, to mark the inauguration of the new century, the members should exert themselves individually and raise the membership of the association to 200 by the end of the present year. The committee had decided that the Annual County Show this year should be held at Moorgreen in September next.

The County Council had during the past year continued their grant of £30 for Technical Instruction in bee-keeping and by that aid NBKA had been enabled to give demonstrations in seven centres, besides lectures at Chilwell, Upper Broughton, Hucknall, Watnall, and Balderton. The City Council had again renewed its grant of £2 2s which enabled them to extend their aid to the bee-keepers in Nottingham.

The report and Balance Sheet were adopted unanimously.

The remaining business was the election of officers for the ensuing year, Viscount St. Vincent being again chosen President. Mr. G. Hayes was re-appointed secretary and Mr. Scattergood auditor. Messrs. Hayes and Pugh were re-elected delegates to the BBKA and Messrs. WS Ellis, AE Trimmings, P. Scattergood, SW Marriott, R. Mackender, C. Forbes, GE Puttergill, GE Skelhorne, AG Pugh and TN Harrison constituted the committee.

Leicester Chronicle 1 Mar 1902
Under the auspices of the NBKA, aided by a grant from the Technical Committee of the County Council, a lecture was given in the National School, Sutton Bonington, on Monday evening entitled 'Bees and Beekeeping' by George Hayes of the NBKA. The lecture, which was freely illustrated with lantern slides, proved very interesting to the

large audience present and showed how bees could be managed to profit. Mr. MJR Dunston occupied the chair.

MJR Dunston worked for the Midlands Agricultural and Dairy Institute and later became Principal of Wye College. In October 1912 he addressed a meeting of 600 fruit growers in Maidstone Town Hall on the subject of 'The scope of scientific research in fruit growing'.

BBJ 21 Mar 1901
Leicestershire BKA. Mr. P. Scattergood of Stapleford gave a practical and helpful lecture on "Foul Brood," and afterwards an interesting, homely talk, illustrated by coloured slides, on the "Relation of Bees and Bee-keeping to Flowers and Fruits."

BBJ 28 Mar 1901
The monthly meeting of the BBKA Council was held at 105, Jermyn-street, SW. on March 21st, Mr ED Till occupying the chair. Letters apologising for enforced absence were read from Messrs. G Hayes, AG Pugh and P Scattergood.

The election of Council for the year 1901 resulted in the following selection amongst others - Dr. Elliot and P Scattergood.

Thomas Stokoe Elliot was a Medical Practioner born in Southwell in 1873.

Minute Book – Quarterly Committee Meeting held in the Peoples' Hall on April 6th, 1901, at 3pm. Present: Messrs Faulconbridge, W Herrod, Hardy, Harrison, Marriott, Pugh, Puttergill, Swann, Skelhorne, Scattergood, Rawson, Hesslewood, McKinnon and secretary.

Mr Pugh having been voted to the chair, the minutes of the previous meeting were read when Mr Scattergood proposed and Mr Hardy seconded that the same be adopted.

Mr McKinnon took objection to the second paragraph in the Annual Report stating the same was untrue and that he did not object to Mr Ellis occupying the chair and that he was not ruled out of order. The secretary explained that he was obliged to mention the matter in the Annual Report and he did so as briefly as possible without any wish to create ill-feeling.

The quarterly Balance Sheet was next read which shewed a falling off in the amount of subscriptions at the Annual Meeting. Mr Scattergood proposed and Mr W Herrod seconded that the same be accepted as read.

The following gentlemen were elected on the sub-committee for dealing with the

County Council grant - Messrs Pugh, Scattergood, Ellis, Marriott, Skelhorne, Puttergill and McKinnon. The secretary read a letter from Mr Dunston stating that the grant had been increased from £30 to £40.

Mr Scattergood proposed and Mr Hardy seconded that we join Southwell (Show) on the same terms as previously.

Mr McKinnon proposed and Mr Faulconbridge seconded that if the Notts. Agricultural Society grants us £5 as before that we accept the same and offer prizes as follows:

	First Prize	Second Prize	Third Prize
Appliances class	30/-	15/-	
Best 6 sections, any year	5/-	3/-	
Best 6 liquid honey, any year - light	5/-	3/-	
Best 6 liquid honey, any year - dark	5/-	3/-	
Best 6 granulated	5/-	3/-	
Best 6 blocks wax, each block or piece to approx. 2oz	2/6	1/-	
Best single frame of bees	15/-	10/-	7/6
Best pint vinegar	NBKA Certificate		

Mr Pugh proposed and Mr McKinnon seconded that the secretary be empowered to arrange any other local shows where they would agree to pay three-quarters of the prize money, NBKA to pay the other quarter.

A question of establishing an apiary for Technical Instruction was considered and it was referred to the sub-committee as a matter for them to deal with.

A unanimous vote of thanks was accorded to Mr W Herrod for his kindness in giving the very interesting and instructive paper at the Annual Meeting.

BBJ 2 May 1901.
WH Wood (Nottingham). Joining Bee-Keepers' Associations. The Secretary of the NBKA is Mr. G. Hayes, 48, Mona-street, Beeston, from whom you may obtain all information as to membership and securing "expert" assistance.

NEP 15 May 1901
Beautiful weather was again experienced to-day. The proceedings in connection with the Newark Agricultural Show were continued and there was a prospect of beating previous records, as far as attendance was concerned. Mr. Scattergood continued his demonstrations and lectures on bee-driving, under the auspices of the NBKA.

NEP 15 May 1901
Notts. Agricultural Show. There will be bee-driving demonstrations and lectures in connection with the NBKA.

East Bridgford Parish magazine July 1901
The day was delightfully fine and the attendance was in excess of last year. The show was declared by competent judges to be better than of late years despite the late Spring and continuous drought.

The special prizes given by Mr WF Fox JP, included:
Honey – 1st, Mrs Bouverie; 2nd, Miss Fox

The annual extracts from a local church magazine are included not just for their beekeeping interest but also for the comments on the weather which obviously affects horticulturists and beekeepers alike.

Minute Book – Committee Meeting held in the Peoples' Hall on July 7th, 1901 at 3pm. Mr Ellis in the chair. Messrs Faulconbridge, Gray, Hardy, Hesslewood, Marriott, Pugh, Puttergill, Skelhorne, Swann, Turner and secretary.

The minutes of the last meeting were read when Mr Skelhorne proposed and Mr Swann seconded that the same be confirmed by signature. The quarterly account was presented by the secretary and Mr Turner proposed and Mr Gary seconded that the same be adopted.

Mr Turner proposed and Mr Hardy seconded that Mr Scattergood be asked to judge at the Southwell Show. Mr Pugh proposed and Mr Skelhorne seconded that the judge for the Annual Show at Moorgreen be one of the following and in the following order as the secretary may be able to arrange: Messrs CN White, FJ Cribb and Dr Sharpe. Mr Pugh proposed and Mr Skelhorne seconded that Mr Marriott be assistant judge at Moorgreen at the usual fee of 7/6d.

BBJ 25 Jul 1901
The Annual Show of the Lincolnshire BKA, in conjunction with that of the Lincolnshire Agricultural Society, was opened at Brigg on July 18th, and so far as the quality and amount of honey and the number of exhibits of hives and appliances on the show-bench, it will easily rank as a "record." The rest was light honey of such an even colour and consistency.
Collection of Hives and Appliances - vhc, Varty & Co., Colwick, Notts.

Gerald H Varty was a joiner born in 1874 in Burnaston.

BBJ 15 Aug 1901
The Leicestershire BKA usually hold their Annual Show in connection with that of the Leicestershire Agricultural Society held in June, a date too early for showing honey of the current year. Mr. AG Pugh of Beeston, Notts, acted as judge, ably assisted by Mr. Riley of Leicester. Both gentlemen gave addresses in the bee-tent, Mr. Faulkner of Market Harborough acting as manipulator

BBJ 22 Aug 1901
Honey from Deadly Nightshade. Can you inform us whether honey collected from the deadly nightshade *(Atropa belladonna)* is poisonous to human beings?

On August 5th I noticed hive-bees at work on the flowers of this plant, but whether they were collecting honey as well as the pollen with which they were covered I could not tell. It is well known that the rabbits in the neighbourhood feed on the leaves of the deadly nightshade without ill effects to themselves, although I am told the rabbits which have so dined make a dangerous dinner for human beings afterwards.
                                                                                                HC Wallis, Old Colwick
Reply. No alarm need be felt with regard to honey from deadly nightshade. The nectar gathered from that plant is so small as to be perfectly innocuous to human beings and at the worst it may, like other poisons, be useful medicinally. Anyway, no one need fear its effects in this country.

BBJ 22 Aug 1901
Seven Bar-Frame Hives, four Skeps, Extractors and sundries, Bees healthy, good condition, £10 or offers.   Herrod, Trentside Apiary, Sutton-on-Trent

BBJ 29 Aug 1901
Honey Show at Ammanford, South Wales. The show of honey introduced this year for the first time in connection with the Ammanford Horticultural Society took place on August 17th. Exhibits were staged from the best honey-producing counties in the kingdom, all of which were excellent in quality. The following was awarded:
Single 1-lb Jar Extracted Honey - 4th, AG Pugh, Beeston

BBJ 12 Sep 1901
NBKA held their Annual Show at Moorgreen on September 3rd in conjunction with that of the Greasley, Selston, and Eastwood Agricultural Society. Regarding the honey section, it may be said that while fully up to the standard in point of entries, the exhibits were considerably superior to any the society has staged for some years past, the strongest classes being those for extracted honey. Mr. CN White, St. Neots, was appointed judge and made the following awards:
Collection of Bee-Appliances - 2nd, GH Varty and Co., Colwick (As only part of goods

were staged the 1st prize was withheld.)
Complete Frame-hive - 1st, GH Varty and Co.; 2nd, JT Faulconbridge, Bulwell
Honey Trophy - 1st, Geo. Hayes, Beeston
Six 1-lb Jars Light-coloured Extracted Honey - 1st, J. Herrod, Sutton-on-Trent; 2nd, AE Trimmings, Gedling; 3rd, R. Mackender, Newark; hc, JT Faulconbridge
Six 1-lb Jars Dark-coloured Extracted Honey - 1st, G. Marshall, Norwell; 2nd, R. Mackender; 3rd, AE Trimmings
Six 1-lb Sections - 1st, GH Varty; 2nd, JT Faulconbridge
Six 1-lb Jars Granulated Honey - 1st, JT Faulconbridge; 2nd, J. Herrod; 3rd, AG Pugh, Beeston
Shallow-Frame of Comb Honey for Extracting - 1st, G. Marshall; 2nd, W. Swann, Eastwood; hc, J. Herrod
Six 1-lb Jars Extracted Honey (novices) - 1st, Dr. A. Gregor, Sutton-on-Trent; 2nd, AE Trimmings; 3rd, J. Brumby, Newark
Six 1-lb Jars Extracted Honey (local) - 1st, W. Brooks, Eastwood; 2nd, GM Bolton, Eastwood; 3rd, W. Swann
Honey Vinegar - 1st, Geo. Hayes, Beeston
Honey-Cake - 1st, Mrs. G. Hayes
Observatory Hive, with Bees and Queen - 1st, R. Mackender; 2nd, G. Hayes; 3rd, AE Trimmings; hc, G. Marshall
Beeswax - 1st, AE Trimmings; 2nd, G. Marshall

Mr. White also lectured and gave demonstrations of bee-driving in the bee-tent.

John Thomas Faulconbridge was a house painter born in 1868 in Hucknall.

BBJ 19 Sep 1901
The Annual Show of the Derbyshire BKA was held on September 11th and 12th on the show ground of the Agricultural Society.
Collection of Bee Appliances - hc, GH Varty, Colwick

Minute Book – Committee Meeting held in the Peoples' Hall on October 5th, 1901 at 3pm. Present: WS Ellis in the chair, Messrs Trimmings, Faulconbridge, Harrison, Pugh, Puttergill, Skelhorne, Swann, Turner, McKinnon, Mackender.

The minutes of the previous meeting were read and it was resolved that the same be accepted and signed. The quarterly Balance Sheet was next read and after going into some details as regards the deficit it was resolved the same be adopted.

The secretary then gave a brief report on the Southwell and Moorgreen shows which, on the whole, appeared to be more satisfactory than last year's shows both as regards number of entries and quality of exhibits.

Some correspondence received from Mr G Marshall about the awarding of prizes at the Southwell Show, particularly as regards the lacing of sections, was read and also Mr Scattergood's (the judge) reply thereto and after considerable discussion Mr GE Puttergill gave notice that he would bring the matter forward at the next Annual Meeting with a view to having the schedule framed to (if possible) meet these contingencies.

BBJ 24 Oct 1901
The monthly meeting of the BBKA Council was held on the 10th inst., at 105, Jermyn-street, SW. Mr. TI Weston being voted to the chair. Apology for inability to attend was received from P Scattergood.

Conversazione.
The Chairman, having learned that Mr. Trimmings, one of the victims of the claim for damages to horses in what will be remembered as the "Gedling Bee Case" was present, called upon that gentleman to explain the details of the accident. Mr. Trimmings then described at considerable length the circumstances. The bees in question, he said, were kept in a "spinney" (a narrow space of land running between two hedges) and in the adjoining field were two farm labourers mowing. They had occasionally to sharpen the knives for the machine, which they, without thinking, did close to the hive entrances and here they also halted for rest. The horses were in a perspiration, as it was a very hot day and the noise and smell probably irritated the bees, though not sufficiently to start them stinging had they not received further cause.

Anyway, it appeared that the bees flew round the heads of the horses and men, the latter endeavouring to knock them away. No doubt in this commotion one of the horses was stung, for it began to rear and plunge. Then the mischief was irreparably made worse, coats, hats, sticks and the whip being freely used to drive away the bees, with the inevitable result that the horses kicked and became entangled in the reins, then fell into the hedge-bottom, while the men had not the courage to try and liberate them from the machinery, but ran away and stood looking on at the mischief for nearly an hour and a half before they sent for him (the speaker). When he and his friend (Mr. Mackinnon) arrived they went and cut the harness away from one of the horses, the other having got clear and run away. The poor brute was nearly covered with angry bees and was so fixed that in order to release it the pole had to be sawn in half; ten minutes or more being occupied before the animal could be released.

A little courage and common sense when the horses were first stung would, no doubt, have prevented the disaster, because at that time few of the bees were irritated; but apart from that it was clear the accident arose entirely from ignorance and stupidity on the part of the labourers. One of the men said "the bees came by hundreds of thousands," the other attributing the catastrophe to the fact that he (the speaker) "had been taking the kings and queens away the day before."

Readers of the BBJ very kindly subscribed, as did also the NBKA, towards defraying the claim of about £70 made against him which was paid; but notwithstanding that, some malicious persons during the following week actually put gas-tar under the hives and set fire to them. With regard to the accident the owner was informed that whatever loss he had sustained, if it could be legally or only morally substantiated, should be made good. The actual loss according to his (Mr. Trimmings') calculation was about £28 and yet £70 had been paid in settlement thereof. Both horses died.

Minute Book – Committee Meeting held at the Peoples' Hall on January 4th, 1902. Mr WS Ellis in the chair. Present: ?

The minutes of the previous meeting were read and Mr Turner proposed and Mr McKinnon seconded that the same be confirmed.

Mr Herrod proposed and Mr Pugh seconded that a note of sympathy and condolence be accorded to Mr and Mrs Mackender in the loss of one of their sons in the South African War. This was carried with a request that the secretary convey this resolution to Mr Mackender.

24387 Private (Trooper?) Henry Mackender

Henry Mackender was born in Staunton-in-the-Vale in 1876 so would have been 25 years old. He served in the 34th Company, 11th Battalion, Imperial Yeomanry. He was killed in action on Christmas Day, 1901 at Groenkop in Orange Free State along with 67 others. His name does appear on the Boer War memorial on the Forest Recreation Ground in Nottingham (but is recorded as Mackinder). This was originally unveiled on Queen Street (near the Brian Clough statue) on 26th March 1903 and later reclocated to its present site.

The Imperial Yeomanry was a volunteer mounted force of the British Army that mainly saw action during the Second Boer War. Created on 2 January 1900, the force was initially recruited from the middle classes and traditional yeomanry sources, but subsequent contingents were more significantly working class in their composition.

The annual Balance Sheet was next read by the secretary and explained in detail, the same shewing a deficit of £1 11s 4d.

After some discussion pro and con it was arranged
    that the Annual Meeting should take place on February 15th at 3pm in the Peoples' Hall and tea at 4.30pm.
    that failing the attendance of his Lordship one of the following should occupy

the chair – the Mayor (EN Elborne), Sheriff (Edward G Loverseed) or one of the Vice-Presidents

that the competition should be as last year

that Mr J Gray (who promised to do so) should give a short paper and that there should be a question box and that Mr Gray should answer the questions therein.

that there should be the usual prize drawing, arrangements for which were left in the hands of the secretary.

The circular about insurance was deferred to the Annual Meeting.

The secretary was instructed to write to the chairman of the British (Beekeepers' Association) complaining of the way the business with this association was conducted.

BBJ 23 Jan 1902
The monthly meeting of the BBKA Council was held at 105, Jermyn-street, SW. on the 16th inst, Mr. FB White occupying the chair. There was also present amongst others Dr. TS Elliot. A letter apologising for inability to attend the meeting was read from P Scattergood. A letter was received from the Nottinghamshire Association and after discussion; the secretary was instructed to reply thereto in accordance with the decision of the Council.

Minute Book – Annual Meeting held in the Peoples' Hall on February 15th, 1902. The chair was occupied by the Mayor of Nottingham, EN Elborne, who stated he was there that day not as a beekeeper, but as one who wished to show his sympathy with the association in its work. He did not know much about bees as yet but hoped to learn about them. He, however, thought that amongst the three or four thousand garden holders in the City of Nottingham, bees would be very helpful. He also hoped that at the show at Colwick Park this year the beekeepers' association would be represented.

The minutes of the previous meeting were next read when Mr J Herrod proposed and Mr Trimmings seconded the same be confirmed.

The secretary read several apologies from members who were unable to be present through illness and Mr Scattergood proposed that as a printed copy of the Balance Sheet had been sent to each member it should be taken as read and be received and adopted. He had audited the accounts and felt although the balance was still a little to the bad, it was very satisfactory. Mr J Herrod seconded and it was carried.

The secretary then read the Annual Report after which Mr Pugh proposed and Mr Gray seconded that it be printed along with the Balance Sheet and list of members.

Mr Scattergood proposed and Mr Pugh seconded that our thanks are due and are hereby tendered to Viscount St. Vincent for his services in the past and that he be elected President for the ensuing year. Although his Lordship was unable to be with us as of old, we should be sorry to lose him as our President, for although he was not present with us that day in body, he was in spirit as evidenced by the two hives he had sent for the prize drawing. Carried with acclaim.

Mr Turner proposed and Mr Bolton seconded that Mr Scattergood be thanked for his services as auditor and elected for the ensuing year.

Owing to the resignation of two members of the committee Messrs Smithurst and Randle were elected to fill their places. Messrs Hayes and Pugh were re-elected as representatives to the BBKA. Mr Hayes was re-elected as secretary and treasurer.

Mr Scattergood then rose to propose a vote of sympathy to an old and valued friend of the association viz. Mr Marriott, who was unable to be present through illness. Mr Skelhorne seconded and Mr Pugh supported.

A vote of thanks was accorded to all who had given prizes to this association this year.

The meeting then adjourned for tea which, unfortunately, was not served in very good style. At 6pm the meeting was resumed with Mr Ellis occupying the chair. Medals, etc. were awarded as follows:
>Silver Medal to Mr W Herrod, Sutton-on-Trent
>Bronze medal to Mr G Marshall, Norwell
>Silver Pendants to Mr G Marshall and Mr G Hayes, Beeston
>Certificates to Mr G Hayes and Mr JT Faulconbridge, Bulwell

The question of the Annual Show was considered and it was eventually resolved to leave the matter to the committee to settle.

Mr Puttergill brought forward a discussion on glazing or lacing and edging of comb honey and it was decided that a condition to the following effect shall appear in the schedule:
> "that all comb honey must be glazed on both sides and edging must not exceed $3/8$ of an inch in width on the face."

Mr Gray next gave a very interesting description of an expert's tour and some difficulties he had met and how he had surmounted them. He also answered a number of questions which had been placed in the question box, after which Mr

Pugh proposed and Mr Trimmings seconded a hearty vote of thanks to Mr Gray for his services that evening.

The winners in the class for one jar of honey for the hospital were:
- 1st   Mr Pugh, Beeston
- 2nd   Mr Mann, Stragglethorpe

The meeting then concluded with the usual prize drawing which was as follows:

		Given by	**Won by**
1st	Hive	President	Mr W Scrimshaw, Nottm.
2nd	Hive	President	Mr J Gray, Long Eaton
3rd	Hive	GH Varty	Mr Mackender, Newark
4th	Useful articles value 10/-	AE Trimmings	Mr W Swann, Eastwood
5th	Cottagers home ripener	Mr Meadows	Mr T Manchester, Nottm
6th	Jam jar, value 5/-	TN Harrison	Mr JT Hempsall, Ollerton
7th	"Weed" foundation	Mr Gray	Mr J Rawson, Selston
8th	"Progressive beekeeping"	Mr Gray	Mr J Ackland, Thoroton
9th	Miller American feeder	Mr Deacon	Mr G Hardwick, Ollerton
10th	200 labels, value 2/-	NBKA	Mr JC Wadsworth, Collingham
11th	100 labels, value 1/-	NBKA	Mr J Breward, Staythorpe

NEP 24 Jun 1941
Edward Newcombe Elborne of Hoveringham passed away yesterday in his 87th year. He was Mayor of Nottingham in 1901/2 and a JP. He spent his working life as a solicitor

NEP 25 Feb 1902
Lecture at Sutton Bonington. Last night a lecture on "Bees and Beekeeping" was delivered in the National schoolroom under the auspices of NBKA and the County Council by Mr George Hayes of Beeston. The lecture was illustrated by lantern slides and was much appreciated by a good audience. Mr MJR Dunston, of the County Council, occupied the chair.

BBJ 27 Feb 1902
The Annual Meeting of the NBKA was held on February 15th at the Peoples' Hall. The Mayor (Edward N. Elborne, JP), presided over a numerous attendance. The Mayor said he wished to show his sympathy with the association which was doing good work. He did not know much about bees but he thought that, in the three or four thousand gardens in the city, bees would be very helpful. He hoped that at the show at Colwick Park next year the beekeepers' association would be represented.

The Secretary read the Annual Report from which we learned that sixteen members had been enrolled and the association had come out of the year better in regard to finances than they had expected. The debt of £3 had been reduced to £1 11s 4d.

Last season had been a fairly good one all round for the yield of honey and, for quality, it had been excellent. The shows held during the year had been very successful. In regard to instruction in beekeeping the County Council had increased their grant from £30 to £40 and by its aid the association had been enabled to give lectures at various places in the county. The City Council had revived their grant of £2 2s, and it was proposed to have a lecture and a demonstration.

The Report and Balance Sheet were unanimously adopted and the officers for ensuing year were elected as follows: President, Viscount St. Vincent; Vice-Presidents, the Duke of Portland, Earl Manvers, WS Ellis, SH Sands, JP, Mrs. J. Hind, Rev. HL Williams, AE Trimmings, HW Cooper, WD Warwick and J. Bowes. Committee, Messrs. G. Smithurst, TN Harrison, T. Carlin, SW Marriott, AG Pugh, GE Puttergill, GE Skelhorne, W. Swann and JC Wadsworth. Hon. Auditor, Mr. P. Scattergood. Secretary and Treasurer, Mr. Geo. Hayes, 48, Mona-street, Beeston, Representatives to meetings of the BBKA - Messrs. AG Pugh and G. Hayes.

The first President of NBKA was Charles William Sidney Pierrepont, 4th Earl Manvers (1854-1926). He was styled by the courtesy title of Viscount Newark from 1860 until he succeeded to his father's peerage in January 1900. He served as NBKA President from 1884 until 1890 and died suddenly on 17 July 1926 at his house in Tilney Street, London,

BBJ 6 Mar 1902
T. Marshall (Nottingham). Honey Sample. Your sample is clover honey and is good in colour and flavour, but shows slight indications of incipient fermentation. We fear it will be "out of condition" for the show-bench this season, as the earliest show is some months off.

Minute Book – Quarterly Committee Meeting held in the Peoples' Hall on March 8th, 1902. Present: WS Ellis in the chair, Messrs Harrison, Marriott, Pugh, Puttergill, Skelhorne, Swann, Randle, Turner, Wadsworth, Scattergood, J Herrod, Rawson, Gray and Trimmings.

The minutes of the previous meeting were read and Mr Scattergood proposed and Mr Pugh seconded the same be confirmed by the signature of the chairman. The quarterly Balance Sheet was next submitted. Mr Pugh proposed and Mr Scattergood seconded.

The secretary read a reply from the chairman and also from the secretary of BBKA about the complaint referred to at the last meeting, apologising for neglect.

After considerable discussion, Mr Scattergood proposed and Mr Skelhorne seconded that the Annual Show be held with the Moorgreen Society as per their invitation. It was also proposed that we hold a show at Southwell on the same terms as previously. The schedule was next revised as set forth in the printed report and schedule supplied to each member. Mr Scattergood proposed and Mr Pugh seconded that the secretary be empowered to arrange for any other local shows, as far as the funds will allow, and print a special schedules for the same.

The Technical Instruction sub-committee were re-elected "*en bloc*" except Mr J McKinnon who could not get.

NEP 12 Mar 1902
Newark Borough Police Court. Charlotte Norwell, was charged with stealing fowl and four hens' eggs, of the value of 3s, the property of Mr. Joseph Price, Bathley on March 5th. Complainant said that on Wednesday week he went to fasten up some bees and saw a fowl under a cover dead but warm, and with four eggs. He left it there, and the next night it was gone. Pc. Johnson said that on March 6th, he was instructed by Sgt. Deacon to watch the top of a beehive on Mr. Price's premises. At about eight o'clock at night he saw Charlotte Heaton, Norwell, lift it up and take away a fowl and four hen eggs. He let her get 150 yards with it and then signalled to Sgt. Deacon, and he stopped her and brought her back to where the witness was. He charged her with stealing the hen and four eggs value 3s, the property of Joseph Price. She said. "I do not know what made me do it. It is the first time I have taken anything." Mr Platt commented that the hen was dead and he did not know who killed it. Prisoner elected to be dealt with summarily and was fined 10s or 10 days hard labour.

BBJ 13 Mar 1902
The Annual Meeting of Leicestershire BKA was held at the Victoria Coffee House, Leicester, on March 6th. There was a good gathering. After tea, Mr. Peter Scattergood, of Stapleford gave a couple of short lectures on "A Year's Work in the Apiary" and "Notes by the Way," illustrating his remarks by a series of lantern slides.

BBJ 27 Mar 1902
The monthly meeting of the BBKA Council was held at Jermyn-street, SW. on March 20th, Mr. TI Weston being voted to the chair. Letters of apology for non-attendance were read from Dr. TS Elliot and P. Scattergood. The following, amongst others, were duly elected as members of the Council for the year 1902-3 - Dr. Elliot and Mr. Scattergood.

BEEKEEPING BETWEEN TWO QUEENS

BBJ 15 May 1902
JT, Junior (Notts.). Foul Brood is developing in comb sent and, as drone-brood is being reared in some of the worker cells, we advise entire destruction of the stock. The combs are old, the queen worthless and with seventeen healthy colonies liable to infection to attempt a cure would be folly. We therefore say stamp the mischief out at once.

NEP 15 May 1902
Newark Agricultural Show. Second Day. An attractive programme had been arranged with the second day's proceedings of the Newark Show, but unfortunately for the society the fine weather of the previous afternoon was not maintained. Rain during the early morning, and though the conditions improved as the day wore on there can be no doubt that some people kept away, and the attendance was consequently not as large as it would have been. Demonstrations and lectures, under the auspices of the NBKA, were given as on the first day by Mr. W.Herrod, the College, Swanley.

BBJ 5 Jun 1902
Good natural swarms for sale, 10s each.
H. Holleworth, New Inn Farm, Widmerpool

Minute Book – Quarterly Committee Meeting held in the Peoples' Hall on July 5th, 1902. Present: Messrs Pugh (in the chair), Smithurst, Skelhorne, Swann, Randall, Turner and Wadsworth.

The minutes of the previous meeting were read and Mr Smithurst proposed and Mr Swann seconded that the same be confirmed. The quarterly Balance Sheet was next gone through after which Mr Skelhorne proposed and Mr Swan seconded the same be received and adopted.

Letters were read from Messrs Ellis and Rawson stating their inability to be present, also from the secretary of the BBKA asking for no further insurance policies to be issued until further instructions. It was also thought that the British (Beekeepers' Association) should bear the expenses incurred in the matter of insurance and Mr Skelhorne proposed and Mr Wadsworth seconded that the secretary write to Mr Young (Secretary, BBKA) on the matter and report back at the next meeting.

Resolved that the following gentlemen, in the order given, be asked to judge at the Annual Show at Moorgreen – Mr WF Reid, Mr WB Webster and Mr CN White. Proposed by Mr Skelhorne, seconded by Mr Wadsworth, that Mr Scattergood be asked to judge at Southwell Show.

Mr Wadsworth complained of the delay of experts visits to the members in his locality to which a satisfactory explanation was given by the secretary.

NEP 14 Jul 1902
W. Doleman, Hickling Pastures writes: In your issue of July it is stated the Commonwealth Postmaster-General has decreed that live bees, if properly packed, may be transmitted within the Commonwealth and to and from the United Kingdom. The paragraph adds: "In what sort of case could live bees be properly packed as to survive a five weeks' voyage? Only we fear in a case approved by the SPCA." He appears to doubt the possibility of bees travelling safely, but I may say it is possible and easy to "properly pack" live bees so that they will survive the long journey. I could explain the process, but it would not be understood other than by practical beekeepers. Also, it would be interesting know if a bee is an animal "strictly within the meaning of the Act," Does the society extend its operations to all insects?

BBJ 17 Jul 1902
Referring to Tunisian Bees, Mr. AE Trimmings, Woodside Apiary, Gedling writes as follows: "I notice in the Journal for July 10th, a Mrs. Tomlinson asks where she can get some of these bees? If she will send me a ready addressed envelope I will send a queen, free of charge, as soon as I get a few more on hand than I require."

The Tunisian honeybee (Apis mellifera intermissa) is a breed local to northern Africa, also called the Tellian or Punic honeybee. It is a strong, rather aggressive, highly reproductive, and incredibly active bee. Tunisian honeybees are distinguished for their hardiness and extraordinary ability to reconstruct their hives in favorable years after the drier ones, and are quick to defend those hives. This race is particularly adapted to cold and dry climates.

East Bridgford Parish magazine August 1902
The show was pronounced to be an excellent one considering the backward season, but the gate proceeds were considerably reduced, the rain, which commenced soon after the opening, continuing till 6pm and thus preventing the appearance of many visitors from the surrounding districts.

Special prize given by Mr Fox: Honey – 1st, Miss Fox

NEP 1 Aug 1902
On Tuesday next the Annual Show of the Welbeck Tenants' Agricultural Society will be held on the usual site adjacent to the estate offices at Welbeck. The show becomes increasingly popular year by year, and this year with the visit of Lord Kitchener, the guest of the Duke of Portland, another record attendance is anticipated. Demonstrations will be given in bee driving and lectures in bee management, under the auspices of the NBKA.

BEEKEEPING BETWEEN TWO QUEENS

Horatio Herbert Kitchener, 1st Earl Kitchener (1850-1916). As Chief of Staff (1900–02) in the Second Boer War he played a key role in Lord Roberts' conquest of the Boer Republics.

William John Arthur Charles James Cavendish-Bentinck, 6th Duke of Portland, (1857-1943), was a British landowner, courtier, and Conservative politician. He inherited the Cavendish-Bentinck estates, based around Welbeck Abbey in 1879.

Sheffield Independent 18 Aug 1902
It is some years since a flower show was held on the beautiful Lawn Grounds and the exhibition was resuscitated on Saturday. There was also an exhibition by NBKA, the lecturer being Mr P Scattergood.

BBJ 21 Aug 1902
Xtractor, "Little Wonder," one frame little used, working order, 5s 6d. E. Eddison, Shireoaks

BBJ 28 Aug 1902
Shropshire BKA Show. The entries in the honey section of the show were less numerous than last year, owing to the unfavourable season, but the exhibits were of excellent quality, both of comb and extracted honey.
Honey Vinegar (open) - hc, P. Scattergood, Stapleford

BBJ 4 Sep 1902
Extractor (two-frame), good working order, cost 22s 6d, 17s 6d. White Clover Honey in 28-lb. tins, $6\frac{1}{2}$d lb, on rail. Light colour, very thick.   Ernest Eddison, Shireoaks

BBJ 11 Sep 1902
Wanted, Well-Filled Sections, quote lowest, carriage paid.
                                                        Summers, Broadgate, Beeston

BBJ 18 Sep 1902
The Annual Show of the NBKA was held at Moorgreen on September 9th, and as far as the weather (the first consideration) is concerned was all that could be wished for. The exhibits staged were rather more numerous than usual and the honey of excellent quality generally. Mr. WB Webster, Binfield, Berks, officiated as judge, assisted by Mr. GE Skelhorne of Notts. And made the following awards:

Display of Honey in any Form, not to exceed 80 lb. of Honey (5 entries) - 1st, G. Marshall, Norwell; 2nd, G. Hayes, Beeston; 3rd, J. Gray, Long Eaton
Six 1-lb Jars Extracted Honey (light) (14 entries) - 1st, AG Pugh, Beeston; 2nd, J. Breward, Staythorpe; 3rd, W. Brooks, Eastwood
Six 1-lb Jars Extracted Honey (dark) (9 entries) - 1st, AG Pugh; 2nd, GM Bolton, Eastwood; 3rd, W. Brooks
Six 1-lb Sections (6 entries) - 1st, G. Marshall; 2nd, D. Marshall, Cropwell Butler

Six 1-lb Jars Granulated Honey (8 entries) - 1st, H. Merryweather, Southwell; 2nd, AE Trimmings, Gedling; 3rd, G. Marshall
Shallow-Frame of Comb Honey for Extracting (7 entries) - 1st, G. Marshall; 2nd, GH Pepper, Farnsfield
Six 1-lb Jars Extracted Honey (novices only - 7 entries) - 1st, H. Mackender, Newark; 2nd, L. Walker, Ruddington; 3rd, C. Markham, Retford
Six 1-lb Jars Extracted Honey (local class - 5 entries) - 1st, W. Brooks; 2nd, G. Smithurst, Watnall; 3rd, GM Bolton
Honey Vinegar - 1st, J. Gray; 2nd, G. Hayes
Observatory Hive with Bees and Queen (9 entries) - 1st, R. Mackender, Newark; 2nd, H. Mackender; 3rd, G. Marshall;
4th, Geo. Smith, Bradmore
Beeswax in 2-oz. Cakes (8 entries) - 1st, AH Hill, Balderton; 2nd, R. Mackender

During the day, Mr. P. Scattergood held an examination of candidates for the third class certificate of the BBKA, three candidates presenting themselves.

Henry Mackender born in Staunton-in-the-Vale in 1876 but was killed in the Boer War in 1901 as reported earlier. Herbert Mackender, born in Staunton in 1878, married ? Booth. These two boys were the only children of Robert Mackender and Emma Eliza Cock who married in 1874 in Bromley. Herbert had a son, Herbert J in Newark in 1911. It must have been Herbert who exhibited the Observatory hive.

BBJ 25 Sep 1902
The seventh Annual Exhibition of the Grocery and Allied Trades opened on September 20th and continues the 27th. As was the case last year, the Honey Section was transferred from the ground floor to the gallery where, if less prominently before the bulk of visitors, it afforded bee-keepers a better chance of inspecting their own portion of the show in comfort and without crowding.
Twelve 1-lb sections (17 entries) – 4th. 10s, A. Hunt, Newark
Beeswax in Cakes, Quality of Wax. Form of Cakes and Package, suitable for retail counter trade (7 entries) – 1st, £1, A. Hunt

Minute Book – Quarterly Committee Meeting held in the Peoples' Hall on October 4th, 1902. Present: Messrs Ellis (in the chair), Messrs Harrison, Pugh, Puttergill, Skelhorne, Randle, Turner, Wadsworth, Mackender and secretary.

The minutes of the previous meeting were read and dealt with *serratum* and adopted. The quarterly Balance Sheet was next gone into which shewed that we should be encumbered with a heavier debt at the years end. Mr Pugh proposed and Mr Skelhorne seconded that the same be passed.

Mr Harrison gave notice of bringing to the next meeting a proposal to improve the next Annual Meeting.

BBJ 16 Oct 1902
BBKA Conference of County Representatives.
A conference of representatives of County Bee-keepers' Associations with the Council of the BBKA was held at 105, Jermyn-street, on October 9th, for the purpose of furthering the interests, economical working and general welfare of the various societies, and conducing to the establishment of more intimate relations between the BBKA and its affiliated societies. There was a large attendance, the spacious boardroom being filled by gentlemen representing the various County Associations affiliated to the BBKA.

Mr. AG Pugh (NBKA) thought the Devon BKA was to be congratulated for, although Colonel Walker complained at first of their difficulty in getting money, yet he finished by announcing a very liberal grant from the Devon County Council. Most counties would envy his success. Referring to his (the speaker's) own county, the local Association held a committee meeting on the first Saturday in the present month, when he was deputed to ask at this conference whether the BBKA had any reserve fund from which it could help its struggling affiliated County Associations?

Having been for many years a member of the BBKA, he had told his friends not to think of that knowing as he did the dearth of funds at the Head Office. He rather differed from Colonel Walker regarding the actual value of County Council support, being inclined to think that instead of their being a help to the county BKA, they might possibly be the reverse.

The NBKA had received from the County Council a grant which had gradually increased from £10 to £40, and his Association suggested to the County Council that this sum be spent upon an expert tour. This suggestion being agreed to, his Committee drew up a scheme and made the necessary arrangements. The result, however, was that after the expert had rendered all the assistance required by members, the latter began to ask, "Why should we continue to pay a yearly subscription to the Association for assistance which the County Council is paying for? And this being so, there is no need to subscribe any longer to the County BKA"

It thus became obvious that if those views became general the Association would eventually cease to exist. Some people do not take a broad view of these matters, but were always inclined to ask, "What shall I get for my money?" And it was difficult to convince these people that they were getting anything if the expert service was being paid for by the County Council. If his Association could keep the membership up to 200, it could exist fairly well but it was a difficult matter to keep up their numbers and any suggestions from this meeting in the direction indicated would be welcome to his Committee.

The question was also asked at his meeting, "What does the BBKA do for us in return for the affiliation fee?" He had always supported the payment of this fee and pointed out that the "British" did much more for the cause throughout the country than the local members were aware of. Probably, among other things, they did not know that the BBKA had a large library of bee-books which was at their disposal. He thought it would be a wise thing for the central body to publish a leaflet stating what they had done and were prepared to do for the affiliated Associations. This would open the eyes of many country bee-keepers, who thereby would feel themselves in much closer touch with the parent body than they could be under present conditions.

Mr. Pugh hoped this meeting was the precursor of many others of a similar character, which could not fail to enlighten bee-keepers and advance their industry. He also trusted that the parent body would make the county representatives feel that their attendance was appreciated and would send them the agenda and other papers from time to time.

After a few words from the Secretary, Mr. Reid, and Mr. Meadows on the same subject, Colonel Walker inquired whether any of the counties paid the expenses of their representatives' attendance. Mr. Pugh said that NBKA allowed 10s (about half the train fare to London).

But did he disclose that he was an employee of the railway and probably got subsidised train fares?

BBJ 6 Nov 1902
The monthly meeting of the BBKA Council was held at 105, Jermyn-street, SW. on October 23rd, Mr. Tl Weston occupying the chair. The Chairman reported that, in accordance with a request of the Council at the last meeting, he had interviews with Mr. Compton-Bracebridge of the Royal Agricultural Society, in regard to the proposed new classification for honey at the 1903 Show. It was thought the suggested changes were likely to meet with the approval of the Society, and it was requested that the proposed alterations should be embodied in a letter to be put before the Stock Prizes Committee on November 3rd. The Council has therefore grouped the counties as enumerated below subject, of course, to such minor alterations as may be considered necessary or desirable.
Group 6. Notts, Lincoln, Cheshire, Derby, Stafford.

BBJ 13 Nov 1902
Deferred Breeding after Queen Introduction. Will you kindly give me your opinion on the following?

BEEKEEPING BETWEEN TWO QUEENS

On September 24th I introduced an imported Italian queen to a black stock after taking away the old queen. On examination a week later I found she had laid a small patch of eggs; so, concluding all was well, I packed them down for winter. Seeing however, last week, that in my other two hives (both young queens) the bees were carrying in pollen freely, while the bees of the hive in question which, with the imported queen, is the strongest stock I have, were not, I examined them to see if the queen was all right. I found no eggs or brood in any form, but the queen was there and seemed lively enough; so I came to the conclusion that the patch of eggs I saw were all she had laid, seeing that, four weeks later, there was no brood, neither have I seen any Italian workers yet.

1. Am I right?
2. This queen was purchased direct from Malan Bros. Are they known as reliable dealers, and do you think I can depend on this queen for next year?
   WHS(toppard?), Nottingham, November 10th

Reply.
1. Probably you are quite right, but it does not follow that the Italian queen is any the worse for the fact. You had better defer judging till next spring, when the Italian may possibly resume the maternal work of egg-laying freely, and perchance shoot ahead of the young ones.
2. We consider the firm referred to quite reliable.

Minute Book – Quarterly Committee Meeting held in the Peoples' Hall on January 3rd, 1903. Present: WS Ellis in the chair, Messrs Harrison, Marriott, Pugh, Puttergill, Skelhorne, Swann, Randle, Wadsworth. Scattergood, Brooks, Mackender, J Herrod, W Herrod, Gray and secretary.

The minutes of the previous meeting were read and confirmed. The annual Balance Sheet which shewed the adverse balance reduced to 16s 2d was next gone into; after which Mr Pugh moved the same be adopted and ordered to be printed on the circular convening the Annual Meeting, and sent to members. Mr Skelhorne seconded.

The secretary read letters from several members of the committee regretting inability to be present also letter from the editor of "Bees" and it was decided that no notice whatever be taken of the latter.

Mr Pugh then gave a very graphic description of what transpired at the BBKA meeting in London to which we sent him as delegate when Mr Herrod proposed and Mr Scattergood seconded that our thanks be accorded Mr Pugh for his report.

The arrangements for the Annual Meeting were next considered and it was decided as follow:

      that it be held on February 7th at 3pm at the Peoples' Hall

that the following gentlemen be asked (in the order given) to preside – the Mayor (Alfred Page), Mr Warwick, Newark, CJ Bristowe, AJ Butler

Mr Ellis kindly undertook to arrange for a caterer to supply tea

that there be two classes for competition viz. one for liquid and one for granulated honey. Prizes to be both alike and honey sent to Children's' Hospital.

It was also decided to have a musical evening and the following promised to assist
- Mr Harrison to provide pianist
- Mr Wadsworth to provide a song
- Mr Skelhorne a Scotch reading
- Mr Mackender to provide a song
- Mr Pugh a Yorkshire reading
- Mr Hayes a singer for two comic songs

Mr Herrod was asked, as promised, to give three ten-minute talks with the lantern.

Mr Herrod proposed and Mr Skelhorne seconded with a view to having the secretary at liberty during the evening and that Mr Pugh and Mr Scattergood be asked to take subscriptions on that day.

Mr Gray also promised to unpack and re-pack the prizes for the annual drawing which was, as usual, to conclude the evening.

Alfred Page was Mayor in 1903. He was born in Nottingham in 1835 and went into the family haberdashery business.

CJ Bristowe was the Director of Education, Nottinghamshire County Council at this time. He was buried in Caunton before 1915.

AJ Blakeman was associated with BBKA.

BBJ 8 Jan 1903
Other shows attended included the Newark Agricultural Show for the NBKA. Trusting my labours of the past season may meet with your approval.
    Your obedient servant, W. Herrod. North Bank, Hextable. December 1st, 1902.

BBJ 12 Feb 1903
The AGM of NBKA was held on February 7th, 1903 in the Peoples' Hall. In the unavoidable absence of the President, Viscount St. Vincent, the chair was occupied by Mr. WS Ellis, Vice-President. There was a large attendance, amongst those present being Miss Bingham, Messrs. P. Scattergood, A Pugh, SW Marriott, W. Herrod, R. Mackender, T. Manchester, G. Smithurst, GE Skelhorne, J. Gray, R. Turner, T. Randall and Geo. Hayes, Secretary.

The Balance Sheet for the past year was adopted. It showed that the receipts amounted to £99 3s 11d including £42 7s 4d from the County Council and £33 5s 6d in subscriptions. The expenses were met within 16s 2d, this being the amount of the deficit on the year.

The Secretary presented his Annual Report, which showed the membership had slightly increased in 1902, and that the adverse balance on the year's working was only 16s 2d, an improvement on the previous year, if a small one. The Annual Show for 1902 was again held in connection with the Moorgreen Agricultural and Horticultural Societies in September last, and was a great success. A show was also held at Southwell, as in previous years, and some splendid honey was staged. The County Council had increased their grant to £50 and lectures have been given at Newark, Mansfield, Kirkby, Cotgrave, Sutton-in-Ashfield, Kingston, Welbeck, Sutton Bonington, Clarborough and Moorgreen. Lantern lectures are also being given at Edwinstowe, Collingham, Besthorpe, Farnsfield, Bawtry and Thoroton. The City Council had also renewed their grant, which enables the association to continue its educational work as before. The Insurance Scheme was now in full working order and some members had taken advantage of it. Of two candidates for the 3rd class Experts' Certificate, one passed, viz., Mr. Horace Mackender, of Newark.

As noted earlier there was no record of a "Horace" in the Mackender family. Was this a mistake and should have been Herbert?

During the spring 174 bee-keepers had been visited. These owned 595 stocks in frame hives and 115 in skeps. Of these stocks 467 were examined, but only twenty-one were found affected with "Foul Brood." Attention was called to the fact that the BBKA had adopted a system of grouping the various counties for competition at the "Royal," and other London shows. This offered a favourable chance to members of winning the prizes offered.

The election of officers resulted unanimously as follows: President, Viscount St, Vincent; committee: Messrs. TN Harrison, SW Marriott, AG Pugh, T. Randall, GE Skelhorne, G. Smithurst, W. Swann, JC Wadsworth, F. Chasteney, together with the hon. secretaries and experts. Messrs. AG Pugh and G. Hayes were appointed delegates to the meetings of the BBKA. Mr. Geo. Hayes was re-elected secretary and treasurer.

After tea, the presentation of prizes, medals and certificates was gone through. Mr. W. Herrod then gave a ten minutes' talk on "The Growth, Progress, and Uses of the BBKA's Apiary," following this with other instructive discourses on "Foul Brood," and "College Life." These "bee-talks" were varied by an interesting miscellaneous programme, consisting of music, readings, etc. contributed by Miss Mabel Smith, Mr.

W. Cartledge-White, Mr. Mackender, Mr. E. Spray, Mr. Skelhorne, and Mr. Wadsworth.

George E Skelhorne was a printer born 1858

John C Wadsworth born 1872 Newark

Frederick Charles Chasteney was a School Inspector born in Great Yarmouth in 1856.

Minute Book - NBKA Report for 1902
The AGM of the NBKA was held on February 7th, 1903, in the Peoples' Hall. In the unavoidable absence of the President, the Chair was occupied by WS Ellis, Esq. one of the Vice-Presidents.

The Minutes of the previous Annual Meeting were adopted. The Secretary read several letters from members stating their inability to be present owing to illness and other causes. Mr. Scattergood then proposed and Mr. T. Manchester seconded, that the Balance Sheet as printed and sent to each member be received and adopted. He had examined all accounts and vouchers and found all quite correct. Carried.

The Secretary then read the Report, as follows:
Mr Chairman, Ladies and gentlemen. It will be noticed that the membership has slightly increased since last year, but in this respect we do not make the rapid strides I should like us to do, but there is one satisfaction that so far, if we have been perhaps somewhat slow, we have at least been sure. But I would again urge all to do their best to bring recruits to our ranks and so swell the membership still more.

It is satisfactory too, I am sure to all, to see we are better financially this year than last and I am sure your thanks are due to your Committee, for the careful way in which they have dealt with the funds of this Association. They would like to do considerably more in the shape of prize giving at shows, both annual and local; but till our funds increase it is impossible to do so.

The past season has been a very precarious one for honey, as we had only (in this county) about a fortnight's real honey flow but it is remarkable how those stocks which were strong enough rendered so good an account of themselves. I am sure many of you must have been surprised at the quantity they gathered and also the quality of it.

To a good few unfortunately (amongst the not over observant beekeepers) the weather was misleading; and a good number of stocks were starved to death, much to the surprise of the owners.

The Annual Show for 1902 was again held in connection with the Moorgreen

Agricultural and Horticultural societies in September last, and was a great success. The Silver and Bronze Medals were won by Mr. Pugh, and Certificate by Mr. H Merryweather; Silver Pendants by Messrs. G. Marshall and J. Gray. Mr. WB Webster, of Binfield, acting as judge. A show was also held at Southwell as on previous years and some splendid honey was staged.

The County Council have again increased their grant this last session from £40 to £50, and demonstrative lectures have been given at the following places: Newark, Mansfield, Kirkby, Sutton-in-Ashfield, Kingston, Welbeck, Sutton Bonington, Clarborough and Moorgreen. Lantern lectures are also being given at Edwinstowe, Collingham, Besthorpe, Farnsfield, Bawtry and Thoroton.

The City Council has again renewed their grant, which will again enable us to take our teaching amongst the apiaries in the city.

The Insurance Scheme is now in full working order and some of our members have taken advantage of its securities, but it has not been patronised by our members as it ought to be, but I venture to hope that our members will join it.

Two candidates offered themselves for examination for 3rd class Experts, but only one passed, viz., Mr. Horace Mackender, of Newark.

During the Spring (which for the most part was unfit for expert work owing to cold and wet) and the Summer which was not very much better, 174 Beekeepers have been visited. These owned 595 Stocks in Bar Frames and 115 in Skeps = 710; of these Stocks we examined 467, but only 21 were found affected with "Foul Brood." These were, (or it was promised they should be) dealt with as the case demanded. If the Stocks were bad, burning was resorted to but I am glad to say as far as I can gather, this was only necessary in four cases.

I would call your notice to the fact that the BBKA have adopted a system of grouping the various counties for competition at the Dairy, Royal and other London Shows, this I believe is a step in the right direction and should give our members a chance of not only competing but also of winning the prizes offered.

Also as to privileges of members of this Association, not only are they entitled to what this Association offers, but they are entitled to exhibit at all above mentioned shows of the BBKA at reduced fees; to avail themselves of facilities offered at their shows for disposal of honey; to attend their meetings and enter into the discussions, but not to vote. I think if this were more thoroughly understood no one could conscientiously say they were not getting full value for their subscription to this Association.

The free use of their Library (paying postage of books required to and fro) containing books as under:
Cook's "ABC"; Quinby's "Beekeeping" (1879); Langstroth's "Honey Bee" (1857); Cowan's "Guide Book" (1901); Neighbour's "Apiary and Bee Culture" (1866); Cook's "Manual of Beekeeping" (1880); Pettigrew's "Handy Book" (1875); Dzierzon's "Rational Beekeeping" (1882); "Nutt on the Management of Bees" (1832); "Thorley on bees"; Key "On the Management of Bees"; "Purchase on Bees"; "Huber on Bees" (1821); Cotton's "Bee Book" (1880); Taylor's "Manual"; Hunter's (Secretary of BBKA) "Manual" (1884); "The Honey Bee" by WH Harris (1884) "Bonner on Beehives" (1795) and a number of works by lesser known writers.

But maybe not so modern ones! However, some of these volumes are being reprinted in 2018 and those dated are in the NBKA George Hayes Memorial library in digital format. Despite extensive research no record of Cook's "ABC" could be found.

Although these volumes were written many years ago they still contain much useful advice to beekeepers. They are also a great source of illustrations to any researcher.

Mr. Pugh then proposed, and Mr. Marriott seconded, "that the Report be printed along with the usual matter and sent to each member." Carried.

Mr Pugh then rose to propose a most hearty vote of thanks to the President, Viscount St. Vincent, for his services during the past years, remarking that although His Lordship was not with us it was of no fault of his own, for he evidently fully intended to be present. His Lordship was one who, though not a beekeeper now, had been one, and an adept one too, and his interest in the industry was not abated, at least not, if we judge from the interest shown in this Association to-day by his gifts to the Prize Drawing, etc. He had, therefore, great pleasure in proposing this vote of thanks to him, and a cordial invitation to accept the Presidency for the ensuing year. This was carried with acclamation. The Committee were re-elected "*en bloc*" with the addition of Mr. F. Chasteney, of Beeston.

The Auditor, Mr. Scattergood, was thanked for his services and re-elected, as were also the Secretary and Delegates to the BBKA.

Mr. Herrod proposed, and Mr. Pugh seconded, "that a suggestion be sent to the BBKA recommending that credentials be supplied to the secretaries and representatives of affiliated associations so that each may have an introduction to those they meet so as to promote a greater feeling of welcome."

The question of judges for each group of counties from one of those counties was discussed but it was thought this was a matter for the BBKA alone, therefore no suggestion was offered.

The meeting was then adjourned for tea, to which 63 members and friends sat down. At 6pm the meeting was reopened by a Pianoforte Solo by Miss Mabel Smith, and the medals and certificates were then awarded by the Chairman to the successful exhibitors.

The winners in classes at Annual Meeting were as follows:
Granulated Honey 1st, Mr. P. Scattergood, 2nd, Mr. AG Pugh.
Run Honey 1st, Mr. Gervaise Smith, 2nd, Mr. AG Pugh.

Mr. W. Herrod then gave a ten-minute talk on "The Growth, Progress, and Uses of the BBKA" followed by others on "Foul Brood" and "College Life." These bee talks were varied by an interesting miscellaneous programme consisting of music, songs, and readings contributed by Miss M Smith, Messrs W. Cartledge, White, Mackender, E. Spray, and GE Skelhorne, and a very instructive and enjoyable evening was brought to a close by the usual Prize Drawing, which resulted as follows:

		Given by	Won by
1st	Meadows Guinea hive	Viscount St. Vincent	Mr EH Allsebrook, Wollaton
2nd	Meadows XL hive	Viscount St. Vincent	Mr Scattergood, Stapleford
3rd	Cheshire Wax extractor	Anonymous	Miss Ede J Morris, Worksop
4th	Cottagers honey ripener	Anonymous	Mr Reynolds, Tuxford
5th	Wanted a frame (50)	Mr AG Pugh	Mr H Mackender, Newark
6th	Bronze Hall mirror	Mr TN Harrison	Mrs Scattergood, Stapleford
7th	Parcel value 5/-	Mr Scattergood	Mr G Reeves, Moorgreen
8th	2 boxes candy	Anonymous	Mr J Horridge, Ilkeston
9th	200 honey labels	NBKA	Mr J Hallam, Orston
10th	100 honey labels	NBKA	Mr Beeton, Mansfield
11th	Useful articles value 10/-	A friend	Mr C White, Hawksworth
12th	"Gleanings" magazines	Mr Gray	Mr Mann, Stragglethorpe
13th	Super clearer	Mr Deacon	Mr S Catton, Lenton

Lincolnshire Chronicle 13 Feb 1903
At the Annual Meeting of the NBKA, held in Nottingham on Saturday, Viscount St. Vincent was re-elected President and the other officers were appointed. The Annual Report disclosed a slightly increased membership and an improved financial position compared with last year.

NEP 20 Feb 1903

Sutton Bonington Horticultural Society. The AGM of this society was held last night the National Schoolroom. It was decided to have an exhibition of beekeeping, bee driving, etc., at the show.

Minute Book – Quarterly Committee Meeting held in the Peoples' Hall on April 4th, 1903. Present: Messrs Ellis, Smithurst, Harrison, Marriott, Puttergill, Skelhorne, Randle, Turner, Wadsworth, Hibbert and secretary.

Mr Turner proposed and Mr Puttergill seconded that Mr Ellis be chairman of this committee for the ensuring year.

The minutes of the previous meeting were read and confirmed, proposed by Mr Harrison and seconded by Mr Marriott. The quarterly Balance Sheet was next gone into when Mr Marriott proposed and Mr Wadsworth seconded that the same be received and adopted.

The question of an advertisement card for sale of honey for members was discussed and it was agreed that members of the committee should embody their ideas on paper and submit to the next meeting.

The secretary pointed out that, owing to pressure, he had been obliged to make arrangements for the Annual Show and Southwell Show and also he had tried to get the Moorgreen people to allow us the entry fees in addition to the £15. He had not been able to do so and had accepted the same terms as last year. Mr Turner proposed, Mr Puttergill seconded that the action of the secretary in this matter be fully endorsed.

The members of the present sub-committee for Technical Instruction were re-elected "*en bloc*".

Mr Wadsworth proposed and Mr Skelhorne seconded after some discussion, that the secretary procure a badge of an oval or round form, with dark red ground and gilt letters for committee men and that they be sold to those committee men desirant of having one.

The question of providing an extractor for members use was discussed but it was decided to be left over till the next meeting.

BBJ 16 Apr 1903
GHP (Notts.) Comb is affected with "Foul Brood". In view of your fourteen healthy stocks, we should remove risk by destroying the weak lot outright by burning.

BEEKEEPING BETWEEN TWO QUEENS

EiDLOG (Mansfield) Feeding Bees in April. Syrup-food will be preferable to candy from mid-April forward.

BBJ 7 May 1903
It is very gratifying to illustrate a "Home of the Honey Bee" like that of Mrs. Wootton, along with the experiences of a lady bee-keeper of a so decidedly practical type as the owner thereof.

The fact that so few of the 'Homes of the Honey Bee,' belong to lady bee-keepers has often been a surprise to me. In looking over last year's volume of the Record, I find the proportion is one lady to twenty-two gentlemen. Why this should be I cannot understand. Are there really so few of us, or do our lady friends not care to let it be known what they can do!

"Amongst the numerous occupations that have of late years been open to women, surely bee-keeping is one of the pleasantest, and to those of us who reside in rural districts, not by any means the least profitable, if we bear in mind the small amount of time necessary for the work required, taking the year through.

If, therefore, a few words from an old hand like myself are likely to be an inducement for any younger member of my own sex to start bee-keeping, or encourage those who have made a beginning, I shall feel only too pleased to have been of some little service, should you think well to publish the following particulars:

I am a gardener's wife, and in the year 1869 I had put into my hands 'Pettigrew's Handy Book of Bees' and this book first brought the subject of bees under my notice. Up to that time I should have hesitated in deciding between a bee and a wasp if shown both, but through studying 'Pettigrew,' along with other bee literature, I was fired with enthusiasm, and having purchased two swarms in straw skeps, I resolved to become mistress of the art.

 My resolve was perhaps a little strengthened because of the jocular predictions of gentlemen friends, who all prophesied the speedy collapse of my new venture. The 'collapse,' however, has still to come, for as you see from the date of my start, more than thirty years have passed, and I have not given up yet.

As a matter of fact, I made steady progress for the first few years, and soon gave up the straw skep for the bar-framed hive and other appliances for modern bee-keeping, and by thus keeping up-to-date I soon became a successful exhibitor at several local shows. I also gradually increased my stocks until they numbered twenty, but for the past few years I find twelve to be quite as many as time or strength allows me to manage properly.

When I say that I have taken as much as 500 lb, and seldom less than 300 lb, of good honey in one season, and sold it all in 1-lb jars, you will, I hope, acknowledge that I have worked successfully, for everything has been done by myself, without the least assistance from any one, except when a swarm issued and clustered high up beyond my reach for - in spite of all my endeavours at prevention - the bees sometimes would swarm. If they settled within reach, I always hived them myself; but they more often chose to fly high in the air, and cluster on a topmost branch of one of the tall trees around us, and, of course, I have been obliged to ask some friend to bring the bees down for me.

Had I my time to see over again, I would adopt the new non-swarming hives, which are so much talked of, and thus remove what I have always regarded as the only drawback to my bee-keeping; but my old hives - all good ones - are still serviceable and will last my time.

I attribute much of my success to the fact of having made it a rule to try and keep the bees healthy by giving every hive and colony a thorough spring-cleaning. I start this about April, and begin by transferring the bees and combs into a clean hive, after scraping top bars of frames and removing all propolis, brace comb, etc.; then the hive from which the first lot were taken undergoes a thorough scrubbing with soft-soap and soda. If time permits, a coat of paint is given to the outside. It is then ready for the next stock; and so I go right through the apiary in this way, not omitting to slip a few bits of naphthaline into each hive, and always clipping a wing of the queen if I chance to see her. I never leave the bees without a cake of candy during the winter months, whether they have enough honey stored or not, and I find they always take it. My bees are invariably healthy, and it has been my good fortune never to have seen "Foul Brood".

As swarming time approaches, any stocks that are not exceptionally strong are united after dredging the bees with flour, which makes the task quite easy. This done, they are soon all ready for supering. When the inflow has started, I chiefly use shallow frames for extracting, but work one or two hives for sections only.

With regard to disposal of the honey crop, I have never had the least difficulty in selling all the honey I could get - indeed, could always sell more; have never kept any from one year to another; but, then, no pains are spared to put the honey on the market in attractive form and thoroughly ripened. Last, but not least, I charge only a fair selling price.

I am not without hope that my experience may be of some service in drawing the attention of ladies to the subject of beekeeping, and, although none of us may reach perfection, we may - if suitably situated, and determined to work earnestly and intelligently - depend on some profit, and will most certainly derive great delight from this soul inspiring and intensely interesting study of that industrious little labourer, the honey-bee. With good wishes for a prosperous season in 1903 to all fellow bee-keepers."

Mrs Wootton was hardly "a gardener's wife" as she and her husband were members of the Nottinghamshire landed gentry and owned Widmerpool Hall.

NEP 15 May 1903
Newark Agricultural Show Second Day. What undoubtedly the finest exhibition ever promoted by the Newark Agricultural Society entered upon its concluding stage this morning, when conditions were of a more pleasant description than on the preceding day. Later in the day the weather became somewhat variable, but success seemed assured. Lectures on beekeeping and demonstrations with live bees were continued, under the auspices of NBKA, and advanced demonstrative lectures on extracting honey and wax and preparing the same for sale, with modern appliances, were also given.

East Bridgford Parish magazine June 1903
The Flower Show will be held on June 30th, when it is hoped we shall be favoured with better weather than was experienced last year. As an additional attraction to the Show, the Committee have arranged with the NBKA for an exhibition of bees, with a lecture on bee-keeping. The following Prizes will be given:

Class		1st	2nd	3rd
1	Best four lb sections of Comb Honey produced in any year	5/-	3/-	2/-
2	Best four lb bottles of run or extracted Honey produced in any year	5/-	3/-	2/-
3	Best specimen of bees, any race, to be exhibited, living with their queen in a Unicomb Observatory hive	7/6	5/-	2/6

Entrance Fee 1/6. Members of NBKA and EBHS. 6d each exhibit. Entries to be made to Mr. H. Goldston on or before June 24th. Exhibits to be delivered on Show ground before 10.30am, and not removed before 7pm.

EBHS – East Bridgford Horticultural Society

BBJ 25 Jun 1903
The sixty-fourth Annual Exhibition of the RASE (Royal Agricultural Society of England) opened auspiciously on the 23rd inst., at the Society's permanent show yard, Park

Royal, London. Messrs. TW Cowan and Walter F. Reid judged the bee-appliance and miscellaneous classes; those for honey and honey trophies being taken by Messrs. Henry Jonas and P. Scattergood.

The following were the awards

(Classes 411 to 414 confined to Notts., Lincs., Rutland, Cheshire, Derbys., Staffs.)

Class 412 - Twelve 1-lb Jars (Light-coloured) Extracted Honey - 2nd, AG Pugh, Beeston

Class 414 - Twelve 1-lb Jars Granulated Honey - r., D. Marshall, Cropwell Butler

Minute Book – Quarterly Committee Meeting held at the Peoples' Hall on July 4th, 1903. Present: Mr Pugh in the chair, Messrs Smithurst, Swann, Skelhorne, Turner and secretary. Letters of apology for non-attendance were read from Messrs Ellis, Scattergood and Rawson.

The minutes of the previous meeting were read and confirmed by the signature of the chairman. The following matters arose out of the minutes:

> Advertisement card for members for sale of honey. Owing to the paucity of members and the fact that no design except the secretaries was submitted, it was decided to leave this matter to a future meeting.
>
> Badges for Committee. The secretary had obtained a number of badges as requested and these were approved by those present, the cost of same being 3d each.
>
> Providing extractor. It was considered that this was scarcely an appropriate time for this and the matter was ordered to be left on the table.

The quarterly account was read and considered very satisfactory and Mr Smithurst proposed, Mr Turner seconding the same be accepted.

Mr Skelhorne proposed, Mr Swann seconded
> that Mr Turner be asked to undertake the judging at Southwell.
> that Mr Scattergood be asked to judge at Moorgreen with Mr Turner as his assistant.

BBJ 23 Jul 1903

The twenty-first Annual Honey Show of the Hunts. BKA was held on July 15th in the admirably appropriate surroundings of Buckden Towers, near Huntingdon, the seat of Colonel Sir Arthur W. Marshall, Chairman of the Association. The exhibits, though not numerous, were of first-class quality. Mr. Peter Scattergood, Stapleford, officiated as judge.

Sheffield Daily Telegraph 28 Jul 1903

The Annual Show of the Welbeck Tenants' Agricultural Society will be held in Welbeck

Park on August 4th. There will be exhibitions of bee driving, and lectures in bee management, arranged by the BBKA and NBKA.

East Bridgford Parish magazine August 1903
The Horticultural Society held its 40th Annual Flower Show on June 30th. The weather was gloriously fine, and a good crowd from all the surrounding villages gathered in the Show Ground. On the whole the show was a good one, considering the cold winds and sharp frosts of early spring. An additional attraction was provided in the shape of a bee tent with a lecture on beekeeping by Mr. Scattergood, of Stapleford, and also an exhibition of honey and bees.

Results of beekeeping classes were:
        Class 1 - 1st, Mr Falconbridge; 2nd, Mr WD Gower; 3rd, Miss Fox
        Class 2 - 1st, Mr Mackender; 2nd, Mr D. Gower; 3rd, Mr Turner
        Class 3 - 1st, Mr Mackender; 2nd, Mr H Mackender; 3rd, Mr. J. Higgs

Derbyshire Times 1 Aug 1903
The Annual Exhibition of Welbeck Tenants' Agricultural Society on the 4th inst., promises to rival its predecessors in point of excellence. The Duke of Portland takes the greatest personal interest in the show and its arrangements, and always makes a point of entertaining distinguished guests for the show. This year General Lord Grenfall will visit their Graces and attend the exhibition. Demonstrations in beekeeping by members of the BBKA and NBKA.

Francis Wallace Grenfall, 1st Baron Grenfall became Governor of Malta with the local rank of General on 1 January 1899, serving as such until early 1903. The 1902 Coronation Honours list on 26 June 1902 included his name and he was created Baron Grenfell, on 15 July 1902.

BBJ 20 Aug 1903
Leicestershire BKA held its Annual Honey Show in connection with the Abbey Park Flower Show at Leicester on August 4th and 5th. The weather was all that could be desired, and the whole exhibition was a great success. Some idea as to its popularity may be gathered from the fact that over 33,000 visitors passed the turnstiles in one day alone.

The honey department was about up to the usual average and reflected great credit upon those who were responsible for the arrangements. Lectures and demonstrations were given on both days by Mr. P. Scattergood. Notts. and Mr. Riley, Leicester, who also officiated as judges of the bee and honey department.

Sheffield Daily Telegraph 9 Sep 1903
The Annual Show of the Greasley, Selston, and Eastwood Agricultural and Horticultural Society and the NBKA, was held yesterday on the show ground, Moorgreen.

Fortunately for the coffers of the society, the inevitable rain did not begin to fall until a large number had passed the turnstiles, thus ensuring success in that branch. The NBKA had 13 classes, and 74 entries. The judges for honey and bees were Messrs. P. Scattergood and RJ Turner.

Derbyshire Times 12 Sep 1903
Moorgreen Agricultural Show. Successful Exhibition. NBKA joined with the societies for the purposes of the show, and an interesting exhibition of bees and honey had been arranged.

BBJ 17 Sep 1903
NBKA Annual Show was held at Moorgreen on September 8th in connection with that of the Agricultural and Horticultural Societies. Unfortunately, the day was a very wet one, but in spite of this some thousands of people were present. Mr. P. Scattergood, Stapleford, was the judge assisted by Mr. RJ Turner, Ratciiffe-on-Trent, and their awards were as follows:
Collection of Hives and Appliances (1 entry) - No 1st prize awarded; 2nd, (£1) GH Varty & Co., Colwick
Complete Frame Hive for General Use (1 entry) – 1st, (10s) GH Varty & Co.
Trophy of Honey in any Form and of any year (2 entries) – 1st, (15s and NBKA silver pendant) Geo. Hayes, Beeston; 2nd, (10s) G. Marshall, Norwell
Six 1-lb Jars Light-coloured Extracted Honey (19 entries) – 1st, (10s and BBKA silver medal) Geo. Hayes; 2nd, (5s) GH Pepper, Farnsfield; 3rd, (2s) Miss H. Bingham, Clipstone; hc, Miss Ede J. Morris, Worksop
Six 1-lb Jars Dark-coloured Extracted Honey (12 entries) – 1st, (10s and BBKA bronze medal) Miss H. Bingham; 2nd, (5s) Geo. Hayes; 3rd, (2s 6d) G. Marshall; hc, J. Bee, Southwell
Six 1-lb Sections (6 entries) – 1st, (7s 6d) Geo. Marshall; 2nd, (3s 6d) GH Pepper
Six 1-lb Jars Granulated Honey (11 entries) – 1st, (7s 6d and BBKA certificate) Miss H. Bingham; 2nd, (5s) AG Pugh, Beeston; 3rd, (2s 6d) H. Merryweather, Southwell
One Shallow Frame of Honey for Extracting (3 entries) – 1st, (5s and NBKA silver pendant) GH Pepper; 2nd, (2s 6d) G. Marshall; hc, Jesse Smith, Kimberley
Six 1-lb Jars Extracted Honey (novices, 3 entries) – 1st, (7s 6d) G. Hopkinson, Newark; 2nd, (5s) RC Cooper, Eastwood; 3rd, (2s 6d) P. Francis, Dunham
Six 1-lb Jars Extracted Honey (local class, 3 entries) – 1st, (7s) GM Bolton, Eastwood; 2nd, (4s) W. Brooks, Eastwood; 3rd, (2s) RC Cooper
Honey Vinegar (1 entry) - 1st, (3s and NBKA certificate), Geo. Hayes
Observatory Hive with Bees and Queen – 1st, (15s) R. Mackender, Newark; 2nd, (10s) H. Mackender, Newark; 3rd, (7s 6d) D. Marshall, Cropwell Butler; 4th, (5s) Cecil Hayes, Beeston; hc, G. Hopkinson, Newark
Beeswax – 1st, (7s 6d) GA Hill, Balderton; nd, (5s), Geo. Hayes

BEEKEEPING BETWEEN TWO QUEENS

Cecil Hayes born in Beeston in 1895. Could this be George Hayes' young son (aged eight)?

BBJ 1 October 1903
As a young bee-keeper and reader of BBJ, would you kindly give me your advice on the following? An old skeppist in the village was going to take some honey by destroying bees over a sulphur pit, so I begged him to let me drive them instead, which I did according to instructions given in the "Guide Book." I brought one lot home and put them in a frame-hive on eight frames, ie. five fitted with full sheets of foundation, the other three filled with built-out combs (clean) taken from a skep belonging to another friend. The bees were then fed on syrup made as per "Guide Book." Three days afterwards I drove another lot, and united them to the first lot in the said hive. I intended to kill the queen as she ran in, but did not see her enter the hive, so have left the question of supremacy to be settled by the bees. This was last week, and I have continued feeding them as above.

Are the bees likely to draw out any comb this year, or will they fill comb given them with syrup, and how can I feed them, and on what, to ensure keeping them through the winter? I do not mind any amount of trouble with them, and I have a good reserve of patience, and am anxious to keep them if possible, having saved them from one certain death. Your advice will be strictly adhered to by (name enclosed for reference)
Worker, Ollerton
Reply. The question of building out combs largely depends on the number of bees in the hive. If the whole eight combs are thickly covered with bees, if feeding is liberal, and a rapid-feeder is used, it is very likely that some at least of the five frames of foundation will be built out. On the other hand, the three combs given will be filled with syrup, and the cells facing the foundation probably lengthened out, before being capped over. This is the one drawback in mixing up combs and sheets of foundation in building up driven bees into stocks in frame-hives. You must feed freely with good warm syrup, and keep a watchful eye on the progress made in comb-building; then when packing down for winter give a good sized cake of soft candy.

Minute Book – Quarterly Committee Meeting held at the Peoples' Hall on October 3rd, 1903. Present: Messrs Ellis (Chairman), Smithurst, Harrison, Skelhorne, Randle, Turner, Hallam and secretary.

The minutes of the previous meeting were read when Mr Smithurst proposed and Mr Turner seconded that the same be confirmed by the signature of the chairman. The Balance Sheet was next gone into, the same being accepted as very satisfactory.

Letters were read from various members of the committee regretting their inability to attend the meeting.

A letter from the "Teachers Charities Bazaar" asking for honey, was read and it was agreed that as this association had no actual dealings with honey, it could not make the grant and, as it was also found that several members of the committee were helping and contributing individually, all they could do was to make the request known among such as were not doing so and the secretary was instructed to suitably reply.

A suggestion from Mr W Bell of Eagle Hall about prizes for best kept apiaries was read and although the committee saw a germ of some advantage in this they felt unable to carry it out for lack of sufficient funds.

Correspondence was read next a complaint emanating from the South..... but as the complainant withdrew his accusations the matter was considered to be ended.

It was also resolved that the matter of advertisement card for honey be left over to the Annual Meeting.

BBJ 8 Oct 1903
GW (Sandiacre) Dimensions of the "WBC" Hive. Correct dimensions and measurements (drawn to scale) of the hive named will appear in a new edition of the "Bee-keeper's Note Book," now being prepared for publication, and will be issued in a few weeks.

BBJ 15 Oct 1903
The second Annual BBKA Conference of representatives of County Associations and the Council of the BBKA was held on October 8th, at 4 o'clock, in the Board Room of the RSPCA. The Chairman of the BBKA (Mr. TW Cowan) presided, and was supported by the following members of the Council and delegates from county associations: Notts. Geo. Hayes and P. Scattergood.

Mr. Scattergood (Notts) said the subject (bee-keeping) was one that could be properly taught in every village in England at evening continuation schools. He held in his hand the Board of Education Syllabus and List applicable to Schools and Classes other than Elementary. It was for the year extending from August 1st, 1903, to July 31st, 1904; and it contained a summary - fairly up-to-date - for teaching bee-keeping in English schools. The remarks in the syllabus were interesting, and if suitable persons to undertake the teaching could be utilised, much good might be done to young people, and in this fact a great future was assured for the bee-keeping industry. The difficulty was, however, that the village schoolmaster generally thought he knew everything. He (Mr. Scattergood) spoke from experience, and was sure that jealousy would be engendered if a stranger were called in who did not happen to be a qualified schoolmaster. He wished it were possible for the subject to be classed as an elementary science, and taught theoretically as well as practically to young men, who would thus be started on the road (if they cared to pursue it) to obtain expert certificates. A great effort should be made to get apiculture included in the syllabus of the counties.

The County Council made a grant direct to the County BKA and had a representative on the executive committee thereof. The amount given was only £50, and it was generally admitted that the Association spent the money more economically than any other institution in the county did its grant.

Mr. Geo. Hayes (Hon. Secretary, NBKA) said that at Kingston Agricultural College they had started a bee-keeping class embracing four counties, which was, he thought, a better plan than the schoolmaster's teaching of young people. He was glad to say they were in more fortunate circumstances as regards County Council work in Notts. than some other counties were. In his county (Notts.) the grant was given to the association. That was satisfactory, as the latter would be responsible for carrying out a work which it was always endeavouring to foster. The best body to dispense the money was undoubtedly the county association.

Mr. Scattergood said that competitors had to consider that, besides entry fees, there was the cost of conveyance to the show, and that expense made them hesitate to enter or when several fruitless attempts had been made, they got disheartened.

On the other hand, some exhibitions were not creditable to the bee-industry, which would be better without them, and at such shows competitors did not secure prizes because they did not deserve them. He had judged at four or five county shows this year, where much of the extracted honey was excellent, while there were some exhibits having 1in. of froth at the top of the jars, an evidence of slovenliness and carelessness in making ready for the show. He recommended that beginners in bee-keeping should be taken to exhibitions like the "Dairy Show," as the splendid specimens on view there would be likely to fill them with ambition to become successful on the showbench.

He was also strongly in favour of county labels, and had received a lot of letters on different occasions asking where honey could be purchased bearing the Notts. County label which showed that the label was a guarantee of quality, and the use of it a good advertisement. The plan could not be carried out in London, but might be enforced at all county association shows. He suggested more varied classes for appliances at county shows, not a huge class like they had at the "Royal"; also that there should be a class for a "Complete outfit for a beginner," not making the cost too high or too low.

In 1900 the Agricultural Department of the University College, Nottingham, moved to Kingston-

on-Soar and combined with the Midland Dairy Institute to form the Midland Dairy and Agricultural College.

BBJ 15 Oct 1903
Your kind replies to my first letter encourages me to ask further questions as follows: A few days ago I drove a skep of bees and found they were queenless. I united them to a stock in a frame-hive by simply throwing them out of the skep on to a board and placing it level with the alighting-board. A few puffs of smoke and the bees ran in, no flour or anything else being used, and they joined up beautifully and peacefully. Was there anything unusual in this?

1. I want to make some "Cowan" hives, but am "fixed" over the dimensions. The "Guide Book" says: "The front and back are reduced to $8\frac{1}{8}$ in. high." Now, if this is so, seeing the measurement of standard frame is (from under top bar to extreme bottom of frame) exactly $8\frac{1}{8}$ in., how do I provide the bee space between floor board and bottom of frame?
2. Which month do you consider would be best to re-queen two stocks of native bees with pure Italians?
3. Could you give me the address of nearest B.K. association I could join? Is there one in Lincoln?
4. How many pounds of candy (for seven frames covered with bees) would last a stock through the winter if packed down now, and which have no stores in brood chamber?
5. I enclose name for reference, and sign as before

Worker, Boughton, October 12th

Reply.
1. Nothing very unusual; but it would have been safer to adopt precautions, as the luck might easily have gone the other way.
2. Keep to the dimensions given in "Guide Book"; but do not omit noticing that the frames hang on metal runners, which raises the bottom-bar from floor-board as required.
3. Autumn is the best time for re-queening.
4. Mr. George Hayes, Beeston, Notts, is the Hon. Secretary of your County Association.
5. Your bees are badly prepared for wintering if they have no stores in the combs of brood-chamber in October. You would need to give cakes of soft candy (weighing, say, 2 lb each), renewing as often as taken down by the bees.

BBJ 29 Oct 1903
Bee-Keeping in Schools. Interesting the Children. After reading through the account in BBJ, of last week's meeting of the BBKA, I thought your readers might be interested in our little effort. Bulwell is an outlying suburb of Nottingham and although within the city boundary, is as it were a town in the country.

Taking advantage of this fact, and in view of the Board of Education's scheme of object lessons, Mr. Barker, the headmaster of the Coventry-road Council School, has arranged a list of object lessons, suitable for the neighbourhood. Among these are three lessons on bees, and as "the bee-man" it fell to my lot to give the instruction in question. The lessons are given on Wednesday mornings to over 100 boys in standards five, six, and seven, and the following Wednesday each boy writes an essay on the previous week's lesson. The first was entitled, "Hive Bees - Their Natural History." A fortnight later we took "Hive Bees - Their Work in Nature," and last week we had "Hive Bees - Commercial Production."

The boys have been keenly interested in these lessons, and the promise of a small jar of honey for the best essay on the last lesson, has spurred them on to great exertions.
W. Darrington, Bulwell

William Darrington, Elementary Schoolmaster, b. 1868, Pye Hill, Notts.

BBJ 2 Nov 1903
In your condensed report of my speech at the delegates' meeting of the BBKA I notice you state that I quite agreed with the member from NBKA in his remarks about schoolmasters. This is not correct. I said I agreed with him in his remarks about children, and emphasised that we had to look to them as the coming bee-keepers. I said that "the schoolmaster would be the best teacher if he could be got to do it, but feared that, with the present demands of the Education Department, he had quite enough to do", etc.

I made these remarks re. schoolmasters after a long experience with many of them. As most lectures are given in a schoolroom, the schoolmaster is nearly always in evidence, and I have always endeavoured to persuade them to take up bee-keeping themselves, but have either been met with the remark "that bees did not like them," or "that their time was so fully occupied that they could not." We have one schoolmaster in our Association with a first-class certificate, and several are local secretaries, and if we could get every schoolmaster to take up the subject in a practical way we should do so. With respect to the two schoolmasters in the last two issues (of BBJ) who have so readily got their skeps in evidence, they certainly do not represent the majority of their profession, who are far more courteous and urbane, even if they do not keep bees.
FJ Cribb, Lincs. BKA

BBJ 5 Nov 1903
Schoolmasters and Beekeepers. Referring to Mr. Scattergood's remarks at the London "Conference" on the "conceit" of village schoolmasters, may I say that, although a co-worker with Mr. Scattergood, I did not agree with him on this point, as

evidenced by my remarks at that meeting? When supporting what was advanced as regards teaching children bee-keeping, I said that the best way of doing this was to get schoolmasters and schoolmistresses interested in the matter, and they in their turn would teach the children.

I do not find schoolmasters so "conceited" as our friend would make it appear; in fact (as I have personally told him) my experience has been quite the reverse. We have in our association and on our various committees several schoolmasters, and have certainly no cause for complaint under that head. But, apart from this, I have had under my tuition in the "Rural economy" courses at the Kingston Agricultural Institute during the last summer, between 50 and 60 schoolmasters and schoolmistresses who never showed this characteristic if they possessed it, but were most interested and anxious to learn what they could of the subject in hand.

I have also done a good deal of lecturing with the lantern, and not seldom has the schoolmaster brought some of his scholars to listen to the lecture, and in some cases, has, to my knowledge, previously instructed them to write an essay on the subject afterwards, thus showing his interest in beekeeping as a subject for children.

In my opinion, to further our cause, we must lay the foundation in the minds of the rising generation, and how can we better do this than through the medium of the schoolmaster or schoolmistress - even though we may come across one now and again who may show a little conceit. George Hayes, Secretary, NBKA, Beeston

Minute Book – Committee Meeting held in the Peoples' Hall on January 9th, 1904. Present: WS Ellis (chairman), Messrs Smithurst, Harrison, Pugh, Puttergill, Skelhorne, Turner, Wadsworth, Scattergood, W Herrod and secretary.

The minutes of the previous meeting having been read, Mr Turner proposed and Mr Harrison seconded the same be confirmed by the signature of the chairman.

A letter was received from his Lordship stating his inability to be present at the Annual Meeting and expressing his satisfaction with the financial state of the association.

Correspondence with the BBKA was read next about the medals and pendants and it was left in the hands of the secretary to deal with. The secretary intimated that he had just received the certificates for third-class experts for Messrs. Turner, Ratcliffe, W Marshall, Cropwell Butler and Mr Darrington, Bulwell and also that our esteemed friend Mr W Herrod had gained his first-class certificate. Mr Scattergood proposed and Mr Pugh seconded that the hearty congratulations of this committee be and are hereby tendered to Mr W Herrod on his obtaining the highest diploma.

BEEKEEPING BETWEEN TWO QUEENS

The annual Balance Sheet was next gone through. This shewed a cash balance to the good of £1 14s 2d (a rather unusual thing) as the whole was considered satisfactory. Mr Scattergood said that having audited the accounts he had found them quite correct and proposed that the same be received and printed on the circular convening the Annual Meeting and sent to each member. Mr Wadsworth seconded.

Mr Hayes, who went as delegate to the BBKA meeting in London in October, gave a short resumé of the two items down for discussion and laid special stress on the matter of showing at the Royal.

The following arrangements were then made for the Annual Meeting:
> that it should take place at the Peoples' Hall on February 20th
> that the afternoon meeting should be held at 3pm presided over by WS Ellis
> that the evening meeting should begin at 6pm presided over by the Mayor (Alfred Page) or Colonel Rolleston, or the Sheriff (Robert Fleeman)
> that the tea should take place at 4.30pm
> that Mr Ellis (who kindly consented) should arrange with caterers for same
> that two competitions should be held as usual
> that if more than 20 lbs of honey were brought it should be divided between the Children's Hospital and the General Hospital. If less than 20 lb all to go to the Children's Hospital
> that the evening meeting should take the same form as last year (Mr Herrod kindly consenting to give three short talks on "Apiaries at home and abroad", "Wax - it's production and preparation" and "Showing")
> that the annual prize drawing take place as usual and that the secretary make the best arrangement he can for it.

The following gentlemen were appointed to arrange the musical portion: Messrs Harrison, Ellis, Skelhorne and Hayes.

Col. Lancelot Rolleston, DSO, 1902; JP.; Colonel comm. South Notts Hussars, Imp. Yeomanry; b. 1847 Greasley; d. 1941 at Watnall Hall. This was built about 1690 and was demolished in 1962.

The Rolleston Window in Greasley St. Mary's church was dedicated in 1960 to the memory of Lancelot Rolleston and his wife, Emma Maud Charlotte. The window poignantly incorporates coloured glass taken from Watnall Hall and an illustration of the family home from around 1690.

Robert Fleeman (1847-1912) was Sheriff in 1903/4. He was born in Eccleshall or Walton in Staffordshire. He was an auctioneer and general dealer but by the 1911 Census he was stated to be a Managing Director (business undecipherable)
BBJ 21 Jan 1904
Shows and Showing. I do not think Mr. Weatherhogg treats our friend Mr. Hales quite

as fairly as he might have done. Mr. Hales stated that "the general public would sooner see an exhibition of honey of the average quality than have the same exhibit (prize) hawked from show to show". I would venture to say that the general public would much rather see a show of the average quality of honey than the "not for competition" class as suggested by Mr. Weatherhogg, unless such class is reserved exclusively for exhibits that have already taken a prize.

I ask: Why should exhibits that have already taken a prize be allowed to compete with those that have not done so? The successful exhibitor receives the reward for which he tried, why, then, should he be paid over and over again for the same exhibit? If the exhibitor must hawk his prize exhibits from show to show, let there be a class for such exhibits; and we shall then find out who shows the best produce during the season.

The entries for the various shows will then increase and such competitors as Mr. Hales will then have a better and fairer chance of finding the "missing link" mentioned by Mr. Weatherhogg, whose remarks are (to put it mildly) a trifle sarcastic. He would, to my mind, have shown better taste by recommending our friend to try, try, and try again, instead of endeavouring to crush him with sarcasm.

I do not know either of the gentlemen referred to, but readers in general will, I fancy, endorse the impressions of Mr. Hales as to the objects of shows and showing, in reference to the suggestions offered by Mr. Weatherhogg, whose prize exhibits must form a veritable El Dorado for him. EG Ive, Ollerton

BBJ 28 Jan 1904
Shows and Showing - Increasing entries. Along with other BBJ, readers, I have been much interested in the correspondence on shows and showing, and in my opinion one of the main reasons why small producers, or owners of a few hives, do not show is the expense entailed. Someone has to go without prize-money, and after paying fees of say 2s or 2s. 6d for each entry, carriage to show and back - the latter a no mean item for long distances - while my experience proves that exhibits are seldom returned at owner's risk rates.

There is also loss of time in preparing sections and honey in jars for the show-bench, and frequently more or less damage, with occasionally entire loss of same. Then, if no prize is secured, the loss is considerable for a cottager or artisan. Personally, I am inclined to agree with Mr. Turner that it would be better in most cases to have exhibits of 6 lbs instead of 12 lbs. This would, at any rate, lower the cost of carriage. Then, again, for the owner of two or three hives, whose output of honey is comparatively small, it is not worth his while to exhibit in order either to sell his produce or attend shows to take orders.

I hoped we were going to have some really useful suggestions as to selling honey when the recent correspondence on "price of honey" commenced, but, so far, the result is disappointing. I have myself sometimes feared that the retail price might come down so low as to make beekeeping unremunerative, partly in consequence of foreign competition, along with that among ourselves at home; but if prices do get as low as this, it will, in my opinion, be the fault of bee-keepers themselves.

If more care was taken generally in grading all our honey, in order to secure a good price for the very best of it, I feel sure a market could be secured, for those able to afford it are generally willing to pay a fair price for a really good article. The lower grades could then be sold at a lower price, and the result would be a fair average.

Another cause of low prices is want of co-operation, and this is evident even among members of county associations, some of whom neglect to take full advantage of membership. To give one instance, I may say, in our own BKA (the Notts.) we have a system of free advertisement of honey, etc., among members; yet one member of our association sold 2 or 3 cwt of good honey at 3d per lb "to get rid of it," while at the same time another, located only about two miles away, had not enough honey to supply his customers, and did not know where to obtain any. He would, no doubt, have been very pleased to purchase some from the other member at double the price it was sold at had he known about it.   J. Herrod, Sutton-on-Trent

BBJ 18 Feb 1904
"Owner's Risk" Rates. Referring to Mr. W. Woodley's "Notes by the Way", I think he has been wrongly informed as to owner's risk rate for honey from shows. I have done practically no "showing" for the last year or two, but I was always able to get honey returned from a show when I was there to pack and book it myself. My honey in this way, has also, as a rule, been returned at owner's risk from shows in connection with the NBKA. I fail to see any reason why the rate should not apply to honey returned from any show. Should I do any showing in the future the return label will be plainly marked "Owner's Risk," and I shall refuse to pay any higher rate.
                                              J. Herrod, Sutton-on-Trent

Minute Book - The Annual Meeting of the NBKA was held on February 20th, at the Peoples' Hall. Mr. WS Ellis presided, and among those also present were: Mr. and Mrs. George Hayes, Mr. and Mrs. RJ Turner, Messrs. P. Scattergood, W. Herrod, AG Pugh, SW Marriott, T N Harrison, T. Maskery, Puttergill, GE Skelhorne, Mr. and Mrs. R. Mackender, Mr. and Mrs. T. Manchester, and many others.

The Balance Sheet showed that the receipts had amounted to £116 3s 4d and there was a balance in hand of £1 14s 2d. The Annual Report of the committee showed an increased membership, which now stood at 185, as compared with 150 a year ago. Another matter for congratulation was the fact that the Balance Sheet showed something to their credit, which it had not done for several years past.

Expert work last spring was delayed by the bad weather but out of 189 apiaries visited, comprising 754 hives, only 25 were found to be infected with "Foul Brood," which was very satisfactory.

The County Council had reduced their grant by £10 last year, owing to increased expenditure in other directions, but the committee hoped the grant for 1904 would be increased to £60. The City Council had resumed their grant for this year. The attention of the members was invited to the grouping of counties at the Royal Show in June next, with the hope that more of their members would enter for the prizes that "had to go begging" in 1903 through lack of competitors. Three members had secured the third-class honours as experts, viz., Messrs. RJ Turner, W. Darrington and D. Marshall.

Votes of thanks were passed to the County Council and to the City Council for their grants for the furtherance of the knowledge in bee-keeping; with the President for the substantial help he had given to the association, to the vice-President, and to the various donors who had assisted them, not forgetting the very valued help of some of the district secretaries.

The Report and Balance Sheet were adopted.

Lord St. Vincent was re-elected President by general acclamation. The Vice-Presidents and committee were also re-elected. Mr. P. Scattergood was re-elected auditor, and cordially thanked for his past services, Mr. Hayes being the recipient of a similar compliment upon his re-appointment as secretary. Messrs. Hayes and Pugh were re-elected delegates to the BBKA.

In the evening a social gathering took place, presided over by Colonel Rolleston, and a very enjoyable evening was spent. The meeting was concluded with the usual prize drawing for hives, etc.

BBJ 25 Feb 1904
Leaky Roofs. One can readily understand the feelings of Mr. W. Loveday, and others, with regard to the spirit displayed by your correspondent, "WH", re "Leaky Roofs" I would call his attention to a previous communication from "DMM" and respectfully ask him to note the difference between his reply to "DMM" and that of Mr. J. Gray. This was not simply a question of hive-covering, but the hive itself, and to his credit Mr. Gray immediately responded by offering the use of it. I am led to ask, did "WH"

send any "salt" with that letter, or must we find our own? For when he tells us that it was out of consideration for the correspondence column, we need it - the salt, I mean. Be honest, "WH" Did you give our Editors a chance of gratuitously publishing that real remedy with full particulars and drawings? I can hardly think so, as it is not a question of space when it comes to anything likely to be beneficial to the readers of the BBJ, proof of which we have weekly.
E. Geo. Ive, Ollerton

DMM was DM Macdonald of Banff. He contributed countless articles for many years to the BBJ as will be seen in these pages. His comments were taken up by members of NBKA.

A public advertised meeting was held in the Agricultural class room, Marischal College, Aberdeen on 4th June 1910 at 2pm, with the object of forming a Beekeepers' Association for Aberdeenshire. Mr DM Macdonald, Morinsh, presided over a large and very representative meeting. After an excellent address from the Chairman, it was proposed and seconded and unanimously carried to form an Association for Aberdeenshire and District.

BBJ 3 Mar 1904
Home-Made Extractors. Referring to the subject of homemade hives, one finds a certain amount of pleasure in reading communications from such generous correspondents as Mr. S. Darlington, and I take it he is only waiting for someone to say they would like to have particulars of how to make an extractor large enough to take four Standard frames, and he is ready to supply us with the necessary information. I should like, with your permission, to suggest to Mr. Darlington that the present time would no doubt be found suitable by most readers for perusing the promised particulars, so that they might have an opportunity of making an extractor for themselves before the busy season demands their spare time in a different direction. It might also console Mr. Darlington to know that I have just finished making four "home-made 'uns" this winter, which are quite ready for either rain or shine to be looked at them. I would also add that I have bought several hives which have given the greatest satisfaction, and they were cheap ones, but then I got them from a reliable dealer.
E. Geo. Ive, Ollerton

BBJ 17 Mar 1904
The Annual Meeting of Leicestershire BKA was held at the Victoria Coffee House, Leicester, on the 5th inst. Among those present were Messrs. AG Pugh and Geo. Hayes of the NBKA.

Minute Book – Committee Meeting held in the Peoples' Hall on April 26th, 1904. Present: WS Ellis (in the chair), Messrs Harrison, Marriott, Pugh, Puttergill, Skelhorne, Hindle, Mackender, Turner and Gray.

The minutes of the previous meeting were read. Mr Pugh proposed and Mr Gray seconded the same be confirmed. The quarterly Balance Sheet was next presented.

Mr Gray proposing and Mr Ellis seconding the same to be received as satisfactory.

Several apologies for non-attendance were read and also a communication from the BBKA asking for some statistics which had been supplied by the secretary.

Mr Harrison next proposed that the sub-committee for dealing with the County Council grant should be re-elected *'en bloc'*. After some discussion Mr Gray amended same by adding the name of Mr Harrison. Mr Turner seconded and the amendment was carried. The secretary here reported that he had received an intimation that the grant this season from the County Council was to be £40.

Correspondence was read next holding the Annual Show at Southwell and Moorgreen and after considerable discussion, it was finally agreed to hold this with the Moorgreen Horticultural Society, they offering the same terms as last year. It was also agreed that we write to the Moorgreen Society to grant the exhibitors free admission into the ground and that members be charged an entry fee of 6d each for the first two entries and 3d each for all further entries, that exhibitors may be allowed (in *sub-camera*) to sell their honey at 4.30pm. Some few alterations were made to the list of prizes including the striking out altogether of the local class owing to the fact that the local subscriptions which previously upheld this class were all withdrawn and Mr Harrison proposed and Mr Gray seconded, that the schedule as altered be adopted.

Mr Pugh proposed and Mr Skelhorne seconded that the secretary be empowered to make arrangements for any other shows he may think fit and to the advantages of the association and as far as the funds will permit.

Mr Harrison, with a view to bettering our Annual Meeting, next brought forward a few points when he considered we failed in some degree. These were:
1. Impromptness in commencing meetings
2. A superior tea
3. That winners in competitions be debarred for a time
4. Bad arrangement of room
5. Missing of an item from programme
6. Rendering of thanks to artists.

Mr Pugh suggested that to save valuable time a quicker mode of drawing be adopted. Mr Gray promised to help.

BBJ 5 May 1904
I commenced bee-keeping at Nottingham in 1894, and like many other beginners I bought my first stock in the autumn instead of spring. The stock in question consisted of about half a pint of bees in an ill-fitting frame-hive, with practically no food in

stores. As might have been expected, the bees died before Christmas, and, having been taken in through ignorance, I determined that this should not occur again.

I therefore began taking the BBJ, and read 'Modern Bee-keeping' and all other bee books I could get hold of. Thus prepared, I made another start the following year on the right lines, and succeeded in getting together a nice little apiary; when disaster overtook me in the shape of a regular flood. My bee-garden was three miles from home, and I had put the hives in a hollow for shelter, and weighted them well down. Then we had a heavy fall of deep snow, followed by a sudden thaw and, later on, heavy rain, which drowned the lot!

Nor did this end my troubles. In the following spring I ordered two more swarms. One turned out queenless, the other arrived when I happened to be away from home and was left next door till my return. My neighbour, seeing a few bees escaping, promptly covered them over with a big hearthrug and stifled the lot! My wife now urged me, with some show of reason, I will admit, not to go in for any more bee-keeping; but being fully conscious that this chapter of accidents was not due to any lack of adaptability on my part for bee-keeping, I felt more determined than ever to have another try. That year I removed to my present home, and laid down the lines upon which I intended to work in the future.   GW Noble

Mr GW Noble, Journeyman tailor, moved to Sandy, Beds. "Altogether I had an apiary of over forty hives in 1901, which yielded 1 ton 3 cwt. of extracted honey, besides two gross of sections. That year I paid to a London firm £10 for glass honey jars."

Lincolnshire Chronicle 20 May 1904
Newark Agricultural Show. A bright morning gave hope for a fine day for the Annual Show of the Newark Agricultural Society, which opened on the Sconce Hills on Friday. There were lectures on bee-keeping, with demonstrations

NEP 25 May 1904
Notts. Agricultural Show. Today's Proceedings. Notwithstanding several slight showers in the early morning, the weather became delightfully fine soon after the proceedings of the Notts. Agricultural Society were again entered upon to-day at Ruddington Hall Park. The outlook, too, was promising, and there was every prospect of a big attendance. The early trains were crowded. Members of the NBKA gave lectures on beekeeping and manipulations with live bees.

Sheffield Daily Telegraph 26 May 1904
Notts. Agricultural Show. Second Day's Proceedings. The weather, though dull, remained fine for the second and concluding day at Ruddington. Members of the NBKA gave frequent lectures and demonstrations.

BBJ 9 Jun 1904
Nearly New Observatory Hive, "Howard's,' complete. What offers?

Miss Cockerham, Basford

BBJ 30 Jun 1904
The sixty-fifth Annual Exhibition of the RASE opened on 21st inst., at the Society's permanent show-yard, Park Royal, London.
Class 434. Twelve 1-lb Jars of Extracted Medium or Dark-Coloured Honey - 1st, AG Pugh, Beech House, Beeston
Class 435. Twelve 1-lb Jars of Granulated Honey - 3rd, AG Pugh

Minute Book – Committee Meeting held in the Peoples' Hall on July 2nd, 1904. Present: WS Ellis (chairman), Messrs Harrison, Pugh, Marriott, Skelhorne, Smithurst, Turner, Windle and secretary.

The minutes of the previous meeting were read and confirmed by the signature of the chairman. Mr Pugh proposed and Mr Gray seconded. The Balance Sheet for the quarter was next gone into, the same being considered satisfactory, was accepted. Mr Gray proposed and Mr Ellis seconded.

The sub-committee for dealing with the County Council's grant were re-elected *'en bloc'* with the addition of Mr TN Harrison. Proposed by Mr Gray and seconded by Mr Turner.

The question of holding our Annual Show was next dealt with and the offers of Southwell and Moorgreen societies considered and after a lengthy discussion, it was finally decided to accept the offer of the Moorgreen society. Proposed by Mr Gray and seconded by Mr Smithurst.

The following alterations were then made in the schedule and conditions. As there were no supporters of Class 10 (Local) it was decided to withdraw that class.
> that only one prize be given in the honey vinegar class and that be 4/-.
> Proposed by Mr Skelhorne and seconded by Mr Harrison.
> that the second prize in Class VI (Sections) be 5/-.
> that a third prize be added to Class VIII and also Class XIII (Wax) of 5/6d each.
> that entry fees be charged as follows – 6d each exhibit for first two entries, 3d each all further entries. Proposed by Mr Pugh and seconded by Mr Windle.
> that members may sell exhibits after 4.30pm. Proposed by Mr Skelhorne and seconded by Mr Pugh.
> that the secretary write to the Moorgreen Society for free admission for exhibitors.

BEEKEEPING BETWEEN TWO QUEENS

A letter from the BBKA and voluminous correspondence, etc. from Mr GM Saunders about the 'Foul Brood Bill' was discussed and it was agreed to write to the secretary of the BBKA and also Mr Saunders and say that, as the matter was one of great importance, it would be best to lay it before a General Meeting so that all might have an opportunity of a larger voice on the matter. It was also ordered that a draft copy of the Bill be sent to the County Council for their perusal to prepare them for any future steps we might take in this matter.

Geo M Saunders, Cumberland BKA

East Bridgford Parish magazine July 1904
At the show held on 28th June, the weather was exceptionally fine, and a good number, especially in the evening, patronised the Show. The exhibition was held in a large marquee and the judges pronounced it a very good show especially considering the late dry season.

Special prizes – Honey
    Sections – 1st, Mr Marshall; 2nd, Miss Fox; 3rd, Mr Gower
    Run honey – 1st, Mr Mackender; 2nd, Mr Gower; 3rd, Miss Fox
    Bees – 1st, Mr Marshall; 2nd, Mr Millington

BBJ 7 Jul 1904
The following is the grouping of counties for the Royal Show as arranged in classes of the published prize-list:
Classes 432 to 435 confined to Notts, Leics., Rutland, Cheshire, Derbys., Staffs., Yorks., Lincs., Northumberland, Durham, Cumberland, Westmoreland, the Isle of Man, Scotland, and Ireland.

Sheffield Daily Telegraph 23 Jul 1904
The Welbeck Tenants' Agricultural Society, the head of which is the Duke of Portland, will hold its Annual Show in Welbeck Park on August 2nd. There will be lectures on beekeeping by Mr. Scattergood, the BBKA and NBKA.

BBJ 4 Aug 1904
Fertile Queens, 1904 tested, 8s; very good strain. Cages free.
                                                        Stoppard, Grove Avenue, Chilwell

Derbyshire Courier 8 Aug 1904
The Welbeck Tenants' Show is always to be regarded as one of the most important events in the agricultural calendar. Lectures on bee-keeping by Mr. Scattergood, of the BBKA and NBKA, explained the best and most profitable methods of bee-keeping and showed appliances of the most modern type and how to use them.

BBJ 11 Aug 1904
During the past week I have destroyed the brood-nests of two stocks in the county belonging to members and one stock owned by a non-member. This would have meant 30s compensation; but instead of the members referred to desiring compensation, they were only too glad to get rid of "Foul Brood". In the case of the non-member, he simply asked me to look at the one suspected stock; and now in view of risks to his other eight hives, he may be glad to rejoin the NBKA, and be certain of help and advice when needed. J. Gray, Expert BBKA, Long Eaton

BBJ 18 Aug 1904
I have been asked to drive a skep of bees for an old lady so that she may take the honey. As there may be a considerable amount of brood in the skep at present, I am at a loss what to do with same after bees are driven. I might say she has no frame hives; she keeps nothing but skeps, so that I could not tie brood in the frames. Your advice would greatly oblige. GW, Sandiacre
Reply. Though no mention is made of the fact, we presume you are to have the driven bees and intend to put them in a frame-hive. This being so, the question is: Can you get the combs containing brood safely home without damage or chilling? If this can be done (and it will need great care in keeping warm and not damaging the cappings of sealed brood), you should either tie the latter in the frames or fix the combs up in a small box - supporting them by wooden pegs - and set the box above the feed-hole in a quilt. The bees will then cover the brood and keep it warm till it is hatched out. It is, however, seldom worth while trying to preserve brood in this way.

BBJ 1 Sep 1904
WD (Notts.). The light-coloured sample is good in colour and consistency, but it appears to have acquired an artificial aroma (like the scent of violets) from the bottle in which it came to hand, otherwise we cannot tell what the aroma comes from. The flavour is also peculiar, and we are at a loss to tell the source whence it is derived. The darker sample is very thin, as if extracted from unsealed combs. The flavour is also inferior. We should not mix the two samples, as the better honey would be deteriorated thereby.

BBJ 29 Sep 1904
The NBKA Annual Show was held in connection with that of the Moorgreen Horticultural Show on September 6th. The duties of judging the bee-exhibits were carried out by Mr. P. Scattergood, assisted by Mr. RJ Turner. This was not a heavy task, as the entries this year were fewer in number owing to the poor season generally in the county. With the exception of the first-prize exhibit, the majority of the honey shown was of a darker colour than usual. The following are the awards:
Collection of Hives and Appliances - 1st, GH Varty, Colwick
Complete Frame Hive for General Use - 1st, GH Varty

BEEKEEPING BETWEEN TWO QUEENS

Trophy of Honey in any form and of any year - 1st, George Hayes, Beeston; 2nd, George Marshall, Norwell
Six 1-lb Jars Light-coloured Extracted Honey - 1st, GH Pepper, White Post, Farnsfield; 2nd, AG Pugh, Beeston; 3rd, AH Hill, Balderton
Six 1-lb Jars Dark-coloured Extracted Honey - 1st, AG Pugh; 2nd, GH Pepper; 3rd, George Marshall
Six 1-lb Sections - Equal 1st, G. Puttergill, Beeston and D. Marshall, Cropwell Butler; hc, George Marshall
Six 1-lb Jars Granulated Honey - 1st, GH Pepper; 2nd, H. Merryweather
One Shallow Frame of Honey for Extracting - 1st, GH Pepper; 2nd, AH Hill; 3rd, D. Marshall; hc, George Marshall
Six 1-lb Jars Extracted Honey (novices) - 1st, Cecil Hayes, Beeston; 2nd, E. Chapman, Hucknall
Honey Vinegar - 1st, J. Gray, Long Eaton
Observatory Hive with Bees and Queen - 1st, D. Marshall; 2nd, W. Darrington, Bulwell; 3rd, George Marshall; 4th, AH Hill
Beeswax - 1st, W. Darrington; 2nd, GE Puttergill; 3rd, George Marshall; vhc, AH Hill

BBJ 29 Sep 1904
Leicestershire BKA held its Annual Show in connection with the Loughborough Agricultural Association on September 21st. Splendid weather prevailed, and considering the season a very nice display of honey was staged. Mr. AG Pugh, Beeston, kindly officiated as judge.

Minute Book – Quarterly Committee Meeting held on October 5th at the Peoples' Hall. Present: Mr WS Ellis (in the chair), Messrs Marriott, Pugh, Puttergill, Smithurst, Windle, Bolton, Gray and secretary.

The minutes of the previous meeting were read, Mr Windle proposed and Mr Smithurst seconded that the same be confirmed. The Balance Sheet was next gone through for the quarter. This shewed we were practically in about the same condition as regard funds as in the corresponding quarter of the previous year. Mr Gray proposed and Mr Windle seconded that the same be passed as satisfactory.

The secretary then gave a short resumé of what transpired at the BBKA meeting in London at which he attended as delegate about the 'Foul Brood Bill' and explained that he had stated there that he believed this committee generally were in favour of the Bill providing it was attempted under the auspices of the BBKA and that they would be willing to contribute to the cost of the same as far as their means would allow them to do.

A report as to the various shows was also given and accepted as satisfactory by the committee.

Mr Gray offered a sum up to 10/- towards a class for an observatory hive to shew the bees at work to educate the people to the fact that the bee makes the honey and not the bee-keeper. He was thanked for his generous offer and the matter was to be considered when framing the schedule for another year. It was pointed out that the Bath and West of England show was coming to Nottingham in 1905 and the secretary was instructed to get into communication with its secretary and arrange as best he found he could for any display in connection with apiculture.

BBJ 13 Oct 1904
In accordance with the notice issued by the BBKA, a meeting of representatives of county bee-keepers' associations was held at Jermyn-street, on October 6th, to discuss the question of promoting legislation for the better prevention of "Foul Brood" or bee-pest. There were also present the following delegate of county associations and other friends - Geo. Hayes (Notts.)

Mr. Hayes (Notts.) said his association had not yet definitely replied to the BBKA nor to Mr. Saunders either; but it was in favour of legislation against "Foul Brood", which in their opinion should be brought about by the parent association. The County Council, which granted £40 or £50 annually for educational work in bee-keeping, had been approached, but no reply was yet forthcoming. Nevertheless he thought they would be willing to help. His committee was unable to judge what the expenses would be in connexion with the Bill, and, as the association was only a poor one, they could not ruin it by promising any large sum, but he was sure, without making a pledge, that they would be willing to assist to the best of their power.

BBJ 27 Oct 1904
The monthly meeting of the BBKA Council was held at 105, Jermyn-street, SW. on October 19th, Mr. TW Cowan occupying the chair. There was also present Dr. Elliot. Reports upon the examination in Notts. was received, and in accordance with the recommendations of examiners, it was resolved to grant a certificate to WH Stoppard. (Chilwell)

BBJ 17 Nov 1904
Parthenogenesis. Most bee-keepers know that eggs laid by a fertile worker, or an unfertilized queen, produce drones, and drones only, but few, I should imagine, know how and when the germ which produces drones from eggs so laid enters the "ovarian follicle" especially in the case of a fertile worker, seeing that we are told only a vestige of an appendicular gland is present.

I, for one, should be very glad to have it explained how the germ of life is introduced into the drone eggs, or how it is acquired by the workers or unfertilised queen bees. Again, why designate queens that lay eggs which only produce drones as unfertile, and workers which occasionally do the same thing as fertile, seeing they both produce life in the form of drones? We are dealing with a matter of parthenogenesis, I admit, but anything that is capable of producing life is, to my mind, necessarily fertile, inasmuch as it produces life. E. Geo. Ive, Ollerton, November 14th

NBKA Report for 1904
The AGM of the NBKA held on February 20th, at the Peoples' Hall. The chair in the afternoon was occupied by WS Ellis, Esq. (Vice-President), and there was a fair gathering of members. Letters were read from several members stating their inability, through illness and other causes, to be present.

The Secretary read the minutes of the last Annual Meeting, which were confirmed and signed.

Mr. Scattergood proposed that the Balance Sheet as printed and sent to the members be taken as read, and that it be received and adopted. Mr. Harrison seconded, and Mr. Pugh supported, the same being carried.

The Committee's Report was next read by the Secretary, which was as follows:
We are glad to be able to show an increased membership which now stands at 185, as compared with 150 a year ago. Another matter for congratulation is the fact that the Balance Sheet showed something to the credit, which it had not done for several years past. The season of 1903 was a disastrous one for beekeeping, and especially in Notts. with the exception of a few isolated cases, but beekeepers were again full of hope for the coming season. Expert work last spring was delayed by the bad weather, but out of 189 apiaries visited, comprising 754 hives, only 25 were found to be infected with "Foul Brood," which was very satisfactory, considering the prelevance of this disease in many other counties.

The Annual Show for 1903 was again held in connection with the Moorgreen Societies in September last, Mr. Scattergood officiating as Lecturer and Judge. The Silver Medal was won by the Secretary (Mr. Hayes), and the Bronze Medal and Certificate by Miss H Bingham of Clipstone (this latter fact would rouse up the beekeepers in that locality) and pendants by Messrs. H. Pepper and G. Hayes. A show was also held at Southwell, and one at East Bridgford in connection with the relevant Horticultural Societies.

The County Council had, unfortunately, reduced their grant by £10 last year owing to the increased expenditure in bringing about the change in the Education Act, but the Committee hoped the Council will this year be able to grant the full amount, or perhaps even to increase it to £60, as this Association had more applications than it could meet with the present grant; a healthy sign. The City Council had resumed their grant for this year for work within the limits of the City.

They would again call the attention of the members to the grouping of counties at the Royal Show, which was so badly patronised last year, with the hope that more of their members would enter for the prizes that last year had to go begging through lack of competitors. It might not be generally known that special low rates for honey (as an agricultural product) could be obtained from the Railway Companies.

Three members passed in third-class honours as experts, Messrs. RJ Turner, W. Darrington, and D. Marshall. They had also added to the list another first-class expert in the person of Mr. W. Herrod, who was also to be congratulated upon having become a Fellow of the Entomological Society.

They desired to express thanks to the County Council and to the City Council for their grants for furtherance of the knowledge in beekeeping, to their noble President for the substantial help he had given to the Association, to the Vice-Presidents, and to the various donors who had assisted them, not forgetting the very valued help of some of the district Secretaries.

The Report was adopted, and ordered to be printed.

Mr. AG Pugh proposed the re-election of Lord St. Vincent as President of the Society, referring to the interest which his lordship had always taken in the work of the Association, and expressing the gratification which it afforded the members to find that he was again willing to accept the position. Mr. Peter Scattergood seconded, endorsing all that had been said in regard to their President. Other Committeemen supported this, and the resolution was carried with acclamation.

The election of Vice-Presidents and Committee was then proceeded with. Mr. P. Scattergood was re-elected Auditor, and cordially thanked for the services which he had rendered to the Association for many years past. Mr. Hayes was re-appointed as Secretary, and Messrs. Hayes and Pugh were re-appointed Delegates to the BBKA.

The Committee, on the motion of Mr. Scattergood, seconded by Mr. W. Herrod, were thanked for their services during the past year, and were re-elected *"en bloc"* with the exception of Mr. Chastenay removed, whose place was filled by Mr. WH Windle, of West Bridgford.

Mr. Turner proposed, and Mr. Mackender seconded, a vote of thanks to and the re-appointment of the Auditor, Mr. P. Scattergood. Carried.

The question of Show Cards was brought before the meeting, and after a lengthy discussion it was decided to leave the matter to the Committee.

A letter was read stating that Mr. G. Smith, one of our members, was lying at the point of death, and Mr. Manchester proposed that a letter of sympathy be sent from the Association by the Secretary. Mr. Pugh seconded. Carried unanimously.

The meeting was then adjourned for tea, to which 60 members and friends sat down. At six o'clock the meeting was resumed with Colonel Rolleston in the chair, who presented the Medals and various Certificates to the winners. Mr. Herrod, FES, gave two lecturettes on "Apiaries at Home and Abroad," and "Wax, its Production and Preparation." Misses M. and N. Smith and Messrs. Wilson and White rendered vocal and instrumental items which were warmly appreciated by those present.

The meeting closed with the usual Prize Drawing, with results as under:

		Given by	**Won by**
1st	Meadows Guinea hive	President	Mr J James, Bradmore
2nd	Meadows XL Hive	President	Mr W Layers, Nottm.
3rd	Marvel honey extractor	Anonymously	Mr J Mann, Stragglethorpe
4th	Frames and ends, value 5/-	Anonymously	Mr Hopkinson, Newark
5th	Hot water jug, value 4/-	Mr Harrison	Mr Rawson, Selston
6th	One year's "Bee Journals"	Mr Marriott	Mr Armstrong, Thurgaton
7th	One year's "Bee Journals"	Mr Marriott	Mrs Hind, Papplewick
8th	200 Honey labels	NBKA	Mr Norman, Mansfield
9th	100 Honey labels	NBKA	Mr Hill, Balderton

Thirty Jars of Honey were brought for the Hospital, and the winners in each class were as follows:
For the best Single Bottle or Jar of Granulated Honey of any year – 1st, 2/6, Mr. Pugh, Beeston; 2nd, 100 Honey Labels, Miss Ede Morris, Worksop.
For the best Single Bottle or Jar of Liquid Honey of any year – 1st, 2/6, Mr. AG. Pugh, Beeston; 2nd, 100 Honey Labels, Mr. Scattergood, Stapleford.

BBJ 19 Jan 1905
Cracked Honey Jars. I notice in the issue of January 5th an inquiry as to glass jars cracking after honey has granulated. I think there may be another possible explanation than improper handling of jars, as the following may show:

About four years ago I was asked if I could say why a vessel containing honey had cracked. In this case it was a new, glazed, earthernware "cream pipkin," containing forty or more pounds of honey, and nearly the whole of one side could be lifted away, leaving the honey (granulated) exposed.

In 1902, I filled a twenty-eight-pound tin with liquid honey, the tin being perfectly ply soldered. The honey had granulated, and wishing "to jar it off" later in the year, I picked up the tin, intending to immerse it in a boiler in order to re-liquefy the honey, but I noticed that the joint was broken and separated about $3/8$ths of an inch. There was also a perceptible bulge in the tin.

Again, last summer, while "touring" in Essex, I was shown a number of jars of granulated honey, the greater part of them cracked or broken. In none of these cases had the vessels been handled after the liquid honey had been put in them. The explanation seems to be that in granulating the honey expands and thus cracks or breaks the vessel containing it.   J. Herrod, Trentside Apiary, Sutton-on-Trent, January 14th

Minute Book – Quarterly Committee Meeting held in the Peoples' Hall on February 7th, 1905. Present: WS Ellis in the chair, Messrs Pugh, Puttergill, Smithurst, Windle, Turner, Bolton, Gray W Herrod, J Herrod and the secretary.

The minutes of the previous meeting were read. Mr Smithurst proposed and Mr Windle seconded that the same be confirmed.

Letters expressing inability to be present were read from Messrs Marriott, Scattergood, Rawson, etc. and correspondence about the Bath and West of England Show were read and this matter was then referred to the sub-committee.

Attention was drawn to the death of Mr JH Howard, of Holme, Peterborough, a prominent beekeeper and one well known amongst us. Whereupon Mr Pugh proposed that a note of condolence and sympathy be accorded Mrs Howard and family in their sad loss and that the secretary convey this to her. This was seconded by Mr W Herrod and supported by Mr Ellis.

The annual Balance Sheet was next gone into which shewed a balance in hand of ??? Mr Pugh remarked that it was regrettable to note that some of the 10/- and 10/6d subscriptions had fallen off and expressed the hope that each one would locally do what lay in his power to enlist the sympathy of such a class of subscribers to our cause. He proposed the Balance Sheet be accepted as satisfactory. Mr Gray seconded.

A draft Bill of the Bee Pest Prevention Act together with a circular from the BBKA asking whether our association approves of the Bill as amended and also for any evidence we may be possessed of shewing the need of same was next considered and after considerable discussion a vote was taken on the matter. Mr Gray proposed and Mr W Herrod seconded that this association give its support to the Bill by signing the form accompanying it. Mr Puttergill proposed as an amendment that the matter be left over to the Annual Meeting. This was seconded by Mr Smithurst. The amendment being put to the meeting – 3 votes for it, the original proposition was then put 6 members voting for same, which was carried.

It was next agreed
> that the AGM should be held in the Peoples' Hall on February 18th
> that the same competition be held and the usual prize drawing
> that Colonel Rolleston be asked to again preside at the evening meeting, failing him Mr PJ Allsebrook or Mr Hind.
> that we have a musical programme similar to last year.
> that we have a discussion on the Bee Pest Bill led off by Mr Scattergood
> that we should have explained to the meeting the Swarthmore method of queen raising.
> that we should have a question box on the table, the same to be answered by Messrs Herrod and Gray.

As regards the prize drawing it was considered the old method of drawing was too long and tedious and it was decided to have the numbers of all paid-up members in one receptacle and that the numbers drawn first receive the prizes in consecutive order.

BBJ 5 Jan 1905
It is with the sincerest sorrow that we announce the passing from our midst of one of the most genial and popular men ever connected with the bee industry in the death of John H. Howard, which took place suddenly from heart failure on December 27, at his residence, Holme, Peterborough.

Though keeping fairly well, and attending to business as usual, it has been known among his family for some time past that our friend suffered from a weak heart, and no doubt it was from this cause that he had been more seldom seen at shows and meetings of bee-keepers than formerly. Probably no man in the appliance trade was favourably known to so wide a circle of bee-keepers as John Howard. His business methods were so straightforward, prompt, and thoroughly honourable as to command respect and approval from all who dealt with him.

NBKA Report for 1905
The AGM of the NBKA was held in the Peoples' Hall on February 10th. Owing to the unavoidable absence of Colonel Rolleston, the chair was occupied by P. Scattergood, Esq., and there was a good gathering of members.

The minutes of the previous Annual Meeting were read. Mr. Pugh, proposed, and Mr. Gray seconded, that the same be confirmed by signature of chairman. Carried.

Mr. J. Herrod next proposed that the Balance Sheet, as printed and sent to members, be received as satisfactory and adopted. Mr. WH Dickman seconded, and it was carried.

The Committee's Report was next read by the Secretary and was as follows:
On taking a first survey of our Balance Sheet you may perhaps think the deficit is somewhat large, but on looking at the Liabilities and Assets, and feeling as we do that the amount owing from the Southwell Society is sure - you will see we stand in about the same position as last year. Bearing in mind the advances we made in the value of the prizes at the Annual Show, and the fact that we have several outstanding subscriptions, we trust you will consider the same satisfactory.

The membership has slightly increased this year, but not at the rate we think it should have done in this county, and as it has so often been said, there is still room for the development of our Association; and work for those who are sufficiently enthusiastic to undertake it in connection therewith; and we venture still to hope that new members may be drawn to the brotherhood and benefits of this Association.

30 jars of honey were brought for the children; 20 were sent to the Children's Hospital and 10 to the children in the General Hospital, Nottingham.

Mr. Scattergood next gave his paper on "Leaves from my Note Book", in which he brought before us the inauguration of our Association, and would, had time permitted, given points to help those starting in the craft.

Some questions had been sent up to the table amongst them being:
    Do Bees use pollen in comb building?
    Why do Bees need water?
    Will spores of *Bacillus alvei* germinate without a medium?
    Would it be practicable, or possible, to abolish the '[Beekeepers'] Record' and substitute the 'British Bee Journal'?

These and others were answered, and as regards the latter it was pointed out that many difficulties lay in the way which made it impracticable.

Mr. Meadows next exhibited some celluloid queen cages, which he considered would enable beekeepers to utilize the whole of the queens reared, or at any rate those desired, instead of allowing them to be killed. He also exhibited a cross-slatted honey board which he maintained would prevent brace combs and give winter passages to the bees.

Mr. W. Herrod told us of the relation of the bee to the flower and *vice versa*, and the comparative yields in honey and nectar of the same, illustrating his remarks by the aid of the lantern with some very good slides, and although very much hurried over his discourse owing to lack of time, he nevertheless gave gratification and satisfaction to those who listened to him.

The annual prize drawing took place at 7.45pm for the following articles:

		Given by	**Won by**
1st	Meadows Guinea hive	President	Mr GE Skelhorne, Sneinton
2nd	Meadows XL hive	President	Mr F Burley, Nottingham
3rd	Honey ripener	Harrison and Sons	Mr GW Halstead, Newark
4th	3 queen cages	Mr WP Meadows	Mr P Scattergood, Stapleford
5th	Candy feeder	Mr SW Marriott	Mr GM Bolton, Eastwood
6th	Bee journals	Mr Sw Marriott	Mr E Chapman, Hucknall
7th	200 labels	NBKA	Mr JR Almond, Cotham
8th	100 labels	NBKA	Mr GA Hill, Balderton

BBJ 16 Feb 1905
Enquirer (Notts.) Your sample is, in our opinion, a South American honey. It may be genuine honey, but decidedly not English to our mind. The flavour reminds us of Chilean honey as imported.

NEP 18 Feb 1905
Beekeepers from all parts of Nottinghamshire assembled this afternoon in the Peoples' Hall on the occasion of the Annual Meeting of the NBKA. Mr. WS Ellis, of Nottingham, presided.

In the Annual Report the committee emphasised the necessity for members being more enthusiastic in making known the value of the association to beekeepers and others in their locality. Regret was expressed at the fact that the season 1904 was anything but good, and that only in a very few isolated cases was any great surplus gathered. And most of that, it was mentioned, was spoilt by the so-called honey-dew.

Allusion was also made to the various shows held in conjunction with horticultural shows, and it was said the one at East Bridgford caused quite a little flutter in bee circles. The committee, however, would like see more entries if that show was to be continued.

During the year 188 apiaries were visited by the experts, and of the 694 stocks of bees there, 561 were examined, only 34 being found diseased. On that fact, bearing in mind what was heard from other counties, the committee thought the Notts. beekeepers were to be congratulated. The Balance Sheet showed credit balance £1 2s and both report and financial statement were adopted.

Viscount St. Vincent was unanimously re-elected President; Mr. P. Scattergood (Stapleford), auditor; and Mr. G. Hayes (Beeston), secretary. Mr. AG Pugh (Beeston) and Mr. Hayes were selected as representatives to the BBKA; and the committee was constituted as follows: Messrs. TN Harrison, SW Marriott, GE Skelhorne, WH Windle (Nottingham), AG Pugh, GE Puttergill (Beeston), T. Randall (Cotgrave), G. Smithurst (Watnall), JC Wadsworth (Collingham), FG Vessey (Balderton) and HW Dickman (Babbington).

BBJ 2 Mar 1905
The Annual Meeting of NBKA was held in the Peoples' Hall on February 18th. Mr. WS Ellis presided, and there was a large attendance of members.

In the Annual Report it was stated that the County Council had renewed their grant of £40 for Technical Instruction in bee-keeping and demonstration lectures had been given at the following centres:
>Newark, Ruddington, Welbeck, East Bridgford, Kingston,
>Sutton Bonington, and Moorgreen.

During the year 188 apiaries had been visited by the experts. These apiaries consisted of 694 stocks of bees, 561 of which were examined. Only 34 of these were found to be diseased.

Two candidates offered themselves for examination for expert certificates during the year, viz., Mr. J. Gray for second class and Mr. WH Stoppard for third class, and both gained their certificates.

Viscount St. Vincent was thanked for his past services, and unanimously re-elected as was also Mr. P. Scattergood, auditor, and Mr. Geo. Hayes, secretary. Mr. Pugh and the secretary were appointed to represent the association at the meetings of the BBKA. The following were elected to form the committee: Messrs. TN Harrison, SW Marriott, GE Skelhorne, WH Windle, AG Pugh, GE Puttergill, T. Randall, G. Smithurst, JC Wadsworth, FG Vessey and HW Dickman.

On terminating the more formal business of the meeting the members and their friends partook of tea, after which a musical programme contributed to the evening's enjoyment. There was also an exhibition of bee-appliances, with prize competitions for extracted honey, the exhibits being afterwards presented to the Children's

Hospital. The proceedings were varied by a discussion on the Bee-Pest Bill, and concluded with the annual prize drawing and presentation of prizes.

BBJ 9 Mar 1905
Mr Arthur Judd, Chilwell. "In the year 1897 a friend of mine brought his bees to my father's orchard, and for a time I was simply a spectator. Ultimately I began to help him in his manipulations, and gradually became bolder, and more indifferent to a few stings. In the year 1900 my friend presented me with a swarm, and having bought a hive I made my start as a bee-keeper. In the autumn of the same bee-notes read as follows:

In the photo above, the figure on the left is our village schoolmaster, who is an enthusiastic bee-keeper. He very kindly gave a look round my apiary each day in the swarming season, which was very helpful to me, as I am away in Nottingham all day. The 'Wells' hive I have kept unoccupied since the first year I had it, as I found it unmanageable.

Though my experience has carried me through some depressing seasons, and, in consequence, I have not had that success with bees which has been the good fortune of some BBJ readers, I have found beekeeping an interesting, instructive, and profitable 'hobby,' and with more favourable climatic conditions I hope to do better than hitherto. I have read the 'Guide Book' and the B.K. Record, and I am greatly indebted to Mr. Peter Scattergood, of Stapleford, for advice always given freely. I also have to acknowledge expert help and advice from Mr. Puttergill, of Beeston."

Minute Book – Quarterly Committee Meeting held in the Peoples' Hall on April 1st. Mr P Scattergood presiding. Present: Messrs Dickman, Harrison, Bolton, Pugh, Puttergill, Skelhorne, Smithurst, Randle, Turner, Windle, Vessy and secretary.

The minutes of the previous committee meeting were read and confirmed by the signature of the chairman. The quarterly Balance Sheet was next presented shewing a balance of £14 3s 11d against £19 1s 5d the previous year.

A letter was read from the Moorgreen society stating they could not offer us any terms this year and that they intended to have classes of their own for honey, etc.

Mr Vessey proposed and Mr Dickman seconded that the sub-committee for Technical Instruction be re-elected and the Messrs RJ Turner and T Randle be added to the same.

The secretary explained that he had been in communication with Mr Merryweather about the Annual Show being held at Southwell this year and it was resolved that if the Southwell Society will offer the same terms as last year we have the Show there and discretion was given the secretary to arrange this if possible.

The prize schedule was next gone through class by class and the following alterations ordered to be made:

Class I.	Cut out. Discretion was given to the secretary to try to arrange for a dealer to exhibit appliances, whom he might assist up to 10/-
Class II.	Best Outfit for a Beginner
Class III.	Weight of honey on Trophy not to be stated.
Class IV.	Four prizes – 1st Silver medal and 10/-; 2nd Silver pendant and 7/6d; 3rd 5/- and 4th 2/6d
Class V.	Four prizes – Silver pendant and 10/-; 7/6d; 5/-; 2/6d;
Class VI.	Three prizes – 1st Bronze medal and 7/6d; 5/-; 2/6d
Class VIII.	Four prizes – Silver pendant and 10/-; 7/6d; 5/-; 2/6d

Honey vinegar – one prize 2/6d

Referring to Mr Gray's offer it was resolved that an additional class be inserted in the prize schedule for the best observatory hive shewing that the bees make the honey and not the beekeeper for which one prize of 10/- is kindly given by Mr Gray of the White Apiary, Long Eaton.

The secretary again mentioned that several members asked for classes for mead and for heather honey, but it was not considered advisable to have these.

In pursuance of the resolution at the Annual Meeting with regard to legislation, a copy of the draft Bill with a voting card had been sent to every member with a request to reply for or against by March 31st. The replies to hand numbered 66 out of a total membership of 172.

        For   51 members owning 350 stocks
        Against   15 members owning 88 stocks

The secretary was instructed to report these figures to the BBKA but they were considered unsatisfactory and it was suggested the matter be allowed to rest for a time.

A few comments were made on the last Annual Meeting amongst which the following appeared most important

>   less time to be occupied for tea
>   more attention to be given to the question box
>   the need for a change from the musical programme.

The secretary was empowered to arrange with local horticultural societies for honey, etc. classes as and when he could best do so, special mention being made of the Horticultural and Botanical Society Show in Nottingham.

BBJ 11 May 1905
Bonus (Newark) Comb sent is so badly diseased that we should destroy the whole contents of hive, bees and all, unless the latter are strong in numbers, which is more than doubtful.

East Bridgford Parish magazine June 1905
The horticultural show will be held on June 27th. The usual exhibition of bees and honey will be given.

BBJ 19 Jun 1905
The monthly meeting of the BBKA Council was held at 105, Jermyn-street, SW. on June 21st, Mr. TI Weston occupying the chair. There was also present Dr. Elliot. A letter explaining absence was received from AG Pugh. Arrangements were made for examinations of candidates for third-class certificates in Nottinghamshire on July 20th.

Minute Book – Committee Meeting held in the Peoples' Hall on 1st July, 1905. Present Mr AG Pugh in the chair, Messrs Ellis, Harrison, Skelhorne, Smithurst, Windle, Dickman, Turner, Bolton and secretary.

The minutes of the previous meeting were read and confirmed by the signature of the chairman. The quarterly Balance Sheet which shewed that we were in much of the same condition as for the corresponding quarter in 1904 was gone into and passed.

The question of a judge and assistant for the Annual Show at Southwell was next discussed and it was eventually agreed to appoint Mr Scattergood as judge but, seeing that no bee tent was to be sent and consequently no lectures would have to be given, no assistant judge would be necessary.

The secretary reported that owing to the East Bridgford Show falling before the committee meeting he had asked Mr RJ Turner to officiate as judge at that show. It was resolved that the action of the secretary be endorsed.

Mr Skelhorne mentioned that he would perhaps be able to arrange for a demonstration of beekeeping at Sneinton and asked if the association would assist him by the loan of the tent, etc. and it was decided to do so.

NEP 10 Jul 1905

The village of Weston-on-Trent was on morning disturbed by a swarm of bees. In their onward progress through the main street they attacked a number of people, and caused much consternation. Eventually they discovered some hives fully occupied, and attempted to take possession. There was consequently much irritation among the local bees, and very soon the whole neighbourhood was alive with them. The insects as they darted about seemed quite mad, and endeavoured to sting anyone who happened to be about. People in the locality had to close their windows and doors to keep them out. Women and girls seemed the greatest sufferers, as they attached themselves to their hair and clothes.

For several hours it was almost impossible to walk down the street, and frequently those who wilfully neglected the salutary advice to reach their destination another way had to suffer for their boldness. Even animals and fowls did not escape. Horses were attacked, as well as their drivers, and in one case at least about half a dozen fowls were completely overwhelmed and died as a result of the stings. The bees disappeared towards the evening.

"It is surely proper that bees should associate in a Bees-town, and the more so as the Urban District Council has adopted a bee-hive as a part of the inscription on its seal. It is, therefore, not surprising to find as reported by Mr George Hayes, the well-known lecturer on bee-keeping, that "in 1905 there were approximately 100 colonies of bees in Beeston" who having the range of the Trent valley, with its domestic orchards and gardens and large nurseries, providing for fruit-growing and afforestation, the conditions are such that a stock of bees under up-to-date management well yield a good surplus to the bee-keeper, so that one stock in recent years gave in one season 105 lbs of honey. Unfortunately, an incurable disease has made havoc among bees throughout the kingdom, reducing the number kept locally by one half but as bees are indispensable to the fruit crop, it is hoped the disease may soon disappear."

Beeston UDC information

East Bridgford Parish magazine July 1905

The Show, in spite of a dry season and late severe frosts, was pronounced to be a good one. Gate proceeds exceeded £9 which wasa very satisfactory considering the uncertainty of the weather.

No prizes for beekeeping were awarded (due to the death of the regular sponsor Mr WF Fox, JP?)

BEEKEEPING BETWEEN TWO QUEENS

BBJ 20 Jul 1905
BBKA and Legislation for the Better Prevention of Bee-pest. Schedule of Information collected to July 12th, 1905. The NBKA Committee were in favour, but their decision was overruled in a General Meeting. An appeal to the members by circular resulted thus: fifty one members, owning 350 stocks, voted in favour and fifteen members, owning 88 stocks, against. Number of members - 172.

BBJ 27 Jul 1905
The NBKA Annual Show was held with that of the Southwell Horticultural Show on July 20th. The duties of judging the bee-exhibits were carried out by Mr. P. Scattergood, who also examined a candidate for third-class expert. This was the best show for several years past, and all the honey was excellent, compared with that shown of late years. The following are the awards:
Beginner's Outfit – 1st, R. Mackender, Newark
Trophy of Honey in any Form and of any Year – 1st, D. Marshall, Cropwell; 2nd, G. Marshall, Norwell; 3rd, W. Lee, Southwell
Six 1 lb Jars Light-coloured Extracted Honey - 1st and silver medal BBKA, W. Herrod, Sutton-on-Trent; 2nd, GH Pepper, Farnsfield; 3rd, JR Almond, Cotham; 4th, J. Breward, Rollestone; hc, B. Bowyer, Swinderby
Six 1 lb Jars Dark-coloured Extracted Honey – 1st, R. Mackender; 2nd, J. Breward; 3rd, G. Marshall; 4th, D. Marshall
Six 1 lb Sections – 1st, W. Herrod; 2nd, W. Lee; 3rd, W. Ball, Eagle; 4th, W. Darrington, Bulwell
Six 1 lb Jars Granulated Honey – 1st, AH Hill, Balderton; 2nd, G. Marshall; 3rd, GH Pepper
One Shallow-frame of Honey for Extracting – 1st, GA Hill, Balderton; 2nd, G. Marshall; 3rd, D. Marshall; 4th, GH Pepper
Six 1 lb Jars Extracted Honey (Novices) – 1st, WH Bowman, Wellow; 2nd, W. Sentence, Shelton; 3rd, HM Gabbett, Rolleston
Honey Vinegar - No award
Observatory Hive with Bees and Queen – 1st, G. Marshall; 2nd, H. Mackender, Newark; 3rd, W. Darrington; 4th, R. Mackender; hc, AH Hill
Beeswax – 1st, W. Darrington; 2nd, W. Herrod; 3rd, G. Marshall

BBJ 17 Aug 1905
The Annual Show of Leicestershire BKA was held in connection with the twentieth Annual Flower Show at the Abbey Park, Leicester, on August 8th and 9th. There was a large show of honey, the quality being excellent, owing to the exceptionally good season this year. Very little dark honey was on view. The number of entries was a record one, and the judges were Messrs. Peter Scattergood (Notts.) and HM Riley (Leicester).

BBJ 7 Sep 1905

The thirteenth Annual International Exhibition and Market of the Confectioners and Allied Trades commenced on September 2nd and remains open till the close of the present week. Owing to the increased demand for space this year, the honey competitions were perforce relegated to the North Gallery Annexe, and thus had less prominence than in 1904. The change, however, gave those especially interested in the display of honey and bee products a better opportunity for examining the exhibits than in the crowded avenues below.

Display of Honey (comb and extracted) and Honey Products, shown in suitably attractive form for a tradesman's window (5 entries) - 4th (£1), T. Marshall, Sutton-on-Trent

Twelve 1-lb Jars Medium-coloured Extracted Honey (27 entries) - 4th (10s), T. Marshall

BBJ 7 Sep 1905

Bees in Notts. Price of Honey. The honey-crop here has not been as large as I expected this season, owing to the drought and prevailing rough winds. Although all my hives were crowded with bees on ten frames, only one has succeeded in filling the second super of shallow frames, and even as late as now all is not sealed over. I have been interested in reading the letters dealing with the price of honey, and agree with your correspondent that selling at a low price to the retailer is likely to do serious harm to those who depend on bees for their living. The honey will keep, and there should be no great difficulty in storing it; therefore why not hold out for a fair price? I have never sold any for less than $6\frac{1}{2}$d per lb. in bulk, and 8s per dozen for tie-over 1-lb jars.

At one shop where I sold a lot last year they say they have bought this year 1-lb "screwcaps" at the same price as my tie-overs, and appeared to want mine for less. But I declined to take less, saying "It would keep, and I would go elsewhere." I am, in consequence, now selling at 9s per dozen, and likely to continue at that price.

Then there is that "Scotch honey" - deception, should I call it? In a shop window in Nottingham I notice 2-lb tins Scotch honey for 1s. It is labelled, "Superfine (Scotch) Pure Honey," and on the margin is the legend, "This honey is of uniform quality and specially prepared for table use." Now I ask "What preparation does pure honey require for any use?"

They said it had been specially prepared to sell cheap, it might mean something - but so much for present-day smart-trading methods. Of course, we all have to "live," but I think among the bee-keeping fraternity we ought to make our motto, "Live and let live," and try to do something to keep out of our business methods such cutthroat competition as is too prevalent nowadays.

I enclose two samples of honey. Will you kindly say what you think of them? I enclose

card and sign   Boot, Nottingham, August 23rd.
[Both samples are of excellent quality. - Eds.]

BBJ 21 Sep 1905
The thirteenth Annual Exhibition and Market of the Grocery and Allied Trades, held at the Agricultural Hall, London, was opened on the 16th inst. and continues till the end of the present week. Favoured with fine weather, as it has been so far, we doubt if any of its predecessors has equalled this exhibition for extent or for all-round excellence, while the section in which bee-keepers will feel most interest greatly surpasses anything previously seen at the "Grocers'."
Twelve 1-lb Sections (26 entries) – 4th, (5s), G. Hunt, Newark
Twelve 1-lb Jars Medium-coloured Extracted Honey (43 entries) – 4th, J. Herrod, Trentside Apiary, Sutton-on-Trent
Twelve 1-lb Jars Dark-coloured Extracted Honey (13 entries) – 1st, (£1), G. Hunt
Twelve 1-lb Jars Granulated Honey (10 entries) – 1st, (£1 5s), J. Herrod
Beeswax in Cakes, Quality of Wax, Form of Cakes and Package, suitable for retail counter trade (13 entries) – 2nd, (15s), G. Hunt

BBJ 28 Sep 1905
The monthly meeting of the BBKA Council was held at 105, Jermyn-street, SW. on the 20th. Mr.Tl Weston occupying the chair. A letter regretting inability to attend was received from Dr. Elliot. A report upon examinations in Nottinghamshire was received.

BBJ 5 Oct 1905
The Bee-Season in Notts. I have never been favoured with the big "takes" of honey which we hear of in the BBJ, but having this year secured my record harvest, I thought it might interest and encourage some at least of the many readers of your valuable weekly to have a few details of what has been done.

I have been a "modern beekeeper" ever since the year 1877, and the converts I have made to the "better way" are so numerous that I have lost count of them; one is a first-class expert known to most bee-keepers in the British Isles.

But about the bee-season. Let me say I have two colonies of bees in frame-hives, standing in my sister's garden at Mildenhall, Suffolk, and these two have yielded a gross weight of 295 lb. of beautiful, light-coloured honey, including forty-two sections. I pay an annual visit to Mildenhall - which is my native place - and have a young friend there who was only a skeppist when I induced him to start on modern lines about ten years ago. He still has some skeps, but now owns eighteen or twenty stocks in frame-hives. I might say, in addition to being a fruit and seed growing district, the farmers there grow an abundance of sainfoin.

I cannot boast of any very large "takes" here in Newark - where I live - although I secured a first prize and silver pendant at our NBKA Show this year. My largest "take" in Notts. is 85-lb. from one hive. Enclosed is a photo of myself and my Newark apiary, and I shall be pleased if you consider it worthy of a place in "Homes of the Honey-bee ". Robert Mackender, Newark, September 30th [Very pleased to get photo, and shall give it a place in our bee-garden pictures in due course (in January 1906). Eds.]

Minute Book – Committee Meeting held in the Peoples' Hall in October. Present: Mr AG Pugh in the chair, Messrs Puttergill, Smithurst, Windle, Dickman, Vessey, Hallam, Turner, Bolton and secretary.

The minutes of the previous meeting were read and confirmed by the signature of the chairman. Mr Smithurst proposed and Mr Dickman seconded. The quarterly statement was next gone into and it was found we were in much the same state as for the corresponding quarter of last year. Mr Turner proposed and Mr Windle seconded.

Messrs Hayes and Pugh gave a short resumé of what transpired at the British Council meeting and conversazione at London on October 5th about the Foul Brood Bill, bee clubs, size of sections and strengthening the standard frame.

The secretary read a report on the two shows held in 1905 at Southwell and East Bridgford and it was resolved this report be accepted as satisfactory and that it be embedded in the Annual Report to be printed later on.

Memo: Mr Puttergill stated that he had heard objections raised to the secretary assisting the judge at the last County Show. Although this was done by the cognizance of the committee, it was seen that there was a real objection to it and the secretary took note for the future. Comment (*in camera*) was also made as to the prize medal honey.

BBJ 12 Oct 1905
The meeting of representatives of County Bee-keepers' Associations, convened by the BBKA, was held at 105, Jermyn-street, SW., on the 5th inst. at 4pm for the purpose of furthering the interests and general welfare of the bee industry and the maintenance of friendly relations between the parent Association and its affiliated societies. Among those present were Geo. Hayes, AG Pugh and P. Scattergood,

The Chairman said the division of opinion among County Councils was about equal regarding the desirability of legislation.

Mr. Scattergood (Notts.) supported the resolution most thankfully, and said that although legislation could not be secured, he thought it might be possible to deal with the disease in another way, viz., through the County Councils being approached with the view of obtaining pecuniary support for the county bee associations.

Mr. Geo. Hayes (Notts.) advocated local meetings of members of County Associations in preference to bee-clubs, which would, in his opinion, damage associations by drawing away people who might otherwise support them. A good local secretary would be glad from time to time to organise these meetings in various centres, which would bring new life into the County Association.

Mr. Scattergood (Notts.) thought the object of bee-clubs was not to teach beekeeping, but more to promote mutual helpfulness, members lending each other extractors or other appliances, and cooperating towards mutual support; but he quite agreed that these bodies should in no way be allowed to detract from County Associations. It was a *sine qud non* that the latter must be the prime authority in a county, and their influence should not be weakened. He might say that he was in a village club himself, for the local cottagers came to him if they wanted foundation, or the use of an extractor, etc.

Mr. Pugh (Notts.) thought they could only judge by experience in these matters. Many would remember the existence of the Wootton-under-Edge Club, which did valuable work so far as its individual members were concerned, but had never been affiliated nor of any help to the County Association. There were many of these little district associations in Yorks. but they never came together; even Mr. Grimshaw, the county secretary, had not been able to secure that any assistance they might have rendered the cause was marred by disunion.

He agreed with Mr. Hayes that a society might have its subsidiary branches, and considered the need well met by district secretaries calling their members together on afternoons, by which probably more subscribers would be obtained for the County Associations. A stipulation should be made that if help was required applicants must first join the latter. He was himself distinctly in favour of district associations as being preferable to village bee clubs.

BBJ 26 Oct 1905
The monthly meeting of the BBKA Council was held on 18th inst., at 105, Jermyn-street, SW. Mr. E. Till being voted to the chair. A letter of regret at enforced absence was received from AG Pugh.

Conference of Bee-Keepers. (Continued)
Mr. P. Scattergood, in opening the discussion on the third item on the agenda, ie.,

"the strength of the standard frame," said that personally he did not think the top-bar needed much strengthening. He preferred frames with a solid top-bar, free from saw-cut or wedge arrangements of any kind, but having a bottom-bar $3/16$ths or a $1/4$in. thick. If such frames were properly wired and the foundation fixed to the top-bar with molten wax, there was not much danger of any deflection or sagging.

His friend, Mr Geo. Hayes, secretary of the NBKA, however, took a different view and - being compelled to leave the meeting at an earlier stage - had requested him (the speaker) to show to those present the frame he held in his hand as a suggested improvement on the "standard," as he considered that the top-bar of the latter needed strengthening. Mr. Hayes said that if the top-bar was $9/16$ths of an inch thick instead of $3/8$ths as now, the frame would be much more rigid, and if extra strength was added on the underside of the top-bar it would be interchangeable with the present standard frame and, consequently, not interfere with hives in use, nor would the cell-capacity of the frame be reduced to any appreciable extent, as only $3/16$ths would be added to the top-bar, and the inside of the frame correspondingly reduced,

Other advantages which Mr. Hayes claimed for this alteration were that the frame would bear a greater strain, being stronger at the shoulders, and as he proposed that the frame-ends or lugs up to the side-bar should be reduced to the thickness of the standard, there would be no alteration of the existing "metal ends" now in use. Those were Mr. Hayes's suggestions, and he (Mr. Scattergood) begged in his name and on his behalf to submit them to the meeting together with the frame made by that gentleman according to the particulars already given, and as a decided improvement on the one at present in use.

BBJ 26 Oct 1905
WB Munro (Notts). Chapman Honey Plant. The seedlings raised in a box out of doors from seed sown in August should have been planted out at end of September, but they will take no harm if left in a box and slightly protected from keen frost.

Minute Book – Committee Meeting held in the Peoples' Hall on January 6th, 1906. Present WS Ellis in the chair, Messrs Harrison, Pugh, Puttergill, Smithurst, Windle, Vessey, Turner, Gray and secretary.

The minutes of the previous meeting were read and confirmed by the signature of the chairman.

Correspondence was read about the Southwell Show and the grant from the same, which was fully endorsed.

From the Mansfield Horticultural Society about a show for 1906, and the secretary was empowered to arrange either for annual or local shows according to grant promises.

From the secretary of the Botanical Society, Nottingham – the secretary again being empowered to further the arrangements if possible for a show.

A letter to his Lordship asking him to preside at the Annual Meeting and leaving him to fix a date convenient to him, which was agreed to.

The annual Balance Sheet was next put forward and was passed for printing with the circular convening the Annual Meeting when the quantity of labels in stock had been added.

A rough draft of the proposed report of the committee was put forward and gone through word by word and altered and amended as considered necessary (except the first paragraph) and was afterwards ordered to be read as altered at the Annual Meeting.

Arrangements where then made for the Annual Meeting as follows:
> to take place on February 10th (unless altered by his Lordship) at 3pm in the Peoples' Hall with one of the following as chairman in the order named:
>> His Lordship Colonel Rolleston The Mayor (John Green)
>> Councillors EN Elborne (Mayor 1901/2) AJ Butler Hannah
>
> that tea be at 4.30pm provided by Messrs Ellis and Hayes and that special mention of this be made in the circular.
> that the two competitions be the same as at the last meeting
> that the arrangements for prize drawing be left to the secretary and that the prizes be drawn for same as last year viz. just number drawn first prize and consecutively to the finish.
> that the evening meeting take place at 6pm and be devoted to a general conversazione on debatable points of interest to beekeepers
> that Mr Scattergood be asked to introduce a subject and time to be allowed for questions - 15 minutes.
> that Mr Herrod be asked if he will be present and, if so, contribute
> that it should be stated that members will be given and opportunity to ask questions either in writing or verbally
> that invitations be sent to the three adjoining secretaries and Mr Meadows
> that the prize drawing should take place at 7.45pm.

BBJ 11 Jan 1906
TB Barlow (Notts) Swarthmore Bee-Books. In reply to complaints from readers who have ordered copies of "Increase," by Swathmore, we beg to say that our stock of

this book being exhausted, we ordered a fresh supply over two months ago. We have made enquiries of Mr. Pratt with reference to the delay in forwarding, and he writes to say that the First Edition having been rapidly sold out, a second is being prepared. He regrets the unavoidable delay in getting this out, but promises that no time shall be lost, though some necessary changes are being introduced into the new edition which prevent it being ready as soon as he expected.

BBJ 18 Jan 1906

We are very pleased to illustrate the neatly-kept bee-garden, (along with himself) of so worthy and valuable a helper in the cause we serve as our friend, Mr. Robert Mackender. The modest account, given by himself, of his labours on behalf of the bee industry needs no addition from us, except to express a hope that he may be long spared to make "converts," who will, like the others, prove good and useful recruits to the craft. He says:
"As requested, I will try and give a few details of my bee-keeping, about which much might be written, as my early experience harks back to the seventies. My interest in the craft was first thoroughly awakened at a village flower show, when a lecture and demonstration with bees was given by Mr. Tom Sells.

I was so astonished by the way the bees were driven and handled by the lecturer that I determined to go home and try to do likewise. An old skeppist in the village where I then lived (only seven miles from my present home) promised me his bees instead of resorting to the sulphur pit, and within a month of seeing it done, I successfully drove my first skep. Meantime I had purchased a frame hive, into which I put the bees, thus starting my new-found hobby, and my fondness for it is just as strong with me to-day. (I might here say that the old man mentioned above also took to the bar-frame hives in preference to his skeps.)

But alas, my first lot of bees soon died from exposure, through constantly opening the hive. I determined, however, to start again, and in the following spring purchased a strong skep, and transferred them to the frame hive, from which I obtained my first honey, and from that time my stock increased rapidly till I had got half a dozen hives, which was as many as my small garden would accommodate. Up to this time I knew nothing of bee associations or of a bee journal, so I advertised bees for sale in a Notts. paper, and Mr. AG Pugh, Hon. Sec. to the NBKA at that time answered my advertisement, not to buy bees, but to invite me to become a member of the association.

Eventually Mr. Pugh paid me a visit, and gave me some useful instruction that I have never forgotten. He also induced me to join the association, and for many years I have acted as district secretary. During that time I have made many converts to the craft, some of whom have become experts; one of them, I am proud to say, now holds the first class certificate of the BBKA and is one of the foremost experts of the day. I refer to Mr. Wm. Herrod, who when reading these lines will no doubt remember his first lesson. I have not aspired to a certificate myself, but have a son, Henry, who holds a third-class certificate.

With regard to my success as a honey producer. I have never been favoured in Notts with large takes of surplus. I work mainly for extracted honey, and this year in class for Dark-coloured Extracted honey at our Annual Show held at Southwell I took first prize (as well a First for Beginners Outfit, and Fourth for Observatory Hive).

The number of hives in the accompanying photo, is fourteen, but sometimes I have twenty or twenty five. I am frequently starting fresh recruits in the pursuit, and this season I have supplied four gentlemen of position with six stocks complete, three of which yielded good results. I still take our old friend the BBJ every week, and should not like to be without it."

BBJ 18 Jan 1906
A Beginner's First Report. I am only a beginner with bees, having started with one hive in March, 1905, and I thought you might be pleased to know that it has yielded me 52 lb. of surplus honey. I consider this a very good harvest for a novice, especially when compared with my less fortunate neighbour, who owns about a dozen hives, and tells me they have done very badly last year. I have the "Guide Book," and also take the BBJ which I look forward to with pleasure each week, always gaining knowledge from the experiences of others as given therein.   TAB, Awsworth

BBJ 1 Feb 1906
I have read with interest Messrs. Herrod and Stewart's note re "Early Breeding". The question arose in my mind, is it early or late? I put into winter quarters five baby nuclei, consisting of four quarter-size standard frames. These nuclei were wintered in three ways for an experimental purpose, and here are the notes of one of these baby nucleus hives:

December 8th	Eggs, larvae, pupae, one frame full, gave candy
December 20th	Pupae only, comb built in candy box
January 4th	No eggs, no brood, gave pollen
January 19th	Eggs and larvae
January 27th	Eggs, larvae, and pupae.

From the above it will be noted brood-rearing ceased with the closing year. I have two more queens in baby nuclei with eggs and larvae; one died, the other suffered with dysentery; result, queen dead; the bees were joined to another baby nucleus.

Personally, I am well satisfied with the experiment, as it has given me a successful method of wintering baby nuclei.   J. Gray, Long Eaton, January 27th

BBJ 1 Feb 1906

The apiary of our friend Mr. Darrington illustrates how bees may be kept close to dwelling-houses without being voted a nuisance by neighbours. The owner himself also combines in his own person a bee-keeping schoolmaster, who is a certificated expert, a good amateur joiner, and a useful member of the craft. For the rest, we may allow the interesting notes, sent at our request, to speak for themselves.

He says: "The photo of myself (Darrington) and apiary was taken by my father-in-bee-keeping, Mr. H. Meakin, Newthorpe. Owing to the above gentleman's influence, I had a thorough practical and theoretical knowledge for some two years before I owned a hive. He it was who introduced me to the "Guide Book" and the Record as far back as 1892, and gave me my first hive.

I make my own hives, which are of the "combination" pattern and very large. At the back is a small door, and flight board, for keeping a nucleus. You see, my hives are worked on a non-swarming principle, because of the situation, and the strain of bees in No. 1 (the one with a hive-roof on top) have not swarmed for a dozen years at least.

I have never had a swarm, although I have tried various races of bees.

Last year I made a "WBC" hive from directions in the "Note Book," and although the bees in it were crowded on twenty-two standard frames and two supers of shallow-frames, they never swarmed. It will be noticed that the hives are placed very close together, so in order to give the bees a better chance of distinguishing their own domicile, No. 1 is painted a pale blue, No. 2 oak-grained, No. 3 green, and No. 4 red.

The hive uncovered and showing a feeding-bottle had a queen which did not commence to lay until three weeks after fertilisation in the autumn, and then began with a frame of drones in worker-cells. It was my best hive all through the spring, but in Whit-week I found she had disappeared. The photo was taken in the spring, and shows, on the left, a border of crocuses, which are extremely useful.

I was one of the first Notts men to take advantage of the Insurance Scheme for bee-keepers, and in that matter my apiary is an object-lesson. Behind the wall at the back, when I commenced, was a field; now there are two rows of houses and a street, and in the height of the season I have seen horses driven up so close as to poke their noses over the wall. I was very pleased to think I had prepared for eventualities.

In 1903 I gained my third-class expert's certificate, and hope to go still higher, though I do not think the present expert visiting system is effective. A much better plan, to my mind, would be to encourage members to take the "Expert's" certificate, and then divide the county among them, allotting to each the immediate district in which he lives.

With regard to honey selling, this year I tried a wholesale "deal" for the first time. I asked 10s per dozen 1-lb jars, and was met with a decided "No." Then I asked for a quotation. This was the reply: "I lately bought a gross for 8s 6d per dozen, and I have no doubt I can get it for even less than that." Well, he did not get mine. I think the middleman gets far too much of the profit.

I seize every opportunity to visit well-known apiaries, and in this way have seen Mr. Rymer's and the late JH Howard's. I also make a practice of visiting the large honey-shows whenever convenient. In 1903 I saw the "Grocer's and Confectioners'," and last year the "Royal."

In conclusion, allow me to say how much I am indebted to the BKA Journal, and to thank Mr. Woodley and "DMM" in particular. I was delighted to discover, some months ago, that the latter is one of us, for I am another schoolmaster bee-keeper."

NEP 10 Feb 1906
The Annual Meeting of the NBKA was held at the Peoples' Hall this afternoon, Mr. P. Scattergood (Stapleford) presiding, in the absence of Colonel Rolleston. Amongst those present were Mrs. Laws (Car Colston), Messrs. AG Pugh (Beeston), W. Herrod (Luton), WP Meadows (Syston), R. Mackender (Newark) J. Herrod (Sutton-on-Trent), RW Turner (Radcliffe), P. Smithurst (Watnall), J. Gray (Long Eaton), WS Ellis (Nottingham), W. Darrington (Bulwell), and G. Hayes (Beeston), hon. secretary.

The committee's report stated that the past season had on the whole been much better than the preceding one. Although some cases the quantity was not quite up to the average, the quality as a rule far surpassed what they had had of late years. The experts had during 1905 visited apiaries containing 716 stocks of bees, of which 586 were found to be healthy, and 37 diseased in varying degrees.

In regard to the latter there was a slight increase. The committee were doing all they could to suppress the disease in this county by providing the experts with a thoroughly efficient means for the first treatment of any stocks affected, and it was hoped that the members would back up their efforts by closely following the instructions of the experts.

There had been a slight increase in the membership during the year, the financial position of the association being about the same. The report was adopted.

On the proposition of Mr. Pugh, seconded by Mr. Skelhorne, Viscount St. Vincent was re-elected President. The following were elected as committee: Messrs. TN Harrison, SW. Marriott, AG Pugh, GE Puttergill, GE Skelhorne, G. Smithurst, HW Dickman, WH Windle, FG Vessey, W. Darrington, H. Sheppard, G. White, and W Adams. Mr. G. Hayes was re- appointed secretary, after holding office for 11 years, and Messrs. Pugh and Hayes were re-elected delegates to the BBKA.

After the meeting the members partook of tea, and a competition for the best bottles of honey was also held. Subsequently Mr. AG Pugh presided over the meeting, at which addresses were given by Mr. P. Scattergood, Mr. WP Meadows (Syston), and Mr. W. Herrod, FES.

BBJ 15 Feb 1906
Bee-Keeping as an Occupation. I have read with much interest the letter of your correspondent "Sahib" and the conclusion I have arrived at is that a man who could earn £150 a year from bee-keeping in England would not have time to spend it. My view of the "probabilitie" of making the sum stated is based on the following calculation:
1. An average of 100 lbs of honey per hive would be a "record" take in this country.
2. Fifty pounds would be a good take.
3. Thirty pounds is a more likely all-round harvest. We must also bear in mind the seasons of entire failure.
4. The wholesale value of honey does not exceed 6d per lb so that the last-named figure would produce 15s per hive, and therefore, 200 stocks would be required to yield £150 net clear profit. To establish an apiary or apiaries, and purchase the necessary plant in good working order in a good district, would need the expenditure of close on £500, either in time or money, for if the bee-keeper makes his own hives, etc., there is the value of his time gone before his apiary has reached the profitable stage.

To make the matter plain, I ask: Can a boy, as he enters on the first day of his apprenticeship, say I know my trade? Or, to go further, can he, on finishing the last week of his apprenticeship, say I know all about the business? I, therefore, ask your correspondent, who has, maybe, spent the greater part of his manhood abroad in his country's service, could he enter directly into a life requiring much technical knowledge, combined with experience, and at once expect to succeed?

I have often said a man has only mastered the growing of a plant when he has succeeded in mastering its enemies. And it is just so with bees; sooner or later the extensive bee-keeper finds disease about and he must master that trouble or great will be his loss. We hear of very big things being done with bees in that land of big things, the USA, 750 colonies in one yard, with a 5,000 acre range of buckwheat for bee forage, that has never failed a harvest yet.

Mr. WZ Hutchinson says: "Bee-keeping is not an occupation in which one can easily become wealthy; but it can be depended on to furnish a comfortable living. Fortunately, the professional man's happiness bears little relation to the size of his fortune; and the man with the hum of the bees over his head finds happiness deeper and sweeter than ever comes to the merchant prince, with his cares and his thousands."

Merrie England is too "tight" a corner of the world to do such big things, but the hum of the bee can help to give many a pleasant hour to, say, a retired soldier, and add moderately to his wealth; that is, if he loves bees, their stings, and is fond of work!

J. Gray, Expert BBKA and C.C. Lecturer, Long Eaton.

BBJ 15 Feb 1906

The Annual Meeting of the NBKA was held at the Peoples' Hall on February 10th, Mr. P. Scattergood presiding. Amongst those present were Mrs. Laws, Messrs. AG Pugh, W. Herrod, WP Meadows, R. Mackender, J. Herrod, RJ Turner, G. Smithurst, J. Gray, WS Ellis, W. Darrington and G. Hayes, secretary.

The committee's report, in referring to the honey season, stated that the experts had during 1905 visited 194 apiaries, containing 716 stocks of bees, 586 of which were reported healthy, and 37 diseased in varying degrees. In regard to the latter there was a slight increase. The committee were doing all they could to suppress the disease in the county, and it was hoped that the members would back up their efforts by closely following the instructions of the experts. There had been a slight increase of the membership and the financial position of the association was about the same as last year. The report was adopted.

Viscount St. Vincent was re-elected President, and the following were elected as committee: Messrs. TN Harrison, SW Marriott, AG Pugh, GE Puttergill, GE Skelhorn,

G. Smithurst, HW Dickman, WH Windle, FG Vessey, W. Darrington, WH Stoppard, G. White, and W. Adams. Mr. G. Hayes was re-appointed secretary, after holding the office for eleven years, and Messrs. Pugh and Hayes were re-elected delegates to the BBKA.

After the meeting the members partook of tea. Subsequently addresses were given by Mr. P. Scattergood, Mr. WP Meadows, and Mr. W. Herrod, FES. The meeting concluded by the usual drawing for hives and appliances kindly given by the President and other gentlemen.

Minute Book – Special Committee Meeting held in the Peoples' Hall on February 15th, 1906 between the business meeting and tea at the Annual Meeting.

The Secretary read the correspondence from Mansfield Horticultural Society agreeing to our terms for the Annual Show to be held with them on Bank Holiday, August 6th and asking for a schedule of prizes to enable them to print it with theirs.

After considerable discussion it was eventually decided to adopt the schedule of last year with the addition of a class for appliances in which one prize of £2 only be offered, and in Class II Mr Pugh supplement Mr Gray's first prize of 10/- with a second one of 5/- out of his own pocket.

Minute Book – Special Committee Meeting held February 18th during the Annual Meeting to prepare schedule for the Annual Show.

After considerable discussions on various classes in the last year's schedule, it was eventually decided to adopt the same for the 1906 show with the following alterations:
> that a Class for appliances with one prize of £2 be added.
> that for Classes 4 and 5, it should be 12 lbs instead of 6, and for Class 12 Mr Gray promised 10/- for first prize and Mr Pugh 5/- for second prize.

NEP 3 Apr 1906
Mansfield Horticultural Society. The monthly meeting was held last evening in the Old Meeting House schoolroom. Mr. George Hayes (secretary of the NBKA), gave an illustrated lecture entitled "Beekeeping for Gardeners."

Minute Book – Quarterly Committee Meeting held in the Peoples' Hall on April 6th, 1906. Present Messrs Pugh, Puttergill, Skelhorne, Smithurst, Vessey, Stoppard, Scattergood, Darrington, Turner and secretary. Mr RJ Tuner was voted to the chair.

The minutes of the previous meeting were read and confirmed by signature.

Correspondence was read about doing the Horticultural Society, Nottingham Arboretum, Southwell, etc.

The secretary next presented a quarterly statement of accounts which were passed.

The election of the Technical Instruction sub-committee next took place and it was resolved that the old committee be re-elected *en bloc* with the exception of Mr Randall (resigned) and that Mr Darrington be added.

A proposal shewing reduction in all classes at Southwell from their secretary was considered and it was decided to accept, but that the Southwell Society should be advised that we hope the prize list will be increased another year. Mr Scattergood was appointed to judge.
> Resolved that we accept the same terms for East Bridgford and that the secretary be appointed judge.
> Resolved that the secretary be empowered to arrange for any other local shows at a cost to this association not exceeding 20/-.

Mr Pugh, as delegate to the BBKA gave a resumé of the whole of the proceedings at the London meeting, including the strength of standard frames, foreign bees, etc. and the Baroness Burdett-Coutts prize hives and it was decided we should apply for same.

Baroness Burdett-Coutts, Victorian philanthropist, (President of the BBKA from 1878-1906) set up a Prize Hive Competition for deserving cottagers. Some considerable difficulty has arisen as to the meaning of the words "cottage member." The Competition Committee has decided that for the purposes of this competition the words mean, "any member of an association whose house-rent does not exceed 6s. per week."

BBJ 10 May 1906
Bonus (Newark) No. 1 sample shows "Foul Brood" in a very pronounced form. In No. 2 there are slight signs of the disease in two or three cells, but in the remainder the brood is "chilled" only.

NEP 14 May 1906
Every year the Newark Agricultural Society's Show seems to grow in popularity, and the annual gathering to-day witnessed a greater influx of people into the town than on any previous occasion. The exhibition is always held on the days of the May Fair and hirings, and the two events combined afford the country folk, as well those from large centres farther afield, a pleasant opportunity of foregathering. Lectures on bee-keeping with demonstrations with live bees were given by the NBKA.

NEP 4 Jun 1906
The Annual Show of the Notts. Agricultural Society, to be held Colwick Park tomorrow and Wednesday, gives every promise of being a great success. NBKA will give lectures on beekeeping and manipulations with live bees, etc.

NEP 6 Jun 1906
NAS Second day. Over 10.000 attended Colwick Park with the weather again sunny and clear, as the remainder of the programme was carried out under ideal conditions. The lectures by the NBKA added general interest to the proceedings.

BBJ 14 Jun 1906
WHW (WH Windle?) (Nottingham)."Advanced Beeculture". The author of this work is Mr. WZ Hutchinson, editor of the Beekeepers' Review, an American journal of high repute. The book was reviewed in our issue of March 1 this year. It is written on entirely different lines to the "ABC of Bee-culture."

A digital copy of this book can be found in the NBKA library

BBJ 5 Jul 1906
The sixty-seventh Annual Exhibition of the Royal Agricultural Society was held on an admirably chosen spot at Alvaston about a mile outside Derby. A more suitable place could not have been selected, the formation of the ground being such as to keep the various sections of the show (in spite of the enormous extent of the exhibits) compact, and easy of inspection by visitors. Peter Scattergood amongst others judged the exhibits.
Class 434. Six Jars of Heather Mixture Extracted Honey - 3rd, AG Pugh, Beeston,
Class 441 Exhibit of a Scientific Nature - 1st, Geo. Hayes, Beeston (Nature Study of the Honey Bee)

Minute Book – Committee Meeting held in the Peoples' Hall at 3pm on July 7th, 1906. Present: Mr Pugh in the chair, Messrs Darrington, Dickman, Harrison, White, Stoppard, Turner, Smithurst, Vessey and secretary.

The minutes of the previous meeting were read and confirmed by the signature of the chairman. Mr Darrington proposed and Mr Vessey seconded that the quarterly statement was next gone into and this being considered very satisfactory it was passed. Mr Harrison proposed and Mr Turner seconded.

The matter of the judge for the Annual Show was discussed and the following names were put forward – Mr Bray, Mr Carr, Mr W Herrrod, Mr Scattergood and Mr Crawshaw. After some discussion Mr Vessey proposed that they should be asked to undertake the duties of judging in the order enumerated until one accepted. Mr

Turner proposed that they should be taken alphabetically. The latter was agreed to.

Mr Pugh proposed and Mr Turner seconded that the secretary write to the BBKA informing them that it is considered that the insurance year, as it now stands, is most inconvenient and suggest it should be from Lady Day (25th March) to Lady Day in each year.

BBJ 19 Jul 1906
The Season in North Notts. We have had splendid bee weather here. With the exception of a very light fall of rain there has been no break for three or four weeks and no cold winds, so that the bees have held high revels.

I do not remember having ever before seen such a wealth of white clover bloom as there is now in this district, but unless rain comes speedily and in quantity I fear its time is nearly over. With rain we might still have it yielding well for another fortnight or three weeks. One or two indications show that the flow of nectar is diminishing, not the least significant being the disposition of workers to attack the drones. This tendency is very evident today (July 14th).

I had four stocks in prime condition with which to open the season, one of which threw a large swarm on May 29th. The others have been prevented from swarming simply by giving the bees plenty of room. A strong second swarm - or "cast" - followed after the usual interval, but the bees returned to the parent hive after being hived, and this happened three times. Twice they clustered but once they returned without. From what I hear about the doings of stocks in skeps swarming has been a perfect nuisance in this neighbourhood

My best hive is now four storeys high, ie, brood-chamber, set of standard frames and two sets of shallow-frames – and packed with bees from top to bottom, and this from a single stock driven from a big sugar-box on August 19th last year. Two others were built up from driven bees.

My experiences this year have decided me on two points.
First, I shall never again work sections and extracting frames on the same hive.
Second, I do not intend for the future to make any type of hive except those with loose body-boxes. This is not because of anything connected with yield, etc., but simply because of ease in handling and cleaning.

I have already removed two racks of nicely-finished sections and have commenced extracting. The honey is of fine quality, practically pure clover and the lightest in colour I have ever harvested.

One word with regard to section-racks. I have two kinds in use - one fitted with tin girders, and the other with simple strips of wood. I prefer the latter very much, for the sections come out much cleaner owing to the closer fit. Trusting that the honey-yield all round will be as good as mine promises to be, I sign North Notts.

BBJ 19 Jul 1906
Insurance for Beekeepers. Referring to the invitation from the Council of the BBKA set forth in the report of their meeting, I am instructed by the committee of the NBKA to say that they consider the Insurance year as now fixed, to be most inconvenient, seeing that it breaks into two seasons. Our committee therefore suggests that it should run from Lady Day to Lady Day. The reasons for this are:
> 1st, at that period the season is just opening:
> 2nd, Annual Meetings are being held, and at such times members would be more likely to take it up or remember to renew existing policies.

Geo. Hayes, Secretary, July 12th

BBJ 19 Jul 1906
More Swarming Troubles. On May 12th I successfully hived a fairly large swarm, which by the third week in June covered ten frames but the bees did not start work in the super put on early in that month till the 29th when I found they had taken possession. Three days later, however (July 2nd), two swarms issued, one large, the other small. The latter made off and was lost, whilst the larger one was not found until after I had examined the hive and cut out half-a-dozen queen cells. I attempted to hive the large swarm as soon as it was discovered, but being in a very difficult position for hiving, I was unsuccessful as the bees, on being disturbed, returned to the parent hive. I then put two new frames of foundation in the latter and placed an excluder over the hive-entrance for twenty-four hours. Will you kindly tell me:
1. Why two swarms should issue at the same time?
2. If it was right to cut out queen cells so soon after they had swarmed?
3. If it was right to put a queen excluder at the entrance?
4. Is it probable there is a young queen in the hive?
5. The name of the Secretary of the NBKA?

The hive seems to be doing well, and the bees are in the super and covering all the twelve brood frames. I send name, etc., and sign "Notts."
Reply
1. It was not two but one swarm divided into two clusters. The smaller lot evidently had the queen with them and, being lost, the large cluster hung till found when, on being disturbed and queenless, the bees returned to the parent hive.
2. As it turned out your action in cutting out queen cells will have cost the bees a fortnight or three weeks in raising a new queen to replace the lost one.
3. No, it is always bad (sometimes disastrous) policy to attempt to stop

swarming by preventing the free exit of bees from a hive at swarming time.
4. The bees will probably be raising one.
5. Mr. Geo. Hayes, Mona Street, Beeston, is Sec. of NBKA.

East Bridgford Parish magazine August 1906
It was a most delightful day and there was good attendance at the Annual Show. The number of entries was quite up to the average and, considering the late continuous poor weather, it was pronounced to be a very good one.

Bees and honey
   Sections – 1st, Miss Fox; 2nd, Mr D Gower
   Run honey – 1st, Mr W Sentence: 2nd, Mr D Gower; 3rd, Mr D Marshall
   Bees – 1st, Mr D Marshall; 2nd, Mr J Higgs; 3rd, Mr W Millington

BBJ 2 Aug 1906
For Sale, a quantity of hives, full of Bees and Honey, in healthy condition; also Bee Appliances.   Apply to Mr. Walter Stevens, Bath Lane, Mansfield

Sheffield Independent 8 Aug 1906
Welbeck Tenants' Show. Mr. Pugh, of the NBKA, gave a lecture with practical illustrations of the best and most profitable methods of beekeeping, showing appliances of the most modern type and of reasonable cost and how to use them. This was a feature of the show which attracted considerable interest on the part of the tenants who are learning to look upon bee-keeping as a profitable department of farm work.

BBJ 9 Aug 1906
HPD (Nottingham). Comb Foundation Making. The process of preparing wax sheets and converting the same into comb foundation is far too complicated a business for us to give the necessary description in this column, even if we were sufficiently informed on the subject ourselves, which we are not. The "Root" machine you have acquired must be one of the ordinary machines formerly used by that firm, who now manufacture foundation only by the "Weed" process. For those who care to make their own comb-foundation the "Rietche" press is, we think, more suitable than the old-style roller press, as the wax sheet is "cast" between two compressed plates thus dispensing with all the paraphernalia of tanks and dipping.

BBJ 16 Aug 1906
The NBKA Annual Show was held in connection with the Mansfield Horticultural Society, in the beautiful grounds kindly lent to them by Mrs. Clarke, Carr Bank, on Bank Holiday, August 6th, the day being delightfully fine and the attendance a record one. The entries in the Bee and Honey Department was also a record one and the

quality of the honey staged was excellent. Mr W. Broughton-Carr, London, officiated as judge of the Honey Class, assisted by Mr. W. Herrod, Luton, their awards being as follows:

Trophy of Honey in any Form and of any year – 2nd, WL Betts, Mansfield Woodhouse; 3rd, D. Marshall, Cropwell Butler

Twelve 1-lb Jars Light-coloured Extracted Honey- 1st, T. Marshall, Sutton-on-Trent; 2nd, WL Betts; 3rd, W. Sentence, Shelton; 4th, J. Wilson, Shirebrook

Twelve 1-lb Jars Dark-coloured Extracted Honey – 1st, AG Pugh, Beeston; 2nd, G. Marshall, Norwell; 3rd, D Marshall; 4th, W. Ball; hc, T. Hilton, Laxton

Six 1-lb Sections – 1st, T. Marshall; 2nd, G. Marshall; 3rd, EG Ive, Boughton

Six 1-lb Jars Granulated Honey – 1st, J. North, Sutton; 2nd, AG Pugh; 3rd, GH Pepper, Farnsfield; hc, KG Turner, Radcliffe

One Shallow Frame of Honey for Extracting – 1st, J. Wilson; 2nd, G. Marshall; 3rd. J. North; 4th, W. Ball

Six 1-lb Jars Extracted Honey (Novices) – 1st, J Wilson; 2nd, WL Betts; 3rd, J. North; hc, W. Adams

Honey Vinegar - W. Ball

Observatory Hive with Bees and Queen -1st, EG Ive; 2nd, G. Marshall; 3rd, WL Betts

Beeswax – 1st, G. Marshall; 2nd, W. Ball

Mr. George Hayes, of Beeston, hon. Secretary to the NBKA, delivered several lectures during the afternoon to large audiences. Mr. W. Broughton Carr held an examination during the day for the 3rd class experts' certificates of the BBKA, four candidates presenting themselves.

William Broughton Carr (1836-1909) was a business man in Liverpool. He lived on the Wirral and kept bees before he was invited down to London by Thomas William Cowan to set up and help edit the Bee Journal. He was the editor of the "Bee Journal and Record" developing the WBC style of beehive and publishing his design in 1890 as shown.

BBJ 16 Aug 1906

Dealing with Suspected Combs. Would you please say if there is any disease in the enclosed sample of comb? I have been a bee-keeper some time without ever having seen "Foul Brood", but I am anxious to nip it in the bud if present. The perforated cells containing the white grubs made me doubt whether it was altogether healthy.

JM, Mansfield

Reply. There is no disease visible in the comb sent. All the brood in the sealed cells has reached the imago stage, and some of the young bees were alive when the comb was examined, but the very strong odour of carbolic acid makes it probable that the bees have deserted the brood, and that there has in consequence been

insufficient warmth for enabling the young bees to hatch out. Another possibility is that the hatching brood has been killed by the fumes of carbolic acid.
BBJ 23 Aug 1906
JW (Notts.) Both samples are very good. That in the tall bottle is less clear and bright than the one in the small flat phial, and it would be improved by straining to free it from a few small specks. Of the two, we prefer the sample in the small phial for the show-bench.

BBJ 30 Aug 1906
Few 1906 Queens, 3s 3d each.	R. Mackender, Seeds and Bees, Newark

NEP 4 Sep 1906
The first essential of the success of agricultural or horticultural show is fine weather, but it is just probable that the weather which has been with us during the week-end would have been too fine for such an event, so that the promise of a change to a more serene temperature was regarded hopefully by those interested in the Moorgreen Agricultural Show, which opened this morning. A local beekeepers' show was also held in connexion with this exhibition, and there were 35 entries, the quality being very good.

BBJ 6 Sep 1906
Perfect combs, from Foundation and Driven Bees, by using "Nondescript" device. Cannot stretch. Better than wiring. 20 years' proof in large apiary. Sample, 7 stamps. Full set for one frame, 15 stamps.	Palmer, Hayton, Retford

BBJ 13 Sep 1906
The Fourteenth Annual International Exhibition and Market of the Confectioners' and Allied Trades opened on September 8th and remains open till the close of the present week. There was a further increase in the demand for space this year over that of 1905, and in consequence the honey competitions were again relegated to the annexe in the North Gallery. The change, however, gave those especially interested in the display of honey and bee-products a better opportunity for examining the exhibits than in the crowded avenues below.

Display of Honey (comb and extracted) and Honey Products, shown in suitably attractive form for a tradesman's window (4 entries) – 4th, (£1), J. Herrod, Sutton-on-Trent, Newark.
Twelve 1-lb Sections (12 entries) Three exhibits disqualified for overlacing – 2nd, (£1 5s) J. Herrod
Three Shallow Frames Comb Honey for Extracting (8 entries) – 3rd, (10s) T Marshall, Ivy Cottage, Sutton-on Trent
Twelve 1-lb Jars Light-coloured Extracted Honey (44 entries) – 2nd, (£1 5s) T. Marshall

Twelve 1-lb Jars Heather Blend Honey (5 entries) – 1st, (20s) AG Pugh, Beeston
Beeswax in Cakes, Quality of Wax, Form of Cakes and Package suitable for retail counter trade (9 entries) – 1st, (£1) G Hunt, Hawton Road, Newark

BBJ 13 Sep 1906
Bee Notes from North Notts. The honey season here has been a good one on the whole, the yield being very fair in weight, and the quality good. Most of the surplus I secured was gathered during July, chiefly from clover. Some darker stuff was, however, obtained in June from the hawthorn, etc. The hives round about here of which I can get particulars have stored about 30 lb to 40 lb per hive; not bad for the neighbourhood. I have heard of fifty-seven sections and 48 lb of extracted honey respectively being taken from two stocks. Our locality is not now a very good one for bees, owing to the extensive building operations of late years. No surplus at all has been stored since August 1st owing to the want of rain; in fact, the pastures have lost their usual green colour and become a beautiful brown (quite scorched up), the soil hereabouts being very light and sandy.

We have not had a good rain since June 28th and 29th ie. during the Royal Show time; and I have no doubt our Junior Editor, Mr. Carr, well remembers those two days, as I do myself, after being examined by him in the downpour. In consequence of the lack of bee-forage caused by the drought, and the subsequent hot weather which has prevailed ever since, the bees have been very mischievous and "robbing" has only been prevented with great difficulty.

The splendid weather this season has, however, been a good thing for getting young queens mated and I have not heard of many failures. But it has been a bad one for cases of decamping swarms, many bee-keepers having lost both swarms and casts. One skeppist I called upon last week had lost three out of four swarms through the bees not clustering well. This same bee-keeper (although he had kept bees for forty years) had only seen a queen once during the whole time and, from what I could learn from him, he said he does not really know now what one is like.

I find the brood-chambers of stocks are almost bare of stores this autumn and they will require well feeding up in consequence; but, having had a good season, this does not trouble us much. I will now close, hoping that 1906 has been a good year for most bee-men, and that it will be repeated in 1907. I hope also other bee-keepers will let us know how they have fared this season. I should guess that it has been rather a good time at the moors. I only wish I was fortunate enough to be within easy reach of the "ling." I sign as before: North Notts, September 4th

BBJ 13 Sep 1906
Bee Paralysis. Some years ago I had a bad attack of this in my apiary. About fifteen

stocks were affected and from strong colonies the bees went down in numbers to less than weak ones. In fact, with the exception of one hive, they were reduced to one and two frames of bees - and sparsely covered they were at that. I attributed the trouble at the time to the hives being overheated when taken to the heather for not having far to take them I was rather careless as to ventilation.

When the disease broke out and during the whole of the time it continued, I noticed that all the sealed honey was wet and sweating. As the bees decreased in numbers I took the uncovered combs away and melted them down for wax.

The honey in them had a sour smell and taste, but the smell was especially noticeable. Indeed, so long as there was any of the old honey left in the hives the trouble kept on; but as soon as it was cleared out the disease disappeared and I have never seen any trace of it in my hives since. All the stocks managed to recover their strength again by the end of the season.

I wonder if any of your readers have noticed the "sweaty" or damp appearance of combs when the disease was present? Name, etc. enclosed for reference.

Nondescript, Notts, September 3.

"Nondescript" was the nom-de-plume of W Palmer originally from Retford then of Netherfield

BBJ 13 Sep 1906
WHS (Notts.) Exhibiting Bees in Tradesmen's Windows. There can be no valid objection to enterprising tradesmen seeking to attract buyers in the way you mention, but when it comes to their selling such stuff as the sample sent as "choicest strained honey" we should fancy that buyers would themselves protest and ask for the return of their money. Our opinion of the 'choicest' etc., is that it is some wretched foreign stuff entirely unfit for table use.

With regard to its "being a punishable offence" as you say, the remedy is in the hands of any one who buys it by informing the authorities connected with the Pure Food Acts.

WHS was how WH Stoppard of Mapperley signed himself in the many letters he was to write to the BBJ.

BBJ 13 Sep 1906
Experienced Expert seeks situation as Allround or Handy Man. Fair amateur joiner and hive maker. W. Palmer, Hayton, Retford

BBJ 20 Sep 1906

The Confectioners' Exhibition. I read your editorial of last week dealing with the above and beg to send a line or two on the subject by way of reply. Personally I am not surprised at bee-keepers being reluctant to enter their honey at these large shows.

Take my own case. Last year I made an entry in the class for light-coloured extracted honey at the Grocers' Exhibition, in London. The total cost of carriage, etc., to and from the show was 16s 9d - rather a big amount for the honour of staging one dozen jars of honey. Now add to the above the cost of a railway journey of 300 miles, with loss of day's work and something for the inner man, and you will agree, no doubt, that it was a pretty expensive exhibit.

But this is not all. I made that journey for the educational gain which I felt sure must follow but imagine my surprise upon arrival at the hall on the Monday, to find a barrier keeping the public six or eight feet away from the chief exhibits (with a uniformed attendant in charge), and to find that the prize cards had not been affixed to the exhibits, even as late as 2.30pm, though the awards were made on the previous day.

If this is considered grumbling without a cause, please forgive me, but I have felt sore about it ever since, and if other exhibitors of last year were in anything like my position the disappointment complained of in your remarks is partly accounted for.

Further, I consider the standard testing glasses for light honey serves but one useful purpose, that is, to prevent complete disqualification of "class" entry. To my mind, one might as well show a jar of treacle as stage honey in the right class, which only just comes inside the colour test, as this seldom or never takes a prize, therefore it appears a foolish and unnecessary expense to do so.

If the honey - to win - must (besides possessing the necessary good points) be so very light in colour, what chance has any exhibitor showing in the same class with honey of a darker colour, although classified correctly? His only chance is to enter the next season's honey in the "Not for Competition" class. But, what a position! Beautiful honey, too dark for the light class, yet not dark enough for the medium class!

With every good wish to our Editors, allow me to send name, etc., and sign as before
Worker, Notts.
[We were sorry to receive the above singularly unfortunate account of "Worker's" experience of last year's show, but our correspondent cannot justly blame anyone but himself for what we regard as the most serious item on which his complaint is grounded - viz., cost of carriage. If this is explained we will go into the other matter and endeavour to afford a satisfactory explanation . Eds.]

BBJ 27 Sep 1906

The fourteenth Annual Exhibition and Market of the Grocery and Allied Trades, held at the Agricultural Hall, London, was opened on the 22nd inst. and continues till the end of the present week.

Three shallow frames Comb Honey for Extracting (7 entries) – 1st, (£1) J Herrod, Sutton-on-Trent; 3rd, (10s) J Wilson, Shirebrook

Twelve 1-lb Jars Medium-coloured Extracted Honey (39 entries) – 4th, (10s) T Marshall, Sutton-on-Trent

Twelve 1-lb Jars Dark-coloured Extracted Honey (14 entries) – 1st, (£1) G Marshall, Norwell

Twelve 1-lb Jars Granulated Honey (7 entries) – 4th, (10s) J Herrod.

Beeswax in Cakes, Quality of Wax, Form of Cakes and Package suitable for retail counter trade (14 entries) – 2nd, (15s) J Herrod

Beeswax judged for quality of wax only (18 entries) – 3rd, (10s) J Herrod

Minute Book – Committee Meeting held in the Peoples' Hall on October 6th, 1906. Present: Mr WS Ellis in the chair, Messrs Adams, Pugh, Puttergill, Scattergood, Smithurst, Dickman, Windle, White, Vessey, Stoppard, Mackender, Turner, Darrington and secretary.

The minutes of the previous meeting were read and on the motion of Mr Pugh, seconded by Mr Vessey, they were confirmed by the signature of the chairman. The quarterly statement was next gone into and it was considered to be very satisfactory. Mr Vessey proposed the same be passed and this was seconded by Mr Turner and carried.

A report on each show was next given when Mr Darrington proposed and Mr Pugh seconded that a very hearty vote of thanks should be accorded to Mr WB Carr for coming to judge at our Annual Show and also that the particulars of entries, etc., and of new members and prize winners, should be given in the Annual Report. This was seconded by Mr Pugh.

The awarding of the Baroness Burdett-Coutts prize hive was then considered and after some discussion, it was decided on the motion of Mr Vessey, seconded by Mr Mackender, to award it to Mr W Sentence, Shelton.

A letter was read from the President, Viscount St Vincent, asking to be relieved of the office and it was decided to refer the matter to the Annual Meeting.

Mr Scattergood, referring to the late Mr PJ Allsebrook, proposed that a letter of condolence and sympathy be sent to Mrs Allsebrook and family and it was also decided to have the matter of another appointment for the sub-committee of

Technical Instruction to Mr Bristowe.

Mention was made of the fact that a cheap honey called "Strained Honey" was being exhibited in a shopkeeper's window in Nottingham with an observatory hive of live bees and it was recommended that this should be mentioned in the report with a view of discouraging this undesirable practice.

BBJ 11 Oct 1906
Experienced Bee Expert wants situation as Handyman; gardening (no glass), poultry, frame hive, etc., making. W. Palmer, 174, Curzon Street, Netherfleld

BBJ 18 Oct 1906
Bee Paralysis. I note that your correspondent "LSC" has one "cap" at least that does not fit well. I refer to the one on bee paralysis. I can assure him that there was no sign of dysentery in any one of the hives affected. All hives were dry and clean inside, and the combs same, bar the slight sweating. The bees affected seemed to have all the symptoms mentioned in every article I have since seen in the Journal on the subject of bee paralysis. They were twice and thrice the normal size in the abdomen, and could not fly. Some were dragged out, others had just the strength to crawl out, and would make almost an attempt to fly, but fell to the ground instead, where there would be sometimes a cluster or heap of a hundred or more underneath the flight-board with just enough life in them to show that they were alive. On opening a hive they were to be seen dotted on the hive-sides, on the floor-board, and at the edges of combs, distended, but not one burst among the lot.

When the stocks were practically recovered I noticed another curious thing. For weeks during the hottest part of the day those hives that had been affected seemed to be very busy at the entrances. At first I thought it was a case of "robbers." On some of the flight-boards there would be half a dozen or more captive bees at a hive, as many bees as could get round, and on it at a time appeared to be pulling and biting the captive, without it resisting or trying to get away. In the end the poor bee would be gradually bundled off the flight board on to the ground with several bees hold of it, and when even released it flew up again, only to have the same process gone through again. A good many of these odd bees were black and hairless, and shiny as a black glass bead, and the blacker they were the more feeble they seemed. None of them appeared to be killed outright, but seemed at last to become too exhausted to fly up again, and so died. Nearly all the dead bees on ground (and there were many), were black and shiny. It was certainly not a case of "robbers."

I should like to ask our friend "LSC" if he has tried the plan he mentions as to placing alternately frames with starters only between combs full of syrup for bees to winter on? To me the idea seems like inviting disaster. His idea as to covering foundation in the saw-cut with paraffin wax seems one better; but why have a saw-cut at all? Why not use solid top-bars and fasten foundation on with molten wax? I have used thousands in this way, and only had two or three topbars give way, and these contained knots. The standard thickness for top-bar is, I think, quite strong enough, but would be better if a little wider. If a full inch or an inch and an eighth it is stronger and better, for the reason that very few bracecombs will be found between it and section-racks or next tier of frames.

May I also ask him why foundation in brood combs should be wired? I had a few wired combs in use at one time, but discarded them because I found that where the wires ran a number of cells were never bred in; therefore I lost many young bees every time. The combs I refer to were built from and on American wired foundation. I have seen the same objection to wiring raised by others at times.

Harking back to top-bars, I have had hundreds of solid topped frames, standard size, weighing, when full, from six to over eight pounds each, and not a bent top-bar amongst them, and some of them had been in constant use for ten years at least.

Nondescript. Notts, October 15th.

BBJ 18 Oct 1906

Managing Feed-Holes in Quilts. Most bee-keepers use on top of the first "quilt" some few thicknesses of thicker material for warmth and, when feeding through these many will have noticed that a few bees will get hanged, either by the neck or the legs. To prevent this I cut all my top quilts of a size, and with a sharp chisel cut the feed-holes through ($1\frac{1}{2}$ in. x $\frac{3}{4}$ full), so that anyone of them when on the hive would have the hole exactly at centre of frame tops; but before placing on hive I lined the hole in quilts with tin.

To do this we will suppose the several thicknesses of material are, when loose, a good quarter inch thick, they will be, when tightened up, nearer an eighth; so cut a strip of tin six inches by two and in the centre, lengthways; mark two lines an eighth apart. Then, $\frac{3}{4}$ in. from one end (and each side), cut the tin through down to nearest line; same $1\frac{1}{2}$ in from these "cuts," and so on alternately till you have four cuts on each side.

The tin will now bend up into the same shape as a common match-box cover. Now push one end of this bent-up tin through the holes in quilting, bend back the cut pieces each side, tap with a hammer, and you have a quilt that saves time in fixing, saves bees from hanging, and with no frayed edges for bees to be constantly teazing at. A bit of glass under the tin hole, and atop of the bottom quilt, keeps all snug when feeders not on.

Nondescript (Nottingham)

BBJ 6 Dec 1906

Hiving Driven Bees. It seems strange how "doctors differ." We have, "DMM" sticking to his guns by stating that hiving driven bees above frames is reprehensible for many reasons, and then we have one who signs himself "Expert," going just the opposite.

For myself, I may say, having driven and hived hundreds of stocks, I have come to the conclusion that the "top of frame" plan is most to be depended on. If the hive is a "WBC" or similar, one, the space round the body-box should be temporarily filled with some soft material, so as to keep the bees out of it, and in ninety nine cases out of a hundred no difficulty arises, if the instructions given are followed. In all cases I consider it best to have the "lift" on the hive before shaking in the bees, if possible. The cloth, or dust-sheet, should be dropped inside the "lift," so that its weight is on the bees. (I have generally used cheesecloth, or whatever material has been used to fasten the bees in the skep for bringing home.) Then a sack laid over the top of all to keep stragglers in or out, and there you are!

I have often found trouble when running bees in at the bottom, and it has generally been when two queens have been present. One will have run in with her bees, and the other, with hers, would persist in staying outside - under the porch, generally. A case of this kind happened with a neighbour only this autumn, and he called on me to put them right after the bees had been in two lots, one out and one in, all night.

Nondescript, Notts, December 3rd.

[Being one of the "doctors" who agree with "DMM" in differing with our friends "Nondescript" and "Expert," we add a line to say that we should never dream of hiving either driven bees or swarms on any other plan than the orthodox one followed by nearly all the experts and bee-men with large apiaries to deal with we have met during our more than forty years' experience of bee-work in all its forms. Much, of course, depends on the operator himself, seeing that some beekeepers - even of long experience - have not the same methods of handling bees as are now practised. At the same time, it is better that "doctors" should "agree to differ," and allow everyone to follow the plan most suited to his own method of handling the bees. Our own methods are fairly well illustrated in the "Guide Book." Eds.]

BBJ 20 Dec 1906
Clearing Bees from Supers. The articles written by our friend, "DMM" are always good reading, and the advice he gives therein is, as a rule, worth following; it therefore seems almost like heresy to doubt anything he advises but I really cannot see how his plan of clearing bees from supers is a saving of beelife.

In nearly every super - during the reasonably early season—there are always a number of young bees, many of which have, perhaps, never flown (a fact especially noticeable if the day or two previous have not been warm ones). I ask: What becomes of these young bees after they have escaped through "the cones"? They know not where to fly to, so some may get home, some may go into neighbouring hives, some may get down to ground, never to rise again!

Suppose a shower or storm comes on while they are escaping; then the odds are that the majority of them would get chilled and lost instead of saved. If a "Porter" or other good escape is used, I do not see why a single bee need be killed or lost, however adverse the weather may be. It also appears to me that the difference in labour by following "DMM's" plan and by using a "Porter" is so small as to be not worth mentioning. Where does the interference with brood-nest come in? Does anyone interfere with that by taking any super off? Nonedescript, Notts, Dec. 16th

BBJ 27 Dec 1906
The monthly meeting of the BBKA Council was held on December 20th, at 105, Jermyn Street, SW., Mr. TI Weston occupying the chair. Apologies for inability to attend were read from Messrs. Geo. Hayes and AG. Pugh. On the examiner's recommendation it was resolved to award a certificate to Mr. WH Stoppard, Mapperley.

Report for 1906
The AGM of this Association was held in the Peoples' Hall on March 2nd, 1907, presided over by EF Milthorp, JP. of Newark, a good number of members being present.

The Minutes of the previous Annual Meeting were read and confirmed.

It was next proposed that the annual Balance Sheet as per the copy circulated amongst the members be received as very satisfactory and adopted. Carried unanimously.

The Committee's Report was next read by the Secretary, which was as follows:
The past season opened out well and augured a good honey year. Bees increased and did fairly well on the fruit blossom for some days, but about halfway through, it set in cold and wet. Later on it improved, and all during June the weather was everything that could be desired by beekeepers - except for just one thing - a few showers. If we could have had these generally, the season would have been a record one for honey; as it was one of sunshine but lacking these, the resultant harvest was only fairly good. The colour of the honey this year has been much lighter on the whole. Swarms were not very numerous, as the month of May was mostly very cold. The coming season may be the record one, so be ready for it!

It will, we are sure, be gratifying news to all our members to hear that during the past year we have enrolled 42 new members, 1 Vice-President at 21/-, 1 at 10/-, 22 ordinary members at 5/- and 24 cottage members at 2/6, making a total of members paid of 205 against 177 last year. This is our record membership and I trust that we shall not rest satisfied with what we have done, but that each will strive to increase our numbers, for in them is our strength. It may too, be interesting to quote membership for the last 10 years which is as follows.

Year	'96	'97	'98	'99	'00	'01	'02	'03	'04	'05	'06
Members	140	147	155	173	172	156	159	185	172	179	207

It will be seen the greatest number of members before was in 1892 when the membership reached 189. To the above mentioned 205 we may now add 10 new members for 1907 making the total to-day of 215. Some 10 members have not yet paid their subscriptions for 1906 but, as they have sent no intimation of their resignation, it is hoped their subscription will be forthcoming. As a natural sequence to the foregoing it is only reasonable to expect a favourable Balance Sheet, as we trust such you will find in that presented to you.

The Bee Tent with Lecturer and appliances has been to the following centres: Newark, Colwick, Balderton, Mansfield and Welbeck. The lectures at which, this year more than any other, the attendance has been larger and more interest evinced.

Of five candidates who presented themselves for examination for third class certificates, four were successful: Messrs. W. Munro, Mapperley, WL Betts, Mansfield Woodhouse, H. Dickman, Babbington, and G. White, Sandiacre. Only one went in for second class, viz. Mr. WH Stoppard of Nottingham and he too has been successful.

Experts visiting has this year been divided amongst seven Experts and they have collectively visited 229 apiaries, containing 804 colonies in bar and 76 in skeps; total 880. They have examined 650 colonies in barframe hives and 44 in skeps, total 694. Amongst these 49 were found to be affected with Foul Brood. We trust that every member will use their best endeavours to help us to stamp out this deadly foe.

It has been noticed that observatory hives stocked with English bees and shallow frames of honey have been exhibited in shop windows in Nottingham, along with a quantity of cheap foreign honey, the object of this being apparently to attract and force a sale of the foreign honey and to induce people to believe it was all British. Your committee hope that members will discourage as much as possible the use of bees and English honey for this purpose, as it is detrimental to our welfare as British Honey Producers.

Three shows have been held during 1905, at East Bridgford, Southwell and Mansfield. That at East Bridgford, though only a small show, was a success and helps to stimulate the industry in the locality, that at Southwell, where this year the prize value was somewhat reduced owing to the Southwell Society's grant being smaller, was good, the classes being fairly filled, and, as usual, a good quality of honey staged and we venture to express the hope that that the Society will enable us to increase the prizes at the next show.

The Annual County Show was this year held with new friends, viz. the Mansfield Horticultural Society, the entries to which broke the record of the preceding years as will be seen by the following comparative statement.

Class/Entries	1	2	3	4	5	6	7	8	9	10	11	12	Total
1904	1	2	9	7	6	4	6	3	1	8	0	7	54
1905	1	3	19	8	10	5	14	6	0	11	1	8	81
1906	2	3	15	10	8	7	13	11	2	9	0	8	89

A noteworthy fact in connection with this show was that most of the prize winners were new members and some of them new hands in beekeeping; and we trust this may serve as a stimulus to other new or young members, as it shows what may be done even by amateurs given a good district and season. We had the honour and pleasure of W. Broughton Carr, assisted by Mr. W. Herrod, to judge the exhibits, which proved to be no light task owing to the fairly uniform quality of many exhibits. The Silver Medal was won by Mr. T. Marshall, Sutton-on-Trent, the Bronze Medal by Mr. D. Marshall, Cropwell, Silver Pendants by Messrs. W. Ball, Eagle Hall, J. Wilson, Shirebrook, AG Pugh, Beeston, Certificate by Mr. W. Ball.

The year of insurance has now been altered so as to run from March 25th to March 24th. This was considered to be a better period for several reasons, one being that members might pay the premiums at the Annual Meeting or when they were paying in their subscriptions to the Association, for it is believed many do not insure simply through forgetfulness. It is thought we in Notts, should support this scheme a great deal better than we do for it is noticed there is a falling off amongst insured members.

Thanks are given to the County and City Councils for their grants which we trust will be continued as apiculture is assuming large proportions, and is taking a very prominent place amongst our industries, and as it grows requires more support. To the various Agricultural and Horticultural Societies for their encouragement of the pursuit as they see how important it is in relation to their interest. To the Local Secretaries and Experts and all who have done anything to forward the interest of our Association, to whom we feel our indebtedness and trust for the future help of all.

Mr. Pugh proposed, Mr. Darrington seconded, that this Report be adopted and printed with a list of members and a copy sent to each.

A letter was read from Lord St. Vincent stating that as he was going out of England it would be necessary for him to relinquish the office of President of this Association; and under these circumstances the Committee felt bound to accede to his request and a letter had been sent to Her Grace the Duchess of Portland, asking her to accept the position of President and she had graciously complied. Mr. Pugh therefore rose to propose that a letter be sent to Viscount St. Vincent thanking him most heartily for having held the office for so long and for his most generous help in providing annually the hives for the prize drawing, assistance to our prize fund, and for the many other ways in which he had helped forward beekeeping and our Association. This was seconded by Mr. Scattergood and carried with acclamation. He next proposed that Her Grace the Duchess of Portland be President for the ensuing year. Carried amid enthusiastic cheers.

Mr. Scattergood proposed, Mr. RJ Turner seconded, that the present Committee be re-elected *en bloc*. This was carried.

It was pointed out that an old and valued member of this Association, Dr. TS. Elliott of Southwell was with us again, after a long period of absence, and it was suggested that his name should be added to the Committee. He agreed to this and it was carried unanimously.

Messrs. Hayes and Pugh were re-elected representatives to the BBKA meetings. Mr. Scattergood was thanked for his honorary services as Auditor and re-appointed and the Secretary was also re-elected.

Mr. Stoppard next proposed that we have two General Meetings in the year and that the second be held in the summer time and take the form of a picnic or a visit to some Apiary or Works. This was received very favourably and carried unanimously with instruction to the Committee to arrange.

This concluded the business for the afternoon and the members and their friends, numbering about 70 sat down to a very nice substantial tea with which all seemed delighted and a vote of thanks was accorded to Mrs. Ellis and those friends who had prepared it.

The meeting was resumed at 6pm. presided over by WS Ellis, Esq., who presented the Medals, Pendants, and Certificates won at the Annual Show, and for the competitions at this meeting which were as follows:
Granulated Honey – 1st, Mr. Scattergood, Stapleford; 2nd, Mr. G. Hopkinson, Newark.
Run Honey – 1st, Mr. Ball, Eagle Hall; 2nd, Mr. Scattergood

35 Bottles of honey were brought (an improvement on last year) 23 being sent to the Children's Hospital and 12 to the children in the General Hospital, Nottingham.

Mr. Scattergood proposed that a letter be written to Mr. Milthorp thanking him for presiding this afternoon. This was seconded and carried unanimously.

Mr. WH Hoyte proposed "that our heartiest thanks are due and hereby tendered to WS Ellis, Esq., for stepping into the void caused by the resignation of Viscount St. Vincent, in providing two hives for the prize drawing." This was carried with acclamation.

Mr. Scattergood then gave a short paper on "Beeswax and its adulterants." Answers were given to a number of questions which had been sent up, several novelties were shown, and the delegate gave a report, as far as time permitted, of what took place at the meeting of the BBKA which appeared to give satisfaction to the members.

The usual prize drawing concluded the meeting which resulted as follows:

		Given by	Won by
1st	Harrison WBC hive	WS Ellis	Mr Mackender, Newark
2nd	Harrison Holborn hive	WS Ellis	Mr Osborne, Boughton
3rd	"Weed" foundation	Harrison and Son	Mr Proctor, Hucknall
4th	Frames, sections and bottles	Harrison and Son	Mr Marshall, Arnold
5th	"Beemaster of Warrilow"	Harrison And Son	Mr Bryan, Kirkby
6th	200 honey labels	Harrison and Son	Mr Smithurst, Watnall
7th	"Don'ts and Whys of beekeeping"	Harrison and Son	Mr Hallam, Orston
8th	1 doz. Standard frames	Mr D Marshall	Mr Rose, Radcliffe
9th	100 honey labels	Mr D Marshall	Mr Reeves, Moorgreen
10th	100 honey labels	Mr D Marshall	Mr Munro, Mapperley
11th	Collapsible skep	Mr J Gray	Mr Hallam, Orston

Winifred Anna, Duchess of Portland (1863-1954), was the third President of NBKA. In 1889, she persuaded the Duke to use a large portion of his horseracing winnings to build almshouses at Welbeck which he named "The Winnings". She cared greatly for the local miners and supported them by paying for medical treatments, and organising cooking and sewing classes for their daughters. She also sponsored a miner, with an interest in art, to study in London.

In honour of her support, the Nottinghamshire Miners' Welfare Association petitioned the King on her behalf; and in 1935 she was made a Dame Commander of the Order of the British Empire (DBE) on his silver jubilee.

BBJ 3 Jan 1907

Cleaning Up Wet Combs. I think our good friend Mr. H. Potter is a little off the track when, speaking of combs turned bottom-bar up, he says, "therefore they cannot store honey in them without first making structural alterations, etc." That bees will more readily clean out wet combs when so placed is possible, but that they cannot store honey therein when placed in that position is entirely wrong. Bees can, and often do, store honey in inverted combs; in fact, I have known syrup to be stored in the cells on the underside of a comb laid horizontally on more than one occasion.

Bees will, as a rule, not refuse to clean up extracted combs when placed above quilts with the feed-hole in the quilts open, or above a super-clearer with the slide drawn, especially if there is no honey being gathered at the time and plenty of room below. But it is both easier and more convenient to have wet combs cleaned up in supers than having to handle each one separately. If bees are reluctant to do the work required of them put the combs in an empty hive, allowing a very small entrance after the bees have found it, and they will soon be dry. Of course the hive containing wet combs should be placed as far as convenient from the bees.   WHS, Mapperley

Minute Book – Committee Meeting held in the Peoples' Hall on January 5th, 1907, at 3pm. Present Mr WS Ellis in the chair, Messrs Pugh, Scattergood, Harrison, Windle, Turner, Dickman, Darrington, Vessey, Mackender, White and secretary.

The minutes of the previous meeting were read. Mr Darrington proposed and Mr Dickman seconded that the same be confirmed by the signature of the chairman.

A letter was read from Mr W Herrod stating his intention of withdrawing from the association and, after some discussion pro and con, Mr Skelhorne proposed and Mr Darrington seconded that we accept his resignation with regret.

The letter from Lord St. Vincent was again considered and it was decided that the secretary should ask the following, in respective order, to accept the office of

President, so that the appointment could be ratified at the Annual Meeting:
> Her Grace the Duchess of Portland
> AW Black MP
> EF Milthorp, JP, Newark
> JR Starkey, Norwood Park, Southwell
> Sir TI Birkin, Bart.

The annual Balance Sheet was next gone into which proved one of the best on record shewing a cash balance of £2 6s 2d with a membership of 207. As this was considered satisfactory the same was passed and ordered to be printed on the circular convening the Annual Meeting and sent to each member.

Mr Hayes, as delegate who attended the meeting of BBKA in October, gave a brief resumé of the proceedings.

The arrangements for the Annual Meeting were next considered and it was agreed:
> that it be held in the Peoples' Hall on March 2nd, 1907 commencing at 3pm
>> that one of the following gentlemen be asked, in order, to preside:
>>> The Mayor (John Spalding)
>>> Dr Cunningham, Newark
>>> AW Black MP
>
> that the tea will be provided by Messrs Ellis and Hayes
> that the competition for granulated and liquid honey be held before and that the prize honeys be used for tea
> that Mr Scattergood be asked to give us his lecture on "Wax and its adulterants, etc."
> that the secretary give a full and detailed account as possible of what took place on the London meeting as far as time will allow
> that the annual prize drawing takes place punctually at 7.40pm.

A request was made for microscopes and Messrs Windle and Hayes promised theirs.

An application from Messrs Thos N Harrison for exchange of an advertisement in our report and their trade catalogue and after considerable discussion, it was decided the best course would be to leave the matter as previously viz. that each pay for any advertisement they wish to have.

Arthur William Black (1863-1947) was born in Nottingham and was educated there before entering into business as a lace manufacturer in 1888, and later became a director of Turney Brothers. He was elected in 1895 to the Nottingham Town Council, becoming Sheriff from 1898/99 and Mayor 1902/3. He was knighted in 1916.

Edward Fairburn Milthorp, JP, Balderton Old Hall, near Newark, was educated at East Retford Grammar School. He was connected with the old established malting firm of Messrs R. Bishop and Sons, of Newark, for a quarter of a century. Mr. Milthorp was a member of the Town Council and was Mayor in 1897-98.

Sir William Randle Starkey married Irene Myrtle Francklin, daughter of Captain John Francklin, on 25th April 1935. He was High Sheriff in 1954 and lived at Norwood Park, Southwell. Meetings of NBKA for pollinating apple trees were held here for many years.

Sir Thomas Isaac Birkin was born on 15 February 1831 at New Basford. He was High Sheriff in 1892/3, JP and DL for Nottinghamshire. He was created 1st Baronet Birkin, of Ruddington Grange, Nottingham on 25 July 1905. He died on 16 January 1922 at age 90 and is buried in Wilford Hill cemetery.

John Tricks Spalding (1844-1924) was Mayor of Nottingham in 1907 and churchwarden of St. Thomas' church.

BBJ 10 Jan 1907
Cleaning Up Wet Combs. Referring to the letter of "WHS", I am obliged to your correspondent for his correction re. the word "cannot" in the fourth line and will be glad to meet him half-way by substituting the words "will not." At the same time, it might be of interest to your readers if "WHS" would tell us how his bees were circumstanced for room and general conditions in connection with their storing honey in "the under-side of a comb laid horizontally." He also says: "Bees will, as a rule, not refuse to clean up extracted combs," &c. My experience tells me that they will refuse to do this. I have had combs left on in this way for ten days without being cleaned up, and that at a time when the bees were scarcely able to fly for cold. But when these combs were turned the other way up they were clean the next morning.

"WHS" also says: "It is both easier and more convenient to have wet comb cleaned up in supers." Yes, but I add it is easier to get "Foul Brood" through allowing our neighbours' bees to have free access to our comb than it is to get rid of the trouble of "Foul Brood". I also believe in feeding my own bees only, to say nothing of the danger of "robbing" being started or a number of bees being thus taught the bad habit of robbing.

Perhaps your correspondent says too much stress on the words "structural alterations." I do not mean that the comb is broken down close to the midrib; the bees do no more than pare them down about one-eighth to one-quarter of an inch. In this way the cells will have a downward incline, and then I gradually turn upward. The "cappings" are afterwards usually uneven and of different sizes and shapes, so much so that it is possible that "LSC" in his "Cappings of Comb" will be creating a laugh about them, as his racy remarks often do with myself.

H. Potter, New Brompton, Kent

BBJ 24 Jan 1907

The monthly meeting of the BBKA Council was held on January 17th, at 105, Jermyn Street, SW. Mr. TW Cowan in the chair. A letter regretting absence was received from Dr. Elliot.

BBJ 24 Jan 1907

Cleaning Up Wet Combs. I gladly respond to the request of Mr. H. Potter as to how my bees were circumstanced for room, &c, that they stored in "the under-side of combs laid horizontal," etc.

First let me say the bees referred to as doing this were not mine. My own bees store their surplus in combs hung in supers placed on the hives in the orthodox way; but two others as well as myself can vouch for the fact of the storing as stated having been seen. A small comb containing eggs was suspended between others for the purpose of queen-rearing. The stock was very strong in bees for the size of hive, and was fed liberally. On some other occasions these horizontal combs were almost filled on the under-side with syrup food, and in some cases I believe the cells were extended.

When a novice transfers combs from skeps to frames they are often placed the wrong way up, and I think if they were required for honey the bees could, and would, store therein without making any "structural alterations," etc. Of course, if they were extended, the cells drawn out would curve upward in the usual way. This must be the "structural alterations" our friend's bees have made. If, therefore, Mr. Potter will agree to add the word "always" after the words "will not," making his original letter read, "Therefore they will not always store honey in them without first making structural alterations," we may be said to agree on the question he has raised.

Regarding his second paragraph, it only needs a reference to my letter in order to show that in speaking of combs being cleaned up in supers I clearly meant, over the brood-nest. Only as a last resort would I put them in an otherwise empty hive to be "robbed out," and then it will never fail.

Should any of your readers decide to adopt the plan of inverting combs for cleaning up this year, I will suggest an improvement, by way of balancing each separately over quilts, and that is to put the wet combs in supers; make a light frame of anything handy, and fasten this on the top, either with a couple of "Van Deusen" hive-clamps or by some other easy method, when they will be able to invert a super full of combs for cleaning out.

Curious Swarming Incident. Early in June last year one of my stocks swarmed. On examination of the combs in the brood-nest before hiving the swarm on the old stand, I found a virgin queen already hatched. The other queen-cells I cut out and the young queens in two of them hatched in my pocket within half am hour of the hive having swarmed. The weather had been favourable for swarming for more than a week. The queen accompanying the swarm was fertile, for eggs and young brood were to be run in the combs. I think this must be rather uncommon, as I have never known, bees to defer swarming for so long a time with the weather quite favourable for swarms coming off.　　　　　　　　　　　　　　　　　　　　　　　WHS, Mapperley

BBJ 24 Jan 1907
The twenty-sixth AGM of the Derbyshire BKA was held at the Victoria Cafe, Derby, on January 12th. Amongst those present were Geo. Hayes and P. Scattergood of NBKA. Mr. Scattergood read a paper on "Beeswax and Its Adulterants." He also gave a capital lecture on bees, the lantern being ably managed by Mr. JS Simnett, Burton

BBJ 28 Feb 1907
Foundation Stretching Stopped by simple device, better than wiring; every cell free for breeding; successful where tried. Sample, 7 stamps; full set for one frame, 1s 1d. (patent applied for).　　　　　　　　　　W. Palmer, 174, Curzon-street, Netherfield.

BBJ 7 Mar 1907
The AGM of NBKA was held at the Peoples' Hall on March 2nd, Mr. EF Milthorp, of Newark, presiding over a good attendance, which included Messrs. G. Hayes (secretary), W. Ellis, TN Harrison, WH Hoyte, WH Stoppard, AG Pugh, P. Scattergood, R. Turner, R. Mackender, T. Marshall, W. Darrington, Dr. Elliott, WP Meadows, W. Dickman, and WB Adams.

According to the Balance Sheet there was a balance in hand of £2 6s 2d, the receipts having amounted to £107 2s 9d. The accounts were passed.

The Annual Report stated that during the past year the association continued to increase, forty-two new members having been enrolled, and this brought the total up to 205 - a record membership. For 1907 ten new members had already been admitted, which would make the total 215. The bee-tent and lecturer had been to Newark, Colwick, Balderton, Mansfield and Welbeck and the attendances had been larger than usual. The expert visiting was divided amongst seven experts, who had collectively visited 229 apiaries. Of five candidates who presented themselves for examination for third-class certificates, four were successful, as was also the second-class certificate candidate, Mr. WH Stoppard, of Nottingham.

Three shows had been held, at East Bridgford, Southwell, and Mansfield, the last-named being the Annual County Show, the entries for which were larger than in any previous year. A falling-off in insured members was reported, and the dating of policies was now altered so as to run from March 25th, 1907, to March 24th, 1908, this being considered advantageous for several reasons. The report was adopted.

Viscount St. Vincent desired to resign the office of President on account of his leaving the neighbourhood, and the Duchess of Portland was elected to fill the position, his lordship being thanked for his past services. The retiring committee was re-elected with the addition of Dr. Elliott. Mr. Peter Scattergood was re-elected auditor, Mr. G. Hayes secretary and treasurer, and Messrs. Pugh and Hayes delegates to the BBKA.

It was decided to hold two general meetings during the year, one of them to take the form of a picnic, the committee to make arrangements.

Later on in the evening a paper on "Beeswax and its Adulterants" was read by Mr. P. Scattergood, after which Mr. Hayes gave a report of the proceedings at the London meeting of the BBKA, embracing matters relating particularly to the new bee-disease - "black brood" - and the taking of honey. The prize-drawing for a number of bee-keeping appliances also took place.   Geo. Hayes, Secretary and Treasurer

BBJ 12 Mar 1907.
Cobwebs. I hope, "Mr. Nondescript," after a good sleep, you now see more clearly since writing. Why, sir, you and I - as it is - see eye to eye. I, too, have
   1. discarded single-wall hives;
   2. don't like odd hives of all shapes and sizes ;
   3. believe in "WBC" hives;
   4. have no good to say about a "thin single wall";
   5. believe in American cloth quilts in summer,
   6. and hold a more porous one preferable for winter
   7. stick to supering above
   8. believe in excluders where found necessary.
Why there, your cobwebs are all gone!

BBJ 14 Mar 1907
Cobwebs and Questions. After reading "DMM's" article I feel as though knocked into the proverbial "cocked hat" on the subject of "quilts," etc, and when looking round the garden to-day at the hives I was in doubt whether I had any brains left to mix with the "quilts," or whether I had to mix any possible remainder with my usual habits and tastes. At last, however, I gave it up as a bad job, and am now wondering whether I am to fetch out of the lumber-room the old discarded single-walled hives of all shapes and sizes for use again, for if (as "DMM" states) one is as good as another,

why go in for "WBC" hives at a guinea or more apiece? I had some time ago and came to the conclusion that a thin single-walled hive was rather too warm a domicile for bees when a hot summer's sun was shining on it; but it now appears I have to unlearn again. I also fancied that a cover of American cloth next to the quilt was far and away the best for summer use, and a more porous one preferable for winter.

Now, however, it appears that it is a mere matter of fancy, as both are best. Were I at the beginning of the neophytic stage, instead of being perhaps just past it, I might be led, after reading our friend "DMM.'s" remarks, to place surplus-chambers below the body box (which would be sometimes much readier than placing them above); but I am inclined to stick to above and the side, as the case may be. Perhaps I should get the "cobwebs" out of my head if it had been my good fortune to enjoy a season or two in Scotland, working for heather honey only, as I believe "DMM" does.

It is possible I might discard the use of excluders below sections because in the older days my idea was that bees, when gathering heather honey, had no time to bother with rearing more than a few youngsters; but as my bees are located where they have a good fling at the earlier flowers, I have vague remembrances of some seasons when excluders were not used the queen has got into the sections and spoilt a good number of them, and in order to remedy this evil I have been led at one time or another to invest something like four pounds sterling on excluder zinc.

Thus I feel myself saying, "What a pity to have wasted my means so lavishly if excluders are of no use after all." Perhaps after a night's sleep my head may be clearer of cobwebs, and, like many another "seeker after truth," I may see clearer in the morning; as it is, I am fairly mixed up. I send name for reference.

Nondescript, Notts.

BBJ 21 Mar 1907
New Light on Brood Diseases. It has long been haunting my mind that there was more in the black "shiny" bees that have been sent to the BBJ office by numerous querists than was to be accounted for by your usual verdict of "confirmed robbers," etc. If my communication is referred to, it will be found that I mentioned the fact of having noticed numbers of black and shiny bees which I was sure were not robbers.

On the two last mild days of November last I noticed at two hive-entrances a number of bees undergoing the teasing or dressing process; these were not robbers, I am sure. About half a dozen at each hive had, at the same time, a number gathered round them which appeared to be putting them through a strict examination, and nibbling or biting them. This went on for several hours each day, and from each flight-board during the two days fell at least fifty bees, alive, but too "sick" to get up again. These were darker than the others, but had not reached the shiny stage.

BEEKEEPING BETWEEN TWO QUEENS

The two stocks died during the last spell of frost. They had plenty of food, and were well wrapped and covered. Several others near, under same conditions as to food, &c, but much weaker at the beginning of winter, have come through all right, and will make good stocks. Now, is it not possible - and probable - that these bees were affected by the microbe mentioned in the last paragraph of your editorial headed "New Light on Brood Diseases"? If I come across any more affected in similar manner I will contrive to send you a few live specimens, if possible.

Nondescript, Notts, March 18th.

Foundation Stretching stopped by simple device, better than wiring, costs no more after first, every cell free for breeding, successful where tried; set for one frame, 1s 1d PO; no more odd samples, make for your own use, but dealers keep off. Patent applied for.   W. Palmer, 174, Curzon-street, Netherfield

Minute Book - The AGM of the BBKA was held on March 21st, at 4pm., at the RSPCA Mr. TW Cowan (chairman) presiding. Among those present were Dr. TS Elliot and AG Pugh.

Mr. WF Reid proposed a vote of thanks to the RSPCA for the gratuitous use of their Board Room for Council and other meetings. He had the idea that in according this permission the RSPCA were carrying out most fully their objects, because one only had to consider the enormous number of bees that were formerly killed unnecessarily under the old methods. The number of horses, cattle, and other animals that suffered would be infinitesimal compared with that of the bees, all of which cruelty to animals - for bees, too, were animals - the Association had been able to stop.

Mr. Pugh, in seconding the resolution, said he had attended Annual Meetings of the Association for some years, and the kindness of the RSPCA appealed to him every time he had been present. It was an admirable room for their meetings. Mr. Edwards proposed the next resolution, which was: The re-election of the vice-Presidents. Mr. Pugh seconded the resolution, which was carried unanimously.

NEP 26 March 1907
The County Council have decided to reduce the grant to NBKA from £40 to £25.

BBJ 1 Apr 1907
Foundation Fixing. I was pleased to see a good word for my device for prevention of foundation stretching, and especially as I have not the pleasure of Mr. Fraser's acquaintance.

Although "protecting" it, I am only doing so against dealers. I have no objection - even if I had the right - to anyone making it for his own use. If not "grinding" my little axe too much, perhaps you will allow me to give particulars of a severe test I put the device to last September, when I placed between 8 lb. and 9 lb. of driven bees onto eight frames filled with two sheets each of the thinnest super-foundation, the top sheet fastened to the top-bar by melted wax, the bottom one kept in place by the device only. After two days, the bees having had plenty of syrup to go at, I found the foundation well fixed and joined, with the exception of an odd corner, built into half-combs, very little stretched and as full of syrup as the cell-depth would allow. Adams, Harrison, Pugh, Smithurst, Dickman, White, Vessey, Stoppard, Ellis, Turner, Gray and Darrington.

At the end of a fortnight the bees had taken thirty odd pounds of syrup; there were four good slabs of brood, and the rest of the combs were nearly solid with food, mostly sealed. It is now a grand stock. I shall not repeat this experiment, though, for the combs were so very tender that even to me it was a ticklish job taking the device off. Nondescript, March 11.

Minute Book – The Quarterly Committee Meeting was held in the Peoples' Hall on April 6th, 1907. Present: Mr P Scattergood in the chair, Messrs Adams, Harrison, Pugh.

The minutes of the previous meeting were read and confirmed by the signature of the chairman.

A letter was read from Mr WH Stoppard complaining of supposed irregularities in connection with insurance, etc. and after going thoroughly into the matter it was found he was under a misapprehension and he was satisfied with the information given to him.

The quarterly account was next gone into and same was considered satisfactory and passed.

An offer from the Southwell Horticultural Society, for holding the Annual Show with them of £5, tent and staging and as we had no other offer it was decided to accept.

Mr Scattergood proposed and Mr Pugh seconded that Dr Elliott be asked to officiate as judge and examiner at Southwell. Dr Elliott, who was present, consented and this was carried. Mr WH Stoppard was appointed assistant judge and to receive his train fare for the duty. Proposed by Mr Darrington and seconded by Mr Pugh. The stewards to be Mr H Mackender and Mr H Merryweather or who he may appoint.

Resolved
          that the sub-committee for Technical Instruction be elected *en bloc* and that

it reports to the general committee its transactions. Proposed by Mr White and seconded by Mr Dickman.

that a vote of thanks be accorded Mr and Mrs Ellis and those who assisted in providing the tea at the Annual Meeting.

The delegate (Mr Pugh) gave his report of what transpired at the Annual Meeting of BBKA.

The schedule for the Annual Show was gone through and revised and the secretary was instructed to arrange for other local shows where possible and failing any entries for the appliances class to pay 10/- to 20/- towards cost of carriage for a display of appliances.

In the matter of a Summer Meeting referred from the Annual Meeting to arrange, it was decided after considerable discussion to hold it on June 1st, at Syston if not less than 25 responded and the secretary was instructed to send a circular to members and arrange with Mr Meadows for this.

BBJ 11 Apr 1907
Thomas W. Harrison and Son. Cheapside, Nottingham, issue a neat little catalogue of bee-appliances, dairy utensils, garden tools, and furniture of all kinds. It is well arranged and practical in every detail, nothing useful in an apiary or garden being omitted.

BBJ 18 Apr 1907
(Miss) AT (Nottingham). Foul Brood is developing in the comb sent. It seems a recent outbreak.

BBJ 18 Apr 1907
Foundation Stretching Prevented by "Nondescript" device; better than wiring, every cell free for breeding; successful where tried. Palmer, 174, Curzon-street, Netherfield

BBJ 24 Apr 1907
The twenty-fifth Annual Meeting of Leicestershire BKA was held on March 16th at the Highcross Coffee House, Leicester. In the absence of Mrs. Maurice Levy, Miss VR Levy presided over a company numbering nearly one hundred members and friends, including representatives from the Notts and Derby BKAs. After the prizes had been distributed the assembly partook of tea, and subsequently lectures were delivered by Mr. Peter Scattergood on "Wax: Its Production, Uses, and Adulteration," and by Mr. Geo. Hayes on "Honey and Honey taking".

BBJ 2 May 1907

The Bee to Cure Foul Brood. Seeing that Mr. Simmins in his recent articles in the BBJ, insists on vital force, energy, and vigour in bees, I thought the following particulars would interest some of your readers, while from other points of view the details I send are not altogether devoid of interest. In our part of Derbyshire, on April 2nd last, four stocks of bees were examined, with the following result:

No. 1 had ten frames placed at right angles to the entrance, which had been left open 6 in. the whole winter. Their sole covering during the whole period had been a piece of American oil-cloth and a piece of glass 8 in. by 12 in. by $^3/_8$in; yet the bees were fairly numerous.

No. 2 had ten frames placed as above, and when examined there were about two frames well covered with bees. Their covering had been just the same as No. 1.

No. 3 was a square hive holding ten frames hung at right angles to the entrance, which had been left open 6 in. wide during the whole winter; the sole covering, as before, had been a piece of oil-cloth, yet there were nine frames of bees, and brood in plenty.

No. 4 was a hive of seven frames hung as the others. In the autumn two frames of honey had been removed and replaced by empty combs, when the bees were exceedingly vicious. This hive had a span roof, with a cone fixed at the back and front, the entrance open 5 in., and the frames open to the roof, ie. neither quilts nor wraps of any kind during the whole winter; yet there were six good combs of bees.

When I say the situation of the hives is quite open, I am sure you will agree that the above facts are truly remarkable after such a winter.   W. Darrington, Notts.

BBJ 10 May 1907
Bonus (Newark). No. 1 sample shows Foul Brood in a very pronounced form. In No. 2 there are slight signs of the disease in two or three cells, but in the remainder the brood is "chilled" only.

NEP 15 May 1907
Fortunately for the Newark Agricultural Society the downpour of rain which accompanied the opening proceedings was succeeded on the second day by weather of a genial description. The NBKA arranged a series of lectures beekeeping and demonstrations with live bees.

BBJ 30 May 1907
Recent papers by Mr. Simmins on "Foul Brood" call for, and will no doubt receive, comment from practical bee-keepers. I will therefore content myself with a glance at the scientific aspect of some of his conclusions, and with a few further remarks which have some bearing on the question.

The statement of Mr. TJ Burril to the effect that spores diluted with water, and kept at room temperature (65°F/18°C), perished in less than six months, is of no practical value to beekeepers, and the inference drawn by Mr. Simmins from that experiment, viz. that prolonged exposure at blood-heat or hive-temperature (95°F/35°C.) will kill spores is, I think, distinctly erroneous. In the first place, it is not the temperature of the room which directly kills the spores, but this result appears to be due to a combination of circumstances.

A temperature of 65°F is sufficient to allow of germination of spores, but growth of the resulting bacilli will be slow, and death will ensue when the scanty nutrient material in the water is exhausted, the temperature being too low to allow of fresh spore formation. The wonder is that they survived so long, but probably germination of spores would be slow, and the adverse conditions would also hinder rapid multiplication of bacilli. Life would also be longer in sterile water than in ordinary well water, owing to the absence of saprophytic organisms common in water, with which *B. alvei* would have to contend.

Hive temperature, on the other hand, is undoubtedly the most suitable temperature for the development of the vegetative form, and consequently will not kill spores, whose resistance to heat is so well known. Referring for a moment to "A Lanarkshire Bee-keeper's " experience, it is difficult - not knowing the exact temperature of his slow oven - to tell exactly what would happen. The temperature could not, however, exceed 60°F, for then the combs would be almost at the melting-point. Any temperature over 57°F would kill the bacilli, but would not affect the spores.

The self-cure, or apparent self-cure, of colonies affected with "Foul Brood" in a good honey-season has often been remarked, and it is interesting to speculate as to the causes. I say "apparent" self-cure advisedly, for how often does it not happen that colonies apparently cured by natural means are found to have a few diseased cells the following season, which may or may not develop seriously, but which are a potential source of mischief?

Mr. Simmins, I think rightly, lays stress on the vital force and energy developed by the bees in consequence of a spell of fine weather and rapid flow of honey, and in this connection we must not lose sight of the antiseptic and invigorating effect of the formic acid in the honey consumed by the adult bees and fed to the larvae and queen. Recently, preparations of the formates of various salts have come into prominence as muscular tonics, and it seems likely that formic acid has a similar action on bees as well as inhibiting the germination of spores of "Foul Brood."

Professor Harrison, in his clever monograph on "Foul Brood" gives analyses of buckwheat and clover honey, showing that the former contains a far greater percentage of formic acid than the latter. This is highly interesting, but no clue is given as to the source of the formic acid or as to the conditions of temperature or season, &c, which determine the percentage in honey. Muhlenhoff contends that the acid is deposited in honey by the bees to assist its keeping properties, and as there is no evidence to show that it is present in nectar as secreted by flowers, we must accept his contention. On the other hand, it is difficult to see why the bees should add so much more to one variety of honey than to another, unless, indeed, instinct tells them that buckwheat honey requires more for its preservation.

Being myself very susceptible to beestings, I have had occasion to note that stings inflicted in the height of the season are far more painful than those received in the early spring. Whether the new honey is responsible for the activity of the poison glands I cannot say but the formic acid is evidently produced for the double purpose of defending the bees' stores and helping to preserve them and may, for aught we know, be used directly by the bees themselves for application to putrid brood.

The development of vital force referred to above enables the bees to withstand the disease themselves and the access of energy encourages them to clear out "Foul Brood" material from the combs to the outside of the hive, where after some hours' exposure to direct sunlight bacilli and spores are destroyed, this beneficial effect being due to the action of the ultra-violet rays on the oxygen of the air, resulting in the formation of peroxide of hydrogen - a powerful disinfectant - and not to the heat of the sun.

A factor which probably aids the bees in their work as scavengers is the greater rapidity with which the dead larvae dry up in warm, dry weather, such dried-up masses being more readily removed than sticky, ropy material. During a rapid honey-flow the bees will clear out as many cells as possible to make room for the incoming honey
.
My personal experience of "Foul Brood" is not sufficient to warrant a decided opinion on the practical value of Mr. Simmins's methods of treatment, nor does he very definitely lay down his plan of combating the disease. He appears to waver between faith in his bees to cure themselves and a desire to make doubly sure by the use of antiseptics. Were it possible to simultaneously destroy every affected stock and appliance and to prevent the importation of diseased bees and queens, at the same time ridding the country of wild bees and wasps, etc, which might be a further source of infection, the pest would be eradicated; but as this is clearly impossible, I fear we must be reconciled to the more or less constant presence of the disease unless, in course of time, some sort of acquired immunity be developed by our bees.

This being so, the man who runs his apiary for profit is hardly likely to employ radical measures on the first appearance of "Foul Brood" if he has confidence in his ability to keep the disease within such limits as will not prevent him from obtaining a profitable return. If, however, he is a trader in stocks or queens he will be wise, if he wishes to retain the confidence of customers, to at once stamp out disease without waiting for a cure by natural or other means.

Seeing that almost every apiary in this country is surrounded by others which may present a clean bill of health, it is incumbent on the owner of an infected apiary to do his best to at least prevent it from becoming a focus of infection to neighbouring apiaries the owners of which may not be so competent to check the disease.

TS Elliot, Southwell

BBJ 30 May 1907
HP (Notts). Dead brood in all sealed cells of comb sent is like that of Black Brood shown at the BBKA Conversazione some months ago.

BBJ 6 Jun 1907
Good Natural Swarms for sale, 10s each.   H. Holleworth, New Inn Farm, Widmerpool

BBJ 13 Jun 1907

W. P. MEADOWS,
SYSTON, near LEICESTER.
Telegrams: "MEADOWS, SYSTON." Nat. Tele. 2 X.
ROYAL SHOW, LINCOLN, 1907.
NOTE AGAIN to the FRONT.   2 FIRST   2 SECOND   2 THIRD   PRIZES
Send for KAT-A-LOG.

The Summer Meeting of NBKA was held on June 1st at Syston. The members and their friends travelled by train from Nottingham, and on arriving at Syston were met by Mr. WP Meadows, one of the largest makers of appliances used in connection with beekeeping, who conducted them over his extensive works. The party then adjourned to the Village Hall, where they partook of tea.

Amongst those present were Mr. WS Ellis (vice President of the association) and Mrs. Ellis, Mr. G. Hayes (secretary) and Mrs. Hayes, Mr. and Mrs. AG Pugh (Beeston). Mr. and Mrs. P. Scattergood (Stapleford), Mr. and Mrs. TN Harrison (Carrington), Mr. and Mrs. Fidler (Hucknall), Mr. and Mrs. Faulkner (Melton), Miss Broadhurst (Long Eaton), Messrs. G. Scattergood and White (Sandiacre), Messrs. Stoppard and Jacques (Nottingham), RJ. Turner (Radcliffe), J. Mann (Stragglethorpe), F. Burley (Nottingham), etc. The party were afterwards driven to the nurseries of Mr. James Wright, FRHS, and to Abbey Park.

BBJ 27 Jun 1907
The monthly meeting of the BBKA Council was held on the 20th inst., at 105, Jermyn Street, SW. Mr. FB White being voted to the chair. A letter expressing regret at enforced absence was read from AG Pugh. Arrangements were made for examinations in Nottinghamshire.

BBJ 27 Jun 1907
AT (Nottingham) There is no sign of "Foul Brood" (*Bacillus alvei*) in the sample of comb. The dead larvae bear all the characteristics of "Black Brood," as described by American writers.

East Bridgford Parish magazine July 1907
In spite of the cold season, the Annual Show was generally pronounced a very good one but the rain which commenced in the afternoon quite spoilt the attendance.

Bees and Honey
    Sections of comb – 1st, Mr D Marshall; 2nd, Miss Fox
    Run honey – 1st, Mr D Marshall; equal 2nd, Messrs Cragg and Sentence
        Bees - Mr W Millington

BBJ 4 Jul 1907
Not for many a year has the Royal Agricultural Society of England held so successful an exhibition as that which terminated at Lincoln on last. Amongst the judges was AG Pugh.
Class 421. Twelve 1-lb Jars of Extracted Medium or Dark-Coloured Honey (19 entries) - 2nd, George Marshall, Norwell
Class 430 Honey Trophy (11 entries) - r. and hc, Ball and Grey, Long Eaton

Minute Book – Committee Meeting held in the Peoples' Hall on July 6th, 1907. Present: Mr RJ Turner (in the chair), Messrs Pugh, Smithurst, Harrison, Dickman, Darrington, Adams, Stoppard and secretary.

The minutes of the previous meeting were read, when Mr Dickman proposed and Mr Darrington seconded the same be adopted. The quarterly statement of accounts was next gone into. These shewed us to be in much the same condition as for the corresponding quarter last year and were considered satisfactory. Mr Dickman proposed and Mr Darrington seconded the same be accepted.

The secretary then read a list of new members for the year, same numbering in all 27.

Letters of apology for non-attendance were read from several members of the committee.

One also from Messrs TN Harrison and Sons which was referred back to them asking for a definite suggestion as to their requirements and to be considered at the next meeting.

Two letters from Mr W Munro asking whether he, as expert, could not attend committee meetings. The committee substantiated the secretary's statement in reply and resolved that, in future, Rule VI should be read to include all members who had obtained 'experts certificates' whether they were acting as such for this association or not.

On the motion of Mr Pugh seconded by Mr Turner it was resolved that the committee meetings in future commence at 3.30pm.

BBJ 11 Jul 1907
Controlling Swarming. In answer to your correspondent who refers: to "the curse of swarming," I beg to give him the benefit of my experience in controlling swarming. Before the standard frame was fixed upon I had a number of hives on hand which were an inch or two deeper than was required for it, and by the time I made my first standard frame I had about thirty of these deep hives in use and with them I had no trouble, for a swarm was with me a rarity. But as soon as I began to use the standard my troubles began for the bees would swarm from them when working for sections.

After a year or two of this I tried a box of shallow-frames on top of the broodnest after removing two outside frames and filling up the spaces with dummies. This stopped the trouble and in over twenty years since I did not have more than six swarms from these deep hives. I generally ran from thirty to forty-odd of them for honey, about half for sections, the others for extracted honey and a large bellglass or two.

To increase stocks year after year I drove a number of skeps. The frames ran end on to the entrance, which was the full width of the hive and $1/2$ in. deep. I took no honey off till the end of each season as a rule. I usually had five or six racks of sections on each hive, or two or three full-sized boxes of brood-combs. The bees were natives and I always took care to have every brood-frame full of brood before placing on the super.

About half of the deep hives held thirteen frames so there were sufficient bees to swarm if they had any mind to do so. My floor-boards were all fastened to the body-boxes latterly, as I found loose ones a nuisance when moving to and from the heather. Extra ventilation was given when necessary by wedging up the front of the supers a quarter of an inch.

I tried Italians for two seasons and Cyprians for one but the natives beat them hollow. One bad quality of the Italians seems have been lost sight of, and that is their "sticky" feet. They would not be shaken off combs, while the natives drop off like ripe pears from a shaken tree. I often wonder why people buy the foreigners. It must be for their beauty, for I cannot see any other good quality in them.

Reverting again to deep hives - in my hands they almost without exception came out better and stronger in the spring and were ready for supers before those taking the standard frame without any trouble. The latter were weaker in bees and required more "coddling," and often uniting, to get them ready for the honey flow.

How often we have read and heard that where a super has been left on during winter that stock has been best in the apiary in the spring?   Nondescript, Notts, June 24th

BBJ 25 Jul 1907
Mr. Simmins sums up thus:
"The conclusion I have arrived at is that while a short period of heat at boiling-point may fail to destroy the spores of "Foul Brood" there is no reason whatever why a more protracted term at blood temperature should not bring about the end of their existence, where further propagation is out of the question."

Mr. Simmins gives no explanation of the process by which spores kept in water at 65^0F for six months were ultimately destroyed, and I naturally assumed that he thought the process was a simple and not a complex one.

It is well known that adverse conditions usually favour the formation of spores, but too low a temperature will prevent sporulation altogether, and it is this fact which is really responsible for the disappearance of bacilli and spores. In the case of honey, however, the fate of the spores probably depends on the percentage of formic acid present. Samples with a low percentage might allow of their germination, whilst those with a percentage approximating to that of buckwheat honey would inhibit germination, but would not prevent such honey from becoming dangerous when used as bee-food.

I am still unable to see the practical value of Mr. Burrill's experiment. There is no comparison between keeping spores in water at room temperature for six months and intermittently diluting "Foul Broody" material with a watering-can. The conditions in the two cases are entirely different, even if we omit the added antiseptic in the latter case.

Again: "He will carry the process of dilution right into the hive ... by saturating the hive sides and his quilts with strong Izal solution, and may have his combs filled solid with medicated syrup." If this treatment is successful, is it not reasonable to suppose that success is due to the antiseptics used - otherwise, why use Izal and medicated syrup if plain water and simple syrup will suffice?   Thos. S. Elliot, Southwell July 8th

BBJ 8 Aug 1907
The NBKA Annual Show was held at Southwell on July 25th, in connection with that

of the Southwell Horticultural Society. The weather was not conducive to comfort, for all the morning it rained heavily. The entries in the honey department were fairly well up to the usual number, but, alas, the weather generally has been such that very little surplus honey could be stored; consequently, only a very poor proportion of the exhibits were staged. Dr. TS Elliot, Southwell, officiated as judge (assisted by Mr. Stoppard, Notts), the following being their awards:

Collection of Bee-appliances – 1st, TN Harrison, Notts.

Trophy of Honey in any Form and of any Year – 2nd, G. Marshall, Norwell

Twelve 1-lb Jars Light-coloured Extracted Honey – 1st, J. Wilson, Shirebrook; 2nd, G. Hopkinson, Newark; 3rd, John Bee, Southwell; hc, W. Lee, Southwell

Twelve 1-lb Jars Dark-coloured Extracted Honey – 1st, J. Wilson; 2nd, JR Almond, Gotham; 3rd, AH Hill, Balderton; 4th, W. Ball; hc, W. Lee

Six 1-lb Sections – 1st, G. Marshall; 2nd, W. Lee

Six 1-lb Jars Granulated Honey – 1st, John Bee; 2nd, WL Betts, Mansfield Woodhouse; 3rd, GH Pepper, Farnsfield; hc, H. Merryweather, Southwell

One Shallow-frame of Honey for Extracting – 1st, WL Betts; 2nd, G. Marshall; 3rd, H. Merryweather

Six 1-lb Jars Extracted Honey (Novices' Class) – 1st, E. Varty, Pleasley; 2nd, Miss Hedderley, Southwell

Honey Vinegar - W. Ball.

Observatory hive with Bees and Queen – 1st, EG Ive, Boughton; 2nd, AH Hill; 3rd, WL Betts; 4th, G. Marshall

Beeswax – 1st, G. Marshall; 2nd, AH Hill; 3rd, John Bee

BBJ 10 Aug 1907
Few 1906 Queens, 3s 3d each.   R. Mackender, Seeds and Bees, Newark

BBJ 15 Aug 1907
The Annual Show of the North Norfolk BKA was held by kind permission of Lord Hastings at Melton Constable Park, in conjunction with the Melton Constable Horticultural Society's show, on Bank Holiday, August 5th. Dr. TS Elliot, Southwell, Notts, judged the exhibits.

BBJ 22 Aug 1907
For Sale, Twelve Stocks Healthy Bees, on 10 frames each, in joiner-made "WBC" Hives all new last year and well painted. Lots to suit purchaser. Owner deceased.
                Apply, Mrs. Pogson, Holme-road, West Bridgford, Nottingham

BBJ 29 Aug 1907
For Sale, several Strong Healthy Stocks, in Standard Framed Hives, £1 each.
                D. Marshall, Cropwell Butler

BBJ 12 Sep 1907
The fifteenth Annual International Exhibition and Market of the Confectioners and

Allied Trades commenced on September 7th and remains open till the close of the present week.

Display of Honey (Comb and Extracted) and Honey Products, shown in suitably attractive form for a tradesman's window (5 entries) - 4th (£1), T. Marshall, Ivy Cottage, Sutton-on-Trent

Twelve 1-lb Sections (17 entries) - 5th (5s), G. Hunt, Hawton Road, Newark

Twelve 1-lb Heather Sections (2 entries) - 2nd (15s), T. Marshall

Three Shallow Frames Comb Honey for Extracting (5 entries) - 2nd (15s), J. Herrod, The Manse, Sutton-on-Trent

Twelve 1-lb Jars Light-coloured Extracted Honey (27 entries) - hc G. Hunt

Twelve 1-lb Jars Heather-blend Honey (8 entries) - 3rd (10s), T. Marshall; 4th (5s.), G. Hunt

Twelve 1-lb Jars Granulated Honey (17 entries) - 1st (£1 5s), T. Marshall

Beeswax, judged for quality of wax only (14 entries) - 4th (5s), T. Marshall

A profitable day for Thomas Marshall in winning £3 15s but we don't know what expenses he incurred.

BBJ 26 Sep 1907

The monthly meeting of the BBKA Council was held at 105, Jermyn Street, SW. on the 19th inst., Mr. WF Reid. being called to the chair. Letters apologising for inability to attend were received from Dr. Elliot and Mr. AG Pugh. Report upon examination of candidates for third-class expert certificates, held in Nottinghamshire, were considered,

BBJ 26 Sep 1907

The fifteenth Annual Exhibition and Market of the Grocery and Allied Trades, held at the Agricultural Hall, London, was opened the 21st inst., and continues till the end of the present week.

Three Shallow-frames Comb Honey for Extracting (7 entries) - vhc, WL Betts, Mansfield Woodhouse

Twelve 1-lb Jars Medium-coloured Extracted Honey (29 entries) - 2nd (£1), G. Hunt, Hawton Road, Newark; 3rd (15s.), T. Marshall, Sutton-on-Trent; hc, J. Herrod, The Manse, Sutton-on-Trent

Twelve 1-lb Jars Dark-coloured Extracted Honey (12 entries) - hc, G. Marshall, Norwell

Twelve 1-lb Jars Granulated Honey (19 entries) - 3rd (15s), WL Betts

Beeswax in Cakes, Quality of Wax, Form of Cakes and Package suitable for retail counter trade (10 entries) - 3rd (10s), T. Marshall; vhc, J. Herrod

Beeswax judged for Quality of Wax only (16 entries) - 3rd (10s), G. Hunt

Minute Book – Committee Meeting held in the Peoples' Hall on October 5th, 1907. Present Mr AG Pugh in the chair, Messrs Adams, Bolton, Ellis, Gray, Harrison,

Dickman, Mackender, Munro, Puttergill, Smithurst, Stoppard, Scattergood, White, Turner, Windle and secretary.

The minutes of the previous meeting were read and confirmed by the signature of the chairman.

Mr Harrison's letter stating that if the association were prepared to give more benefits to its members he would, in return, give in confidence the names of his customers so that the secretary might try and induce them to become members of this association. After a good deal of discussion pro and con, it was pointed out that members were now receiving great benefits as such and that the financial position would not allow of any further benefits at present, and the matter was allowed to drop, Mr Harrison promising to give the names to secretary.

A letter from Dr Elliott stating his inability to be present and offering to, in some way, return his fee for judging at the Southwell Show, Mr Munro proposed and Mr Harrison seconded that a hearty vote of thanks be accorded Dr Elliott for his promise.

Correspondence about loan of bee tent to the Derbys BKA was read and it was resolved that the secretary's action be confirmed and that a charge of 12/6d be made for loan of and carriage paid for same.

A letter from Thos Newberry, Hucknall Torkard, stating that his bees had been upset and offering a reward was considered. The secretary stated that he had sent out posters offering a reward from this association but nothing further had been heard of the matter. Resolved that this matter lay on the table.

Correspondence between the West Bridgford District Council and this association as to restriction on beekeepers in their area was read and it was decided now that the matter had been taken up, it be allowed to remain and await results.

Correspondence as to removal of exhibits from the Annual Show before the specified time was gone through and it was agreed that nothing of very serious nature had occurred and that the matter be allowed to drop.

The quarterly statement was next dealt with shewing we were about £1 behind the corresponding quarter of last year which was due to the falling off of subscriptions. Mr Scattergood proposed and Mr Ellis seconded the same be accepted.

The names of five more members were read and it was resolved to admit the same as such.

A short report on each show was given and accepted as satisfactory.

Mr Harrison next suggested that this association should approach the Nottingham Horticultural Society with a view to holding a show of honey, etc. with them and after considerable discussion it was proposed by Mr Harrison and seconded by Mr Scattergood that a deputation consisting of Mr Ellis and the secretary be sent and try to arrange this. (Could not get a meeting, the Horticultural Society would not have us)

A report was read of the proceedings of the sub-committee for Technical Instruction.

The awarding of the prize hive given by a vice-President at the Annual Show was next dealt with and after taking all exhibitors into consideration it was awarded to Mr G Marshall, Norwell.

BBJ 17 Oct 1907

At a meeting of the BBKA Council, held on October 10th, preceding the conversazione of members, Mr. Walter F. Reid was unanimously elected to the vice-chairmanship, rendered vacant by the lamented death of Mr. TI Weston.

After a short interval at five o'clock, during which light refreshments were served, the members reassembled. The company present included AG Pugh. After briefly opening the proceedings the chairman called upon Mr. Hayes (who was obliged to leave the meeting early) to open a discussion on the subject of "Pollen - its Advantages." Mr. Hayes thought it a great honour that the Council had asked him to lead the debate on the subject named. He was of the opinion that bee-keepers should always be ready to do their best, however little that might be, to popularise their pursuit and advance the cause generally.

He was not going to propound any new or startling theory, but rather to re-state something old, and endeavour to emphasise its importance, while avoiding any approach to dogmatism. The world was prone to speak of the disadvantages of a thing rather than of its advantages and that was so regarding pollen. It was quite usual to hear pollen spoken of as a nuisance to bee-keepers, as in the case of pollen-clogged combs, and he agreed that in supers any excess of it was very troublesome, but he believed that by a little thought and management of the brood-nests in districts where pollen was known to be over-abundant that might be avoided.

With regard to the importance of pollen to bee-keepers, it was as well that they should remind themselves of what pollen consisted. Was it not the essence of the plant or tree and its fruit? The roots, the bark, the branches and the leaves were only auxiliaries to the flowers that bore the pollen. This fact demonstrated its importance. The fertilising dust of the flowers, as they were accustomed to briefly call pollen, was, as most present were aware, the ovules with a cellulose covering, something like the covering of an egg, the shell. A better name for this dust was pollen-grains, the size of which varied from $1/200$th to $1/2000$th part of an inch. Of course, to see these grains separately a pocket lens was necessary, or, better still, a microscope with a $1/4$in. or $1/2$in. objective would be of great assistance in studying the shape and size of these grains.

Each plant had its own particular coloured pollen. The form of the various grains was most interesting, and many were beautiful, some more or less beautiful than others; and each form and each colour was identical with the tree that bore it. Investigations of this sort formed, in his opinion, a good nature study.

The microscope provided the means of identifying the source and purity of the honey. The bee-keeper, from any sample of honey he took, was able to trace the origin of such produce by recognising the pollen-grains which, as already explained, differed in size, shape and colour according to the plant from which they were obtained; thus he was enabled to say from what source his honey was derived.

Again, as regards the adulteration of honey, the apiarist could take a sample of that known to be pure honey and compute the number of grains he found within a certain area; this could then be compared with another sample supposed to be diluted with glucose, whereupon the difference in the number of grains would be apparent and that would form a clue to what extent adulteration had taken place.

But, after all, the most important point in connection with pollen lay in the fact that it was a bee-food and that was on what he wished to lay special stress. The very "fitting out" of the bee for the easy collection of pollen showed that the same was necessary for its welfare. The hairs on its body, the arrangement of them, the receptacle for bringing it home, all pointed to the importance of it in the hive, so that it became clear to the bee-keeper that he should see that his bees had sufficient pollen; for pollen to the bee was what bread and meat was to man, or, to put it more concretely, it was as oatmeal to the Scotsman. It built up the frame and renewed the tissues, thus making the bee strong and better able to resist "Foul Brood", as well as carry out the work for which it was destined.

Pollen was also used in the cappings for covering the brood. He had noticed this year a great quantity of brood in an advanced pupa state which had not been sealed over properly and he attributed this in a great measure to lack of pollen. It was well known that sealing was often left over till late, when the conditions of heat were favourable, but he had never seen it postponed so long as it had been this year.

Another consideration was in connection with driven bees. In many cases driven bees were put into winter quarters without any thought of pollen for them. They were placed on combs or on sheets of foundation and fed with syrup but the necessity for pollen was generally overlooked or forgotten. Now here was a use for the pollen-clogged combs. If these were reserved from summer or late autumn, or taken from the hives when driven bees were expected, they could be turned to a useful purpose. Driven bees could not gather sufficient pollen for their own needs because the season was usually far advanced.

Pea-flour was the best substitute for pollen. This might be given by dredging it into the cells dry, or by mixing it with a little syrup, or by placing it in another comb on the cluster. It could also be given to the bees as a sort of paste on the top of the hive in a feeder.

These were the most important points concerning pollen that he wished to bring before the meeting, and they might be summed up thus:
   a. the interest of the subject as a nature-study;
   b. a means of telling pure honey from adulterated;
   c. a necessary food for the bees at all times;
   d. a necessity for the capping of brood;
   e. its value to driven bees when put into winter quarters;
   f. its use in resisting disease. Man must keep his body strong and full of tone to banish disease; so with the bee.

Before proceeding further with the discussion, the Secretary read a letter from P. Scattergood (Notts) regretting his unavoidable absence from the meeting.

Mr. Moore then asked if there was any particular breed of bees that was more or less given to gathering pollen and putting it in the supers, to which Mr. Hayes replied that he had not experienced any difference between the habits of one breed and another in that respect. Mr. Herrod attributed the difference - if any existed - to the nature of the district and the Chairman confirmed this by remarking that some flowers yielded much more pollen than others. Mr. Herrod also pointed out that the use of an excluder would, in some measure, prevent deposits of pollen in supers as the bees could not easily carry it on their legs through the perforation in the metal.

Mr. Garratt thought Mr. Hayes had brought out one point strongly which should be very helpful to young beekeepers and that was the necessity for providing driven bees with a sufficiency of pollen. That was often overlooked, and in such cases it was only by accident that the bees survived. Intelligent observers had been struck by it. Bees which had not received an adequate supply of pollen in the spring of the year showed a tendency to dwindle. Mr. Pugh said the fact of the bees not having sealed over the brood so early this year as in former years was an interesting fact and worthy of consideration. It was certain that they had had more difficulty in gathering pollen than usual.

Mr. Meadows wished to ask Mr. Hayes what his experience was in feeding bees in the hive with pollen. Feeding them outside was of course common but he had never heard of the former.

Mr. Hayes replied that he had mixed artificial pollen (pea-flour) to the extent of a 1-lb. packet with a table-spoonful of honey and sufficient syrup to make its consistency like that of good thick cream. He then took an empty comb and ran the liquid into the cells, afterwards placing this comb on the outside of the brood nest. He found the bees very readily cleared the mixture out and took the contents into the interior of the hive. He had also fed them with the same mixture made into a firmer consistency on the top of the frames but they had not taken it quite as freely as they did from the combs.

The Chairman said there was difficulty in making the bees accept this pollen. They would often not take pea-flour. Lentil flour was equally nutritious. The usual plan, when crocuses were out, was to sprinkle a little pea-flour on them or another way was to distribute it in a box of shavings.

Mr. Carr asked Mr. Hayes if he had observed what was the result of giving the bees pollen and honey mixed? Did they separate the two substances, or simply store it as they do honey? Mr. Hayes had not been able to ascertain how they stored it. The Chairman considered the point an important one. The bees were in the habit of covering their pollen with honey. Mr. Carr begged Mr. Hayes to give bee-keepers the result of his further investigations on this subject for publication at some future time.

Mrs. Pearman endorsed the Chairman's recommendation regarding the distribution of pea-flour in crocuses or amongst shavings. She had had personal experience of its effectiveness. Mr. Hayes agreed, but there were no crocuses in autumn and shavings were not always obtainable.

Mr. Bevan believed in the efficacy of the mixture described by Mr. Hayes. A large sheet of foundation containing it and covering half a dozen frames will carry the bees on till February. He did not quite see in what way the pollen could be a guide with regard to the special flower it came from.

Mr. Hayes, in reply, repeated that each plant bore its own particular kind of pollen, which differed as to size, form and colour according to the source from which it was derived; and a study of all this enabled anyone to trace the origin of the honey. For instance, take pollen from hawthorn or lime or clover whichever was prevalent in a given district, place each under the microscope and observe its characteristics. Then take a honey sample, place it in a glass, stir it up, and leave it till the pollen-grains settle at the bottom; then with a pipette put these on a slide and examine them under the microscope, when the different grains from which the honey was gathered could be compared with the specimens taken direct from the plants and identity established.

The Chairman thanked Mr. Hayes on behalf of the company for his valued suggestions and lucid explanations and hoped that he would continue his inquiries regarding pollen and its advantages, and not fail to let his brother bee-keepers enjoy the benefit thereof through the columns of the BBJ.

BBJ 17 Oct 1907
A Beginner's Query. Will you be kind enough to give me a bit of information through the BBJ about my bees? I am a novice at bee-keeping, having bought my first two swarms last June in skeps. I put the skeps into frame-hives on six frames because they had some brood in the combs when I bought them. Having done this, I thought the bees would go down on to the frames below, but on looking at them I find they are still in the skeps.

I bought a copy of the new "Guide Book," but cannot find how heavy they should be to winter in the skeps. But I weighed them on September 19th, one weighing 24 lb. and the other 23 lb. Since then I have given them 18 lb. of sugar syrup. Are they safe for the winter, and should I leave them in the wooden hives?
            Wishing the BBJ, every success, I send name and sign Miner, Notts.

Reply.The time for setting bees in skeps above the top-bars of a frame-hive, for the purpose of transferring themselves to the latter, must always be gauged by the condition or readiness of the skeps for the operation, as directed in the "Guide Book." In your case the needful conditions were evidently absent; hence the failure. For the rest, we may say the skeps will no doubt winter safely as they are, as far as the food-supply goes but we should be inclined to take away the body-box of the frame-hive and set the skeps directly on the floor-boards of the latter if no work has been done on the foundation.

BBJ 24 Oct 1907
Converszione continued.
Mr. Pugh had earlier in the evening referred to the numerous cases in which brood had not been sealed over. That had occurred in some hives where there was plenty of pollen, but although the brood had not been properly capped over, yet it had hatched out and emerged from the cells satisfactorily. Strange to say, this omission to seal would occur unaccountably perhaps in three or four hives in an apiary, while in all the rest brood-sealing had been finished off properly. But for several years past he had seen occasional instances of brood remaining partially unsealed. It had not been a

BEEKEEPING BETWEEN TWO QUEENS

case of overdosing with naphthaline with regard to them.

Another of the difficulties was connected with queen-rearing; he meant with regard to the fertilisation of young queens and in cases of swarming. Again and again the queens produced had turned out to be drone-breeders in cases where two or three small swarms had been joined together. No doubt they were second swarms, and the queens had not been fertilised. Another difficulty had reference to the comparative frequency with which queens had been turned out of hives. One morning he picked up four or five queens thrown out in this way and that from swarms working under normal conditions. Undesirable swarming had occurred to a great extent but it was not owing to the need for room.

He had seen cases in which the bees had been allowed any amount of room and every facility for going to work and yet swarming took place. Then later on, when the nights were cold, the queens ceased ovipositing. There was more brood in some hives three or four weeks ago than in the height of summer, the diminution being due to the cold nights. With regard to the excess of drone breeding, possibly the explanation he had given might in some measure account for it or the queens found their way through the excluder and upstairs into the surplus chambers where they started depositing eggs in drone-cells. Mr. Ellison said there had not been any undue deposit of honey to explain it.

Mr. Bevan asked if any bee-keeper present had experience of bees eating their own eggs. A friend of his had complained of this occurring and he (Mr. Bevan) attributed it to the fact that there was no food in the hive. The Chairman jocosely suggested that this must surely be some new disease! Mr. Herrod, in reply to Mr. Bevan, said that bees would eat the eggs, and even suck the juices from the larvae, when hard pushed for food. Egg-eating also happened sometimes in the case of small colonies headed by very prolific queens.

Mr. Carr said that he could hardly give any practical experience of his own on the difficulties of the past season but only that of others and his conclusion, derived from a considerable amount of information received at the BBJ office during the season, was that a great amount of trouble had arisen in consequence of the failure of bee-keepers to keep an eye on the storehouse.

Thus the bees muddled on and their owners, in some cases, wondered why there was no honey in the supers without reflecting whether there was any honey to be gathered or not. Then, in another direction, they had an instance in which a bee-keeper had lost eight strong stocks out of fourteen or more this autumn and had sent up a frame of comb as a sample, explaining that the bees had deserted the hives, leaving similar combs in each, and asking for our opinion as to the cause thereof, saying: " The bees had gone - he did not know where."

Upon examination of the comb he (Mr. Carr) had no great difficulty in explaining the most probable cause. There was absolutely no sign of food in the comb sent and he therefore regarded it as a clear case of neglect or unintentional carelessness on the part of the bee-keeper. There was no doubt in his mind that the bees left the hives as "hunger" swarms and distributed themselves wherever they could in the contiguous hives of the owner's apiary or in others in the immediate neighbourhood. Mr. Carr then passed round the comb in question for inspection.

Mr. Pugh agreed that many bee-keepers were often wanting in that forethought which would at times avert a catastrophe. He had himself an outapiary and, in April last, visited it for the purpose of putting on supers and seeing whether all was going well with the queens, etc. He at once recognised the necessity of starting to feed and the result was that those bees had got on fairly well all through the bad season thanks to a little heather in the locality whilst other bees in the immediate neighbourhood had died out. The truth is these last-named bees had no food wherewith to maintain themselves and had not a sufficient number left to gather the little nectar available.

In 1888 they had a similar summer to the last except that there was more rain. There had not been an extra quantity of rain this year, the greatest evil being that clouds and dull skies had intervened so largely between the bees and the sun.

On the other hand, 1887 was one of the best seasons for honey he himself had ever known and it looked as though the Giver of all good had taken care to provide the stores for the lean year that was to follow. He thought they might draw some consolation from that, which should fortify their hopes for the future.

BBJ 31 Oct 1907
Converszione continued.
Mr. Pugh said he liked these things (the subject of labels) localised as far as possible, and was of opinion that the counties' own labels had the best chance of success; better than anything that could be decided on in London. In Notts. when a person applied for labels, he had to fill up a form and practically undertake to supply nothing but good honey while being able to trace the source from which the same was derived.

They in Notts. had, by this means, built up a good sale for all local honey and were anxious that that sold in their county should be the produce of that county alone, so far as it was possible to keep it so. He thought the BBKA must be careful not to infringe what he might call the vested interests of the local branches. Anything that had a tendency to interfere with local trade was looked on with a jealous eye. Fifty per cent profit was made on the labels and it was considered that the purchasers received their money's worth in the better sale of their honey. The labels cost about 5d per hundred and were retailed for 10d or 1s; but obviously a deal of work in connection with this matter fell on the secretary. He did not think the members of the NBKA would desire a change.

Mr. Pugh suggested that although complaints have not been made, nevertheless the numbering and registering had no doubt acted as a deterrent against the bottling of bad honey.

BBJ 14 Nov 1907
Plurality of Queens. Around the "Alexander" system of keeping a number of mated queens in one hive, there naturally centres an exceptional amount of interest to all bee-keepers.

The chief points of advantage appear to me to be its help in prevention of swarming – which came as a surprise to Mr. Alexander - and its method of retaining surplus queens. Other advantages claimed for the system are open to question. Dr. Miller mentioned in his "Stray Straws" that Mr. Alexander had successfully wintered five queens in one stock up to date (February, 1907).

We in England have for many years been acquainted with a plural-queen system in the "Wells" hive, which is now falling into disuse, mainly, I think, on account of the heavy, cumbersome hive, the single hive being so much easier to handle. Yet some valuable lessons were learnt from that hive. But the "Alexander" method opens up new ideas worthy of trial and no new hive is required; therefore the outlay is not great. I have wintered a single-comb observatory-hive and it is surprising to see how close bees can cluster, emphasising the saying, "The best packing for bees is bees."

If, then, we could unite the results of two queens of the "Wells" system into one compact cluster, as is done on the "Alexander" method, it is worth trying for wintering if it shows stronger stocks in spring. It is hardly necessary to mention the great advantage that would accrue to the apiarist who is located a distance from his bees if he can completely control swarming. It is clear to most of us that Mr. Alexander's location and the race of bee he works with may lend themselves to his system. Yet, even with our variable climate and varied flora, we may in some measure succeed with his method; so it is worth a trial.

There are those who say it is "contrary to Nature"; which may be true in a limited sense. In the progress of civilisation and in improving breeds by selection the life habits of various members of the animal kingdom are constantly changing and it is quite possible that we may succeed in changing the habits of the bee to a certain extent sufficient for our purpose. It may be that in the very multiplicity of queens there is safety from "balling."

Two years ago I studied the "balling" question closely, and found that in the early spring manipulation, if a stock had been fed a little so that there was some unsealed food in the combs, the bees were unwilling to break into their sealed store. Consequently a stock with a greater number of field workers at home are not able to fill their honey-sacs so quickly as when there are plenty of unsealed stores and there is thus less risk of "balled" queens because filled bees show practically no fight to apiarist or queen. This is in line with Mr. Alexander's method of introducing plural queens. I have introduced two queens into a hive and they have been at once accepted, but in twenty four hours the queens were fighting!

I have also seen two active laying queens in a stock, one above and one below a queen-excluder, making one powerful stock. Again, two queens - mother and daughter - have been found laying side by side in my apiary. I have also seen two queens successfully introduced by mistake.

A singular thing occurred to me the other day. I had by me a laying queen without attendants, when it chanced that a strange bee alighted on the open cage. The piercing cry of terror that came from that queen surprised me; so loud and prolonged was it that my daughter in the next room remarked, "Father, are the bees 'balling' that queen?" It was accompanied by the curling of the abdomen to ward off attack. After a pause this solitary worker showed fight and would soon have killed the queen but for my interference.

In conclusion, I say the "Alexander" method of preserving queens is beset with difficulties, and few will be the number that succeed; yet it is worth the effort.

Joseph Gray, Expert and C.C. Lecturer, Long Eaton

A copy of the writings of EW Alexander, first published in the USA in the magazine "Gleanings in Bee Culture" in 1904, were gathered into a book published in 1907. A copy of this, in digital format, is in the NBKA library.

BBJ 21 Nov 1907

The Importance of Pollen. Referring to the paper read by Mr. Geo. Hayes at the late BBKA Conversazione, may I say that some time ago I carried out a series of experiments with pollen? My object in doing so was to ascertain if bees that have reached the imago stage are physically injured by being debarred from the same. The result of these experiments led me to the conclusion that there are circumstances when bees are much better without pollen and to give it them at such times causes physical suffering. And I asked myself the question, "Does a virgin hatched in a nursery need pollen in her candy?"

BEEKEEPING BETWEEN TWO QUEENS

The first experiment proved that artificial pollen (pea-flour) in candy made on the "Good" plan, ie., honey thickened with caster sugar, caused the virgin and her few attendants to be badly affected with dysentery; while that made with fine oatmeal was not so bad in its effects. But cages supplied partly with pollen candy and partly with plain gave still better results and those cages with plain "Good's," without any pollen at all gave the best results. The query then arises, "Are there not sufficient pollen-grains in honey alone to supply the needs of the adult bees?"

My next experiment was in wintering stocks; and in this direction I found that driven bees placed on combs partly filled with honey, supplemented with plain candy, came out best; those placed on combs of honey with pea-flour candy to make up the shortage wintered the worst, those with oatmeal candy coming out midway between the two. I have also wintered stocks short of stores with pollen candy and each time they suffered with dysentery, while those with plain candy wintered well.

The results of this series of experiments have led me to the following conclusions:
1. Bees in confinement should not be pollen-fed.
2. During the time that bees are required to be kept perfectly quiet and very rarely take flight they need no pollen at all, and if fed on it will suffer in being unable to discharge the faeces, while if cleansing flights are not possible they will suffer in consequence.
3. That great care is required if pollen is given in early spring to cause brood-rearing, or the loss in unnecessary flights will counterbalance the gain in brood.

J. Gray, Long Eaton

BBJ 19 Dec 1907
Natural vs. Artificial Pollen. Bee-keepers generally will agree with Mr. LS Crawshaw when he states that pea-flour and fine oatmeal are only substitutes for natural pollen. I am also well aware that it is difficult to wrest from the hive some of its secrets, yet all my experiments have so far tended to one conclusion, viz., that pollen is not consumed during the resting time of the bees, except for the few grains that are already in the honey when taken as food. The consumption of pollen in quantity causes an accumulation of excreta, which must be got rid of when the bees are in flight; hence a well-made cake of candy is excellent food during the quiet winter-time.

My latest experiment is with a stock of bees covering one comb only. These were fed with heather-honey, which resulted in a certain amount of dysentery, just enough to cause the bees to soil the hive and combs a little. I have now given them a 5-lb. cake of candy, and the unrest has given place to rest and, what surprised me most, they are actually building comb. Heather-honey was given in preference to clover on account of its density but it may also contain more pollen grains.

Joseph Gray, BBKA Expert and C.C. Lecturer, Long Eaton

Minute Book – Committee Meeting held in the Peoples' Hall on January 4th 1908 at 3.30pm. Present: Mr WS Ellis, Chairman, Dr Elliott, Messrs Adams, Bolton, Darrington, Dickman, Gray, Munro, Pugh, Puttergill, Mountney, Smithurst, Stoppard, Turner, Vessey, Windle, White and secretary.

The minutes of the previous meeting having been read and confirmed by the signature of the chairman, correspondence was read about the holding of a show with the Nottingham Horticultural and Botanical Society.

The annual Balance Sheet which shewed a cash balance of £4 4s 10d was next gone into and the same was considered to be satisfactory. Mr Pugh proposed and Mr Vessey seconded the same be accepted and printed with the circular convening the Annual Meeting.

Suggestions were made to the secretary for drawing up the Annual Report of the committee.

Arrangements were next made for the Annual Meeting to take place in the Peoples' Hall on March 7th at 3pm and that the following be asked in this order to preside thereat.
>> Her Grace the Duchess of Portland
>> The Mayor of Nottingham 1908/9 (John Ashworth)
>> County Sheriff 1908/9 (Edwin Mellor)
>> EF Milthorp (accepted)
>> Sheriff 1907/8 (William Henry Carey)
>> Colonel Rolleston
> that the tea and competitions take the same form as in the precious years
> that words be added to the invitation to make it fuller
> that certain words be left out of the portion re. question time
> that Dr Elliott take the chair at the evening meeting
> that the agenda be as follows:
>> The chairman's address
>> Distribution of medals and certificates
>> Answers to questions laid upon the table
> Paper by Mr Thos. N Harrison on "Bee appliances dealing, its difficulties and pleasures."
>> Introduction of any novelties
>> Delegate report of the proceedings at BBKA Meeting
>> 10 minute talk on "Miniature sections" if time permits.
>> To close with the prize drawing at 7.45pm.

Edwin Mellor, JP, lace manufacturer at Hill's factory on Manvers-street, Beardsley and Mellor

in the lace market, lived in Addison-street and represented the Forest Ward. He died in 1927.
The matter of prizes for the drawing was brought forward when Mr Pugh stated he considered that, as the association were bound according to the rules to provide a 'hive' for this purpose, that none of the committee should be asked to contribute anything for this purpose, and this was accepted.

BBJ 6 Feb 1908
The twenty-seventh AGM of the Derbyshire BKA was held at the Victoria Cafe, Derby, on February 1st. Mr. R. Giles occupied the chair. There were also present P. Scattergood (Stapleford), and George Hayes (hon. sec. NBKA). Mr. Hayes then gave a most interesting lecture entitled "Bees and Flowers." which was greatly appreciated, Mr. Hayes being warmly thanked for his address.

NEP 8 Mar 1908
The AGM of NBKA was held at the Peoples' Hall this afternoon. The secretary's and treasurer's reports were unanimously adopted. The Duchess of Portland was re-elected President.

Suggestions were next made to the secretary for drawing up the Annual Report of the committee. Arrangements were made for the Annual Meeting to take place in the Peoples' Hall on March 7th at 3pm

There appears to be some confusion here about the timing of the AGM.

BBJ 19 March 1908
The AGM of NBKA was held on March 7th, in the Peoples' Hall, Dr TS Elliott, of Southwell, presiding, in the unavoidable absence of Mr. EF. Milthorp, JP. (Newark). The attendance was very satisfactory and included members of the Derbyshire and Leicestershire BKAs. Several letters regretting inability to attend were read, among them one from Mrs. P. Scattergood on behalf of her husband, who was too ill to be present.

The Annual Report and financial statement were also presented, the latter showing a balance in hand of £4 4s 10d and both were adopted. The report showed that the total membership was 208. Thirty four new members had joined during the year.

They had to announce with regret that the County Council had reduced their grant for Technical Instruction in aid of bee-keeping from £50 to £25, which had to a great extent minimised their usefulness in the county. It was understood that the Council was quite satisfied with the work done for the money spent but lack of funds had compelled the reduction. They therefore hoped that in the coming session the Council might be enabled to restore the grant to the former sum, especially in view of the experiments in cottage farming which would shortly take place under the Small Holdings and Allotments Act, the cultivation of bees being distinctly advantageous to agricultural and horticultural work.

Lectures had been given at the Newark and Welbeck Tenants' Agricultural Societies shows, and an open-air demonstration at Elston; 148 apiaries had been visited, in which were 558 stocks of bees in frame-hives and 49 in skeps, advice being given as to immediate necessities.

Thanks were passed to the officers for their services during the past year, and the Duchess of Portland was unanimously re-elected President. The retiring committee, together with the district secretaries and experts were re-elected. Mr. Peter Scattergood was also re-elected auditor and Mr. Geo. Hayes, secretary and treasurer. Mr. Pugh and Mr. Hayes were appointed representatives to the BBKA.

There was a competition as usual for the best single 1-lb. jar of honey in liquid and in granulated form, the exhibits being given for the children in the Nottingham hospitals.

In the evening Dr. Elliott presided at a well-attended meeting in the same hall, when medals and certificates were distributed, after which he addressed the members and friends on "Scientific Knowledge in connection with Practical Beekeeping," pointing out its direct bearing on apiculture. A hearty vote of thanks was accorded Dr. Elliott for his valuable and instructive paper, a desire being expressed that, if possible, it might appear in the BBJ. Various novelties and interesting articles were next shown and explained, and finally Mr. TN Harrison read a most interesting and very amusing paper on "Bee-appliance Dealing: Its Difficulties and Pleasures," for which he was heartily thanked. The meeting concluded with the usual prize-drawing.

Report for 1907
The AGM of this Association was held in the Peoples' Hall on March 7th, 1908, and in the absence of EF Milthorp, Esq., who was unavoidably absent, the chair was taken by Dr. Thos. S Elliott, of Southwell.

The Minutes of the previous Annual Meeting were read and confirmed by the signature of the Chairman.

Letters were read from Mr. R. Godson, Secretary of the Lincolnshire Association; Mr. RH Coltman, Secretary of Derbyshire BKA, and Mr. P. Scattergood regretting their inability to be present, the latter owing to a severe illness and it was resolved that a letter of sympathy be sent from those assembled at the meeting to Mr. and Mrs. Scattergood.

BEEKEEPING BETWEEN TWO QUEENS

The Annual Balance Sheet was next considered, and after several questions had been answered concerning the same, it was resolved that it be received and adopted as satisfactory.

The Committee's Report was next read by the Secretary which was as follows: Well! It seems to us the least said about an unpleasant thing the better. Everyone found it a *bad* season, though some found it better than others, and it is sure to have caused in the minds of some a feeling unfavourable to the industry. But it is hoped all will come out good staunch beekeepers, in whom "Hope" is an evergreen; and that in the coming season we shall be well repaid. The good season *will* come so let us be prepared for it.

Last year we beat the record so far as regards membership and we are glad not to have to report a falling off from this high-water mark; but on the other hand to shew we have advanced even slightly further, for whereas our total membership last year was 205 we now number 208. 34 new members have joined hands with us during the past year, and 31 have left us through removal, death and other causes. Amongst those whom time has removed from us may be mentioned one of our Vice-Presidents for some years - Mrs. Jesse Hind of Edwalton. There may also be some whose subscription may yet be forthcoming for 1907, which would of course raise the membership.

Mrs Jesse Hind was, in fact, married to someone of that name but was actually Mrs Eliza Hind (1842-1907). In the 1891 Census she was living with her family at the Grange, Papplewick (demolished 1932) but by 1911 the family was at Edwalton and her husband was recorded, obviously, as a Widower. Her husband was a well-known solicitor in Nottingham.

Perhaps with a better season before us, and a little incentive which will be held out later in the day, we may still go forward increasing in strength and usefulness. Let our watchword be "Advance NBKA" and hope that every individual member and each District Secretary and Expert in particular will assist the advance by putting all the energy into the movement.

Never for the last 20 years, nor do we think, even from the inauguration of the Association, have we been able to shew so large a cash balance as we do at the end of the year; not that it is very great, or that we wish to be considered parsimonious, on the contrary we think the money should be spent in our work rather than to be hoarded up; but it is - we think you will agree - far pleasanter that we have not over-stepped the mark, and we are well on the safe side, than to have a feeling of hovering insolvency and insecurity. We therefore venture to think you will be satisfied with our present financial position; but at the same time hope it will not create too great a feeling of security, to cause any slackening in our interest in the work and its needs.

It is with regret that we have to announce that the County Council have reduced their grant for Technical Instruction to £25, which has to a very great extent minimised our usefulness in the county, for the same amount of work cannot be done as when their grant was £50. We understand the Council are quite satisfied with the work done for the money spent - as those Experts who do the work do it not for gain, but for the good of the industry - but we learn that the reduction is through lack of funds. We therefore hope that in the coming season they may be enabled to increase our grant to £50, more especially is this necessary now in view of the experiments in cottage farming which will shortly take place under the "Small Holdings Allotments Act"; for the cultivation of bees is distinctly advantageous alike to agricultural and horticultural work.

We shall be glad if our members will wherever possible bring the usefulness and the necessity of our teaching before those Councillors they meet.

With the amount granted for 1907-8, lectures have been given at the Newark and Welbeck Tenants' Agricultural Societies Shows, and an afternoon open-air demonstration at Elston. 148 Apiaries have been visited in which were 588 stocks of bees in barframe hives and 49 in skep hives, a total of 607 of which 489 were examined and advice given as to immediate necessities, and answers given to sundry enquiries from each.

Owing to the exceptionally cold and wet season, many beekeepers did not get a visit, as it was unsafe for even Experts to open up stocks in such weather. Certainly there were a few fine warm days intervened, but as soon as the Expert thought it had taken up and sent notice of his coming, so sure would it break up again, and this made the matter of visiting a cause of anxiety to him, and so he waited until he found it was impossible to get at all.

Two members offered themselves as candidates for the examination for Experts: W. Mounteney, Southwell, for 3rd Class, W. Darrington, Eastwood, for 2nd Class and we are glad to state that both have passed successfully. It is hoped that more members will avail themselves of this examination so that we may have a greater number qualified for Expert work.

A case of interference to bees was reported by Mr. Newbury of Hucknall Torkard. A reward was immediately sent out for the conviction of the offenders, but they were not found. It however stopped the practice, as they have not since been molested.

The Annual Show was held at Southwell on July 25th, Dr. Elliott officiating as judge, the number of entries being 74 against 80 in the previous year. Although we had a good entry only comparatively few exhibits in the honey classes turned up owing to the bad season. The principal prizes taken were:

Silver Medal	J. Wilson, Shirebrook.
Bronze Medal	G. Marshall, Norwell.
Silver Pendants	W. Ball, J. Wilson, WL Betts.
Certificates	J. Bee, Southwell; Messrs. Thos. N. Harrison & Son, Nottm., W. Ball.

The prize hive given by a Vice-President to the cottager exhibiting the best sample of comb and extracted honey at this show was awarded to Mr. G. Marshall, Norwell.

At East Bridgford the usual show was held and was about up to the standard of previous shows at that place. Mr. RJ Turner officiated as judge.

At the last Annual Meeting it was decided that the Association should arrange for a Summer Outing for its members and their friends. This appeared to be desirable and the Committee arranged a trip to Syston and Leicester on June 1st. Fifteen members and their friends making the party up to thirty-three, assembled and having been shewn over Mr. Meadow's appliance works partook of a substantial tea in the Village Hall close by. After tea the party were driven to the Abbey Park Gardens, etc., but as the evening turned out wet this portion of the outing was somewhat damped thereby and so marred the enjoyment which otherwise would have been.

Our thanks are due and hereby tendered to the County Council, the City Council and the various Agricultural and Horticultural Societies for their grants. To the District Secretaries and Experts and to all who have in any way helped forward the Association and its work.

Resolved that the Report be adopted and printed with a list of members and copy sent to each.

Mr. Hill proposed and Mr. Hoyte seconded, that a vote of thanks be accorded to all retiring officers for their services during the past year. Carried.

Mr. Pugh then rose to propose a very hearty vote of thanks to Her Grace the Duchess of Portland as President during the past year, and that she be again elected for the ensuing year. This was carried with acclamation.

The Committee were re-elected *en bloc*, and without any addition. Mr. Scattergood was re-elected Auditor. The Secretary and Delegates to the BBKA meetings were also re-elected.

The Committee were instructed to arrange for a Summer Meeting, and the wish expressed, that considerably more members would avail themselves of it so as to make it a success. A vote of thanks was accorded Mr. Meadows for his help at last summer meeting.

Mr. EM Varty, of Pleasley, promised to give a new hive as a prize in the Amateur Class for honey at the Annual Show in the hope that others would follow his example and to encourage those to exhibit who have never done so.

This concluded the business of the afternoon, and members and friends (including Mr. J. Waterfield, of Leicester and Mr. WP. Meadows, of Syston) to the number of 50 partook of a capital and well prepared tea, for the preparation of which our thanks are due to Mr. and Mrs. WS Ellis.

17 bottles of honey were sent for competition, and the prize winners were:
For Granulated Honey – 1st, Adams, Mansfield; 2nd, JT Wilson, Shirebrook.
For Liquid Honey – 1st. Thos. Marshall, Sutton-on-Trent: 2nd, Wm Ball, Eagle.

The honey was sent to the Children's Hospital, Nottingham.

The meeting was resumed about 6pm. and after the distribution of medals, certificates, etc., Dr. Elliott addressed the meeting on "Scientific Knowledge and Practical Beekeeping," pointing out the numerous sciences that were involved in apiculture. His discourse was listened to with rapt attention, and great applause greeted him when he had finished. A hearty vote of thanks was accorded him, and Mr. Darrington desired the paper might be printed in the Bee Journal.

A number of questions on the taking up of pollen from flowers, cell construction, dusting bees with flour, renewing combs, spreading brood, etc., etc., were dealt with from the question box, and as one member expressed himself - these supplied plenty of food for reflection.

It was announced that Dr. Elliott wished our Association to increase in numbers, and as an aid to that end he would offer three prizes to the members of this Association, Secretary excluded, as follows:
For Visiting Experts - A prize of 10/6 cash for the greatest number of new members made from this meeting to August 31st, 1908.
For Other Members:
    1st prize of 7/6 for greatest number of new members made from this meeting to August 31st, 1908.
    2nd prize of 3/- for those bringing the second highest number.

BEEKEEPING BETWEEN TWO QUEENS

(A new Vice-President to count as two members. If a name and address is given of anyone whom the competitor has spoken to and the Secretary is able to follow it up and receives their subscription; this to count to the credit of that competitor.)

Mr. Thos. N Harrison next read a rather lengthy paper on "Bee Appliance Dealing; its difficulties and pleasures," which proved to be most interesting and amusing, and which was afterwards discussed by Mr. Meadows and others. A hearty vote of thanks was accorded to Mr. Harrison for his paper.

Various novelties and objects of interest were shown and briefly explained - New Marvel Honey Extractor, New Wax and Fruit Strainer, Paper Jars, Miniature Sections, Spoon-ledge Jar, Young's Spoon-rest, Rolls of Honey, Gray's New Show Jar, Travelling Box for Bottles, Cardboard Section Case, Queen Cages attached to top bar, Bulldog Smoker, etc.

This was followed by the usual prize drawing for members who had paid their subscription;

		Given by	Won by
1st	XL All hive	NBKA	Mr Bryan, Kirkby
2nd	Volume "Beekeepers' Record"	NBKA	Mr Stoppard, Nottingham
3rd	200 honey labels	NBKA	Mrs Lewis, Ilkeston
4th	Wollert spur embedder	NBKA	Mr Wood, Arnold
5th	100 honey labels	NBKA	Mr Faulconbridge, Bulwell
6th	2 vols. "Record"	Mr Ball, Eagle	Mr Broome, Rufford
7th	1 gross miniature sections	Mr Ball, Eagle	Mr Cragg, Flawborough
8th	2 vols. "Records"	Mr Ball, Eagle	Mrs Blagg, Burton Joyce

A vote of thanks to Dr Elliott for presiding concluded the meeting which was a very successful and enjoyable one.

NEP 17 Mar 1908
Mr. Peter Scattergood, one of the best-known figures in Stapleford, having occupied many positions in the district, died this morning from a heart seizure, after an illness extending over some weeks. Mr. Scattergood was a member of the Stapleford Parish Council, and correspondent to the school managers, and amongst other positions held was the secretaryship of the Stapleford and Sandiacre Water Company. For years he had been a local preacher in the United Methodist Connexion and his loss will be deeply felt in the district. Mr. Scattergood was for long time connected with the NBKA.

BBJ 26 Mar 1908
The AGM of BBKA members was held at Jermyn Street, SW on March 19th, Mr. TW

Cowan, FGS, presiding. There was also present AG Pugh. Mr. Till moved, and Mr. Pugh seconded, a vote of thanks to the RSPCA for the use of their boardroom, which was carried.

BBJ 26 Mar 1908
We deeply regret to record the death of Mr. Peter Scattergood, which took place at his residence, Gladwin House, Stapleford on the 17th inst., from heart failure. He had just past his fifty-third year.

The well-known figure of Mr. Scattergood will be sadly missed from among beekeepers not only in Notts, but throughout the Midlands. He was, we believe, the oldest member of the NBKA. at the time of his death, having been present at the preliminary meeting prior to its formation in April, 1884, and since that time had remained one of the most active members of its Council.

A view of Mr. Scattergood's apiary, with himself and wife, appeared in our issue of January 13, 1898, and from the notes written by himself we gather that he came from a bee-keeping family, an uncle of his having some forty years ago owned about 240 colonies of bees in North Notts. Our late friend was in all things he undertook (and they were many) a busy and prominent worker, and he will be greatly missed in the place where his life was spent; indeed, it may be truly said few men have so long a record of service in the causes of temperance, religion, education, and general philanthropic work among his fellows.

The public offices filled by Mr. Scattergood would make a long list, but readers of the BBJ, know of him mostly in connection with apiculture. He was a successful exhibitor at shows, a first-class expert of the BBKA, while later he judged at many leading shows, and became himself an examiner of candidates for the BBKA. In all of these capacities Mr. Scattergood rendered useful service to the BBKA (of which he was a loyal supporter) and to the craft generally, on whose behalf we offer sincere sympathy to his widow in her bereavement.

BBJ 26 March 1908
The monthly meeting of the BBKA Council was held at 105, Jermyn Street, SW on 16th inst., Mr. WF Reid occupying the chair. Apology for enforced absence was read from AG Pugh. The Finance Committee's report was presented by Mr. BD Till, and gave particulars of receipts and expenditure to date. It was duly approved. The following was elected as a Finance Committee member for the ensuing year.

Dr. TS Elliot,

BBJ 2 Apr 1908
BBKA Conversazione. Mr. Pugh said that the present affiliation fee was anomalous in its working, because the same amount - namely, £1 1s was expected from each

association, whatever its membership - whether twenty or thirty, or 500 or 600. He thought a *pro rata* fee would be justifiable as well as helpful to the BBKA.

Some of the most thriving associations derived a great deal of their present success from holding meetings in the winter so enabling their members to know each other. He recently had the pleasure of attending the meetings of three different county associations, where he found plenty of good fellowship, as well as a desire to learn and exchange ideas. The conference it was proposed to hold would, no doubt, be a stimulus to bee-keeping generally.

Minute Book – Committee Meeting April 4th, 1908, held in the Peoples' Hall. Present Mr WS Ellis in the chair, Messrs Pugh, Turner, Moult, Dickman, Adams, Gray, Vessey, Darrington, Stoppard, Harrison, Smithurst, White and secretary.

The minutes of the previous meeting were read and confirmed by the signature of the chairman. Proposed by Mr Pugh and seconded by Mr Dickman.

The doings of the sub-committee were reported and the members of the same were re-elected *en bloc* adding the name of Dr Elliott. Proposed by Mr Vessey.

The quarterly Balance Sheet which shewed a falling off of subscriptions compared with previous years was next read and passed. Mr Gray proposed and Mr Darrington seconded.

A vote of thanks was accorded those who assisted at the tea, the secretary to write to those not present. Proposed by Mr Pugh, seconded by Mr Turner and supported by Mr Darrington.

A list of the names of 12 new members were read and it was resolved that the same be accepted. Proposed by Mr Pugh.

Mr Pugh, delegate to the BBKA, gave a very full account of the general meeting and conversazione of the BBKA which proved very interesting to those present and raised considerable discussion, Mr Pugh being thanked for what he had recounted. From this emanated the desire that this association should do all it could to help the BBKA.

It was resolved that the Summer Meeting should be to the Conference at the Franco-British Exhibition and that the secretary write to the BBKA stating their intention and also to Mr Young, secretary of the BBKA, stating that Mr W Darrington promised to read a paper at the conference if desired. Proposed by Mr Gray, seconded by Mr Darrington supported by Mr Pugh.

As regards the Annual Show, the secretary stated he had not been able to get any suitable offers from any society and it was resolved to defer the matter to a special committee meeting and, in the meantime, the secretary write to the Hucknall and Kingston Societies and try to get terms from them.

Resolved on the proposal of Mr Gray that a letter be sent from this association to Mr Wm Ball, Eagle wishing him "God speed" on his journey to Canada and success on his arrival.

BBJ 16 Apr 1908
Franco-British Exhibition. It may possess interest for both Editors and readers of the BBJ, to know that the committee of the NBKA have decided that the Summer Meeting this year shall be at the above exhibition instead of, as previously, at some local centre.

Each of our members will be advised of this and of the most economical way of getting up to town, and in this way it is hoped that a good number will respond, so that we may he well in evidence at the convention. One of our committee has also offered to read a short paper on some subject if it is desired.

I hope other county associations will endeavour to do something on the same lines, in order that British bee-keeping may be well represented, and our French brethren may thus be able to see the interest displayed in the craft on this side the Channel.
Geo. Hayes, Secretary NBKA.

In view of the use of the two languages at the conference, may I suggest that some kind friend provide us with a fairly complete English-French glossary of beekeeping terms? I am sure that this would be a great help even to those who speak French fairly, and might even be of interest to our French friends at the Conference.

Minute Book – Special Committee Meeting held May 16th, 1908. Present WS Ellis in the chair, Dr Elliott, Messrs Darrington, Vessey, Dickman, S Adams, Moult, Stoppard, Harrison, Pugh, Smithurst, White and secretary.

The secretary had correspondence from several societies and it was decided to accept the offer of the Kingston Show of £7 10s. NBKA to provide our own tent. It was unanimously agreed that Mr Herrod should be asked to officiate as judge and examiner and that Mr Darrington be appointed as assistant judge and lecturer.

The schedule was gone through and the following alterations made.
> Silver Medal to sections and bronze to extracted honey. Proposed by Mr Vessey and seconded by Mr Darrington

> Class 7 (Frames) – leave out 1st prize making the remaining 1st, 2nd and 3rd.
> Class 8 (Novice) – 1st prize to be hive given by Mr Varty, Pleasley.
> Class 12 (new class) - A hive given by Mr WS Ellis for three bottles and three sections to member not previously getting a prize.

BBJ 11 Jun 1908
Bees and Buttercups. Referring to this matter, may I be allowed to say the bulbous crowsfoot certainly yields pollen in great profusion in a handleless cup or bowl shape, and if any of those interested in the subject will carefully separate the petals from the calyx they will be able to see with the naked eye, but of course better still with a pocket-lens, the nectary at the base of each petal on the inside. Like others of your correspondents, I have not found bees giving very much attention to the flower named above, probably for the same reason that animals avoid them, viz. their very acrid taste. Geo. Hayes, Beeston, Notts

East Bridgford Parish magazine July 1908
The Annual Show was held on June 30th and the produce exhibited was remarkably good.
> Bees – 1st, Mr G Huskinson; 2nd, Mr C Allwood
> Sections – 1st, Mr D Gower; 2nd, Mrs Cartwright; 3rd, Mr T Hill
> Honey extracted – 1st, Messrs Gower and Higgs; 2nd, Messrs Colville and Millington

BBJ 2 Jul 1908
The Conferfence of Bee-keepers was held in the Congress Hall at the Franco-British Exhibition on June 25th, under the presidency of Lord Avebury, PC, FRS, when a distinguished company numbering over 250 assembled including AG Pugh and Geo. Hayes (hon. secretary NBKA),

Minute Book – Committee Meeting held in Peoples' Hall on July 4th, 1908. Present: Mr AG Pugh in the chair, Dr TS Elliott, Messrs FG Vessey, W Adams, W Darrington, WH Moult, RJ Turner, G Smithurst, D Marshall, W Mountney, Thos. Harrison, W Munro, WH Windle, WH Stoppard, G White and secretary.

The minutes of the previous meeting were read and adopted. Proposed by Mr Darrington and seconded by Mr Vessey. The quarterly statement of accounts shewed us to be in much the same position as at the corresponding period last year and were passed as satisfactory. Proposed by Mr Turner and seconded by Mr Stoppard.

The names of 38 new members were submitted and received, making a total of 50. Proposed by Dr Elliott and seconded by Mr Vessey.

A letter from Mr W Herrod was read stating he could not, owing to a previous engagement, accept the offer to judge at Kingston and the following gentlemen, in the order given, were to be written to for this purpose:

    Dr Sharpe   Mr Richard Brown   Dr Elliott

Dr Elliott, who was present, promised to undertake the task if neither of the two foregoing could do so.

Mr Harrison proposed that Mr Darrington be elected auditor for the current year in the place of the late Mr P Scattergood. Mr Munro seconded and this was unanimously agreed to.

It was reported that 7 members and 2 ladies attended the summer meeting at the Franco-British Exhibition.

NEP 22 Jul 1908
August 3rd, Bank Holiday. Kingston and District Shire Horse Society Annual Show Good entries. Trotting Races. Gymkhana Bare. Jumping, etc. Splendid Horticulture, NBKA Annual Show, and the Beautiful Grounds of Kingston Hall thrown open to the public kindness Lord and Lady Belper. Good Band and Great Display of Fireworks.

Henry Strutt (1840-1914) 2nd Lord Belper, JP, DL (Nottinghamshire), JP (Leicestershire and Derbyshire). Between 1901 and 1910 he was Yeomanry Aide-de-Camp to Edward VII.

BBJ 23 Jul 1908
The monthly meeting of the BBKA Council was held at 105, Jermyn-street, SW on the 16th inst. Mr. E. Garcke being voted to the chair. There was also present Dr. TS Elliot. Apology for inability to attend the meeting was received from AG Pugh. Judges and examiners were appointed to officiate at fixtures in Nottinghamshire.

BBJ 30 Jul 1908
Honey Samples. ME Varty (Pleasley) An excellent sample of clover honey; fit for any show-bench. Do not select the darker sample mentioned on any account.

BBJ 6 Aug 1908
Troubles in Re-queening Stocks. I should be much obliged if you would give me any explanation of and advice on the following facts:
On June 23rd I found that the queen was missing (I don't know the reason) in one of my hives, and several queen-cells were found sealed over. I placed a frame containing several of these cells in the queenless portion of an artificial swarm I made, and when examined a week later (June 30th) the queen-cells in both hives were empty. On July 5th I inserted a frame of brood in all stages in each hive, and no queen-cells were formed in either, nor have any eggs been laid since.

Accordingly, on July 17th, I united the two lots and placed a queen I had obtained (in an introducing cage) on the top of the frames, allowing the bees to get to the candy and release the alien queen forty-eight hours later. On the following evening I found the enclosed dead bee, which I take to be a queen - although it looks much smaller than a live one - in front of the hive. I have not yet examined the hive, not feeling quite certain on this point, and thought it better not to disturb them.

I shall, therefore, be much obliged if you can tell me:
1. Is it possible for a queen from a cell sealed over on June 23rd not to have started laying by July 17th, and still be in good condition? The weather has been very fine up till July 4th, and wet since, and there were drones in all my three hives. Of course, I could never find the queen in either hive during this period.
2. If the bees are queenless will they rear a queen on being given a comb of young brood?
3. Does the fact of one queen having been refused diminish the chance of another being accepted, if I try to introduce another? I must apologise for the length of this letter, but thought it better to give full details. Thanking you in anticipation of a reply in the BBJ, I send name, etc. for reference, and sign

                                  Medico, Nottingham, July 22nd

Reply.
1. Without entering into the possibilities of the case, it would appear from the details given that there was already a queen in the hive when the alien queen was introduced, and one or the other has been destroyed. We therefore advise you to examine the hive at once, in order to ascertain if the queen released from the cage on July 19th is there and is laying. If this is so, all will be right. We say this because the dead queen sent has the appearance of a virgin; but has evidently been viciously "balled" and the dead insect is too dry and hard for post-mortem examination.
2. After what has happened to both the stocks in question, there will not be much chance of the bees raising a queen from brood or eggs now given.
3. After killing one alien queen, bees have a tendency to do the same again, and so the risk is increased.

BBJ 6 Aug 1908
The Annual Show of the Lincolnshire Agricultural Society was held at Sleaford on July 16th and 17th, the honey department being, as usual, under the management of the Lincs BKA. Unfortunately the weather was very unfavourable, heavy rain on the first day, transforming the show-ground into a perfect quagmire. Mr. AG Pugh and Dr. Percy Sharp acted as judges.

BBJ 13 Aug 1908
The Annual County Show of NBKA was held in conjunction with the Kingston Horticultural and Agricultural Society in the beautiful park of Lord Belper at Kingston, near Derby, on Bank Holiday, August 3rd, under ideal weather conditions.

The exhibits were of first-class quality, and entries numerous compared with other years. Dr. Percy Sharp, Brant Broughton, assisted by Mr. W. Darlington, judged the exhibits, and their awards were as follows:
Collection of Bee-appliances – 1st, Thos. W. Harrison and Sons, Nottingham (only one exhibit.)
Honey Trophy – 1st, U. Wood, Arnold (only one exhibit staged.)
Honey - 1st, AG Pugh, Beeston; 2nd, WL Betts, Mansfield Woodhouse; 3rd, G. Marshall, Norwell; hc, G. Hopkinson, Newark.
Twelve 1-lb Jars Dark Extracted Honey – 1st, G Marshall; 2nd, GE Puttergill, Beeston.
Twelve 1-lb Sections – 1st, W Lee, Southwell: 2nd, GE Puttergill; 3rd, WL Betts: hc, WH Stoppard, Mapperley.
Twelve 1-lb Jars Granulated Honey – 1st, AG Pugh; 2nd, GE Puttergill; 3rd, ME Varty, Pleasley.
Single Shallow-frame of Comb-honey – 1st, G Marshall; 2nd, WH Stoppard; 3rd, WL Betts.
Twelve 1-lb Jars Extracted Honey (novices only) – 1st, J Wood, Nettleworth; 2nd, Thos. N. Harrison, Carrington; 3rd, WH Stoppard
Honey Vinegar – 1st, GE Puttergill
Observatory hive – 1st, Dr Elliot, Southwell; 2nd. G Marshall; 3rd. EG Ive, Boughton; 4th, C Fincham
Beeswax – 1st, G Marshall; 2nd, AH Hill, Balderton; 3rd, GE Puttergill
Three 1-lb Jars Extracted Honey and three 1-lb Sections (amateurs only) – 1st, E Wood

An examination of four candidates for the third-class certificate of the BBKA also took place during the show.

BBJ 13 Aug 1908
Birds and Bees. Referring to Mr. Newth's letter, "Conscience does not make cowards" of the thrushes here. Since my bees commenced to turn out the drones, a pair of thrushes, who had a nest close by, almost live on the bed in front of my hives. They seem to have an acquired taste for a good fat drone, and only run under the gooseberry bushes when I appear on the scene. They would doubtless like the berries also, but these are protected by a lot of thread.   FJC (FJ Cribb?) Retford.

BBJ 20 Aug 1908
ABC (Worksop)."Mead and How to Make It" and "Honey-Vinegar." Directions for

making the above are published as separate pamphlets, which may be had from this office for 2½d each post free.

BBJ 27 Aug 1908
The Annual Show of Leicestershire BKA was held in connection with the twenty-third Annual Horticultural Exhibition at the Abbey Park on August 4th and 5th. Demonstrations with live bees were given at intervals on both days to large audiences by Messrs. AG Pugh (Notts) and EJ Roper (Birstall). The same gentlemen officiated as judges.

BBJ 3 Sep 1908
Wax Rendering. Several interesting questions are asked by your correspondent Mr. King as to the best manner of treating old frames of comb, and for a tabulation of results to be obtained in dealing with them.

A short time ago I dealt with about thirty old combs, not by the fire, but by the far more economical method of placing them inside a "Gerster" wax extractor, 10s 6d size (the 3s size is no better than an ordinary saucepan), and after removing the debris and refining the wax, obtained over 4 lb of wax, for which any appliance dealer would give appliances to the value of 5s 6d, or, if I could sell it to retail customers, I should obtain a shilling or so more in cash.

As a business man, the argument about time being of value is one that appeals very strongly to myself but, I ask, is not bee-keeping with the great majority a hobby to be followed in our leisure hours for the pleasure of it, as yielding a mine of information, besides the profit to our pockets which it affords?

I enclose a sample of beeswax obtained from the above-mentioned very dark, old combs, and claim that it is of a good colour and of the commercial value named.
Thos. N. Harrison, Yew Tree Apiary, Carrington

BBJ 10 Sep 1908
The sixteenth Annual International Exhibition and Market of the Confectioners and Allied Trades commenced on September 5th and remains open till the close of the present week. The honey competitions were:
Display of Honey (Comb and Extracted) and Honey Products, shown in suitably attractive form for a tradesman's window (6 entries) – vhc, J Herrod, Sutton-on-Trent
Twelve 1-lb Heather Sections (9 entries) – vhc, T Marshall, Sutton-on-Trent
Twelve 1-lb Jars Light-coloured Extracted Honey (43 entries) - 1st (£1 15s and BBKA Certificate), J Herrod; vhc, J Woods, Mansfield and AG Pugh, Beeston
Twelve 1-lb. Jars Medium-coloured Extracted Honey (36 entries) - 2nd, (£1), J Herrod; vhc, G Marshall, Norwell
Twelve 1-lb. Jars Heather-blend Honey (6 entries) - 4th, (5s), J Herrod

BBJ 17 Sep 1908
Mated Queens for sale, 2s 6d each.   Munro, Oxshott, Mapperley

BBJ 24 Sep 1908
The monthly meeting of the BBKA Council was held on September 17th, in the board-room of the RSPCA, Mr. JB Lamb being voted to the chair.

A report upon examination of candidates for third-class certificates held in Nottinghamshire was received, and, acting upon the recommendations of the examiners, it was resolved to award certificates.

BBJ 24 Sep 1908
The sixteenth International Exhibition and Market of the Grocers and Allied Trades was opened at the Agricultural Hall on September 19th, and continues until the end of the present week.
Twelve 1-lb Sections (21 entries) – 1st, (£1 15s and BBKA Bronze Medal), T. Marshall, Sutton-on-Trent
Three Shallow-frames Comb Honey for Extracting (7 entries) – 1st, (£1), T. Marshall
Twelve 1-lb Jars Light-coloured Extracted Honey (47 entries) - 2nd, (£1 5s), T. Marshall; vhc, AG Pugh, Beeston; G. Hunt, Newark
Twelve 1-lb Jars Medium-coloured Extracted Honey (42 entries) - vhc, G. Marshall, Norwell
Twelve 1-lb Jars Dark-coloured Extracted Honey (11 entries) – 1st, (£1), T. Marshall; hc, G. Marshall
Twelve 1-lb Jars Granulated Honey (17 entries) - 4th, (10s), T. Marshall
Beeswax in Cakes, Quality of Wax, Form of Cakes and Package suitable for retail counter trade (11 entries) - 4th, (5s), T. Marshall

Quite a profitable show for Thos Marshall - £5 15s!

BBJ 1 Oct 1908
EW. (Mapperley) Honey-extracting.
1. You cannot possibly "extract" honey from combs without the help of a honey-extractor. To proceed as described is to obtain "run" or dripped honey, and requires the combs to be sliced up and put into a muslin or coarse flannel bag, and hung up before a fire to drip through after being made very thin by warming.
2. You can obtain a cheap form of extractor for 7s or 8s.

Minute Book – Committee Meeting held in the Peoples' Hall on October 3rd, 1908 at 3.30pm. Present: Mr AG Pugh in the chair, Messrs Turner, Darrington, Vessey, Moult, Adams, Dickman, Puttergill, Stoppard, Harrison, Windle, Smithurst, White and the secretary.

The minutes of the previous meeting were read and confirmed by the signature of the chairman. The quarterly statement of accounts was next gone through which was considered to be satisfactory and the same passed. Proposed by Mr Turner and seconded by Mr Vessey.

A report on the three shows which had been held during the summer was submitted and this was ordered to be embodied in the Annual Report.

Nine new members' names were submitted and accepted. Proposed by Mr Stoppard and seconded by Mr Dickman.

A report of the doings of the education sub-committee was next read and accepted.

A letter from the Mansfield Horticultural Society offering us £6 with tent and staging for us to hold our Annual Show with them on August Bank Holiday, 1909, was read and it was resolved that we accept their offer. Proposed by Mr Darrington and seconded by Mr Vessey.

Mention was made about trying to get with Moorgreen Society but the matter was eventually allowed to drop.

Mr Harrison proposed that the price of association honey labels be reduced. After careful consideration of the costs of the same, the benefit accruing to members using them, etc. Mr Darrington seconded and a vote was taken on the matter but, as only three votes posted for this, then the proposition was lost.

The Annual Report – a rough draft of which had been prepared was next gone through in detail and altered to meet the wishes of the committee where necessary and it was resolved the same be printed and circulated to the members in the normal way.

Arrangements for the Annual Meeting on March 6th, 1909, were next gone into and it was agreed that the following gentlemen be asked – in the order given - to preside:
Alderman W Mellors, Annesley Road, Hucknall.

Mr Harry Wyles,The Court, Cropwell Butler that the competition and tea be on the same lines as at the previous meeting but that the prize drawing be divided between members who have paid subscriptions and those members present at the meeting.

BBJ 15 Oct 1908
The monthly meeting of the BBKA Council was held at Jermyn-street, SW. on October 8th, 1908, Mr. TW Cowan occupying the chair. There were also present Messrs Geo.

Hayes and AG Pugh. A letter apologising for inability to attend was received from Dr. Elliot.

BBJ 22 Oct 1908
Fertilisation of White Clover. Are bees necessary for the fertilisation of white clover (*Trifolium repens*)? This is a matter on which I shall be pleased to know the opinion of those of your readers who are botanists and who have studied this question, as I have no doubt there must be many such amongst the number.

I have always understood from what I have read, heard and observed that the bee was absolutely necessary to the flower - in fact, that the one was pre-eminently fitted for and necessary to the other and this is what I have always taught. Last summer, however, my faith in this was severely shaken by a statement made by a professor in botany, who is also a great authority on such matters. He said bees were not necessary to the fertilisation of clover, and that most of the *Leguminosse* were self-fertilising.

That some are so, I am, of course, aware, but I have yet to find that white clover is one of them. The matter is of importance to all teachers of apiculture, and of sufficient interest, I think, to make it worth discussion in our journal, to draw out evidence for or against, and thus enable us to have a right understanding concerning this subject.

Geo. Hayes, Melhurst, Beeston

BBJ 27 Oct 1908
The monthly meeting of the BBKA Council was held at 105, Jermyn-street, SW on October 8th, 1908, Mr. TW Cowan occupying the chair. There were also present: Messrs. Geo. Hayes and AG Pugh. A letter apologising for inability to attend was received from Dr. Elliot,

BBJ 28 Oct 1908
The thirty-third Annual Show of the British Dairy Farmers' Association opened in summerlike weather at the Agricultural Hall, Islington on October 6th. We were glad to see four entries in the class for Interesting and Instructive 'Exhibits of a Practical or Scientific Nature' and the one that took a well-earned first prize was staged by Mr. Geo. Hayes, hon. secretary of the NBKA, and was in the form of samples of pollen taken from all the best-known bee-flowers, which were shown in glass phials preserved in solution. There was also a copious collection of drawings (greatly enlarged) showing the various shapes of the different pollen-grains. Then to complete the whole we had a collection of slides mounted for the microscope for the use of lecturers. Mr. Hayes deserves every credit for his most interesting and useful exhibit, which is of interest to all bee-keepers as affording a test for ascertaining the sources of honey from the pollen-grains found therein.

BBJ 5 Nov 1908
Diary Show. Mr. Pugh went to the Confectioners' Show, and asked Mr. Elliott how it was that several honeys had been staged in the wrong class, and he was told that such exhibits had been disqualified.

He (Mr. Pugh) thought it was only fair to the exhibitor that he should be informed that his particular exhibit, on account of non-compliance with the conditions of the schedule, had not been allowed to enter the competition. It was a great fault at any show, even although the honeys in question might be better than that which took the first prize, if they, competing in the wrong class, were allowed to be staged without a disqualification notice on them.

The Chairman explained that the BBKA had nothing at all to do with the matter complained of, and had no right in any way to interfere with the arrangements. The Dairy Show was held under the rules of the British Dairy Farmers' Association, which appointed its own officials and judges, and settled the amount of entrance fees and regulations as regarded prizes. The judges had books given them with the numbers of the exhibits and stating the number of prizes to be awarded in each class. The exhibitors made their entries knowing that the rules stated that no second or lower prize would be awarded if there were fewer than six entries and no third or lower prize if there were fewer than nine entries. The judges did their best by commending when they did not feel justified in recommending more prizes.

The Secretary said that one rule of the Association authorised the judges to withhold prizes or to recommend additional ones subject to confirmation by the Council. As a matter of fact the judges did recommend an extra prize in one class, and the Council have confirmed it.

Mr. Pugh wished to avoid all personality, but would nevertheless like the feeling of the meeting in regard to this matter put on record, and he would therefore move: "that it be a recommendation from this meeting that at all future shows disqualification cards be put on all exhibits that are shut out of competition owing to non-compliance with the rules, and that such cards be returned with the exhibits."

BBJ 5 Nov 1908
Picture Post-Cards, Bee subjects, 6 for 1s.
                                W. Darrington, School House, Eastwood

BBJ 19 Nov 1908
Wintering Bees. The writer of a query signed "Country Mouse" has kindly drawn my attention to your reply, and, on behalf of beginners in our craft and appliance dealers, may I be allowed to point out a possible misconstruction which might be placed upon your reply, which says: "There should be no occasion for putting candy on the

hive for wintering bees, as they ought to have been fed up sufficiently to supply them with all they require until spring."

Then, I would ask, why do the hive manufacturers who advertise in your interesting journals, directly November arrives, commence their advertisements with "Do not let your bees starve," and why, if you pick up a copy of the November issue ten years old, do you find the same advertisements?

Also, how do you explain the fact that last season our firm, though, comparatively speaking, beginners as bee appliance dealers, sold upwards of a quarter of a ton of candy, of which by far the larger proportion was bought and sold before Christmas? As an unsuccessful candidate for the third-class expert's certificate, may I, in all humbleness, say that in my opinion the whole question turns on the word "ought"?

The weather ought now to be cold and foggy, bees ought now to have commenced their winter sleep, but years of interested observation of the weather proves that here in the Midlands we often get a short sharp spell of cold weather, and then more or less mild weather right up to Christmas; in fact, I am sure it would be an interesting point if those whose memory goes back many years would say if seasons have changed during their lifetime.

I am not a poet, otherwise I should like to write a poem in praise of candy. If weather is cold it assures the bee-keeper that plenty of food is within reach of the bees, without disturbing the hive during the dead season.

TN Harrison, Carrington, November 16th
[We fully appreciate the force of our correspondent's remarks from the advertisers' point of view, and we are always glad to see beginners and careless beekeepers forcibly reminded - in our advertisement pages - of their shortcomings in not seeing that bees go into winter quarters so well supplied with sealed stores that no candy is needed. This does not, however, make it less true that bees "ought to be fed up," etc., as stated in our reply. Eds.]

BBJ 26 Nov 1908
The monthly meeting of the BBKA Council was held at Jermyn Street, SW. on the 19th inst., Mr. WF Reid occupying the chair. There was also present AG Pugh. Apology for enforced absence was received from Dr. Elliot.

BBJ 10 Dec 1908
3 Stocks Bees, on 10 Standard Frames each, 2 Crates Shallow Combs each, one good single walled, 15s; 2 good double walled, 22s 6d each; full Stores for winter; guaranteed healthy. Deposit. Mrs. Williams, Rainworth

Minute Book – Quarterly Committee Meeting held in the Peoples' Hall on January 2nd, 1909. Present Mr AG Pugh in the chair, Dr Elliott, Messrs Darrington, Turner, Vessey, Mounteney, Moult, Dickman, Smithurst, Munro, Adams, Puttergill, Harrison and secretary.

The minutes of the previous meeting were read and confirmed by the signature of the chairman.

Correspondence about the withdrawal of the City Council's grant was read, as was also a letter from the President to say she would not be able to be present at the Annual Meeting.

The annual Balance Sheet was gone through which was considered very satisfactory. Mr Darrington, in proposing the same be passed, suggested a banking account and also receipts for the County Council grant and expenditure but it was explained that the most desirable, it was impracticable. The resolution was then seconded and carried.

Mr Harrison also offered to become agent for the sale of honey labels but it was thought they should remain in the secretary's hands.

Mr Pugh proposed that, in future, all candidates for examinations be instructed to find their own skep of bees for driving and, after considerable discussion pro and con, it was decided to leave the matter to a future meeting.

Minute Book – A special Committee meeting was held in the Peoples' Hall on January 23rd, Mr AG Pugh in the chair. The other members present being Messrs Darrington, Adams, Turner, Harrison, Windle, Smithurst, White and the secretary.

The schedule was gone through and revised for the season 1909 for the shows at East Bridgford, Southwell and the Annual Show in Mansfield. Only dates, etc. were corrected in the first two but for the Annual Show the following more important alterations were made.

- Class I the swarm catcher removed and a second prize of £1 offered.
- Class II add weight not to exceed 100 lb.
- Class III silver instead of bronze medal.
- Class V bronze instead of silver medal.
- Class VIII 1st prize 7/6d instead of hive.
- Class 8 a class for amateur's sections. Prizes 7/6d, 5/- and 2/6d.

BBJ 4 Feb 1909
Bees in Suburban Gardens. The interesting inquiry of "CWW" and your full reply leave

little to be said, but as one who for many years has successfully kept bees "within city walls" I will, with your permission, add a few rules. The experience of keeping bees amongst the chimney-pots has taught me several things which should be borne in mind by all who intend to start bee-keeping in suburban gardens, among which I will venture to name a few.

First, obtain a copy of the "Guide Book" and read it carefully. May I take this opportunity of saying what a help it is to us appliance dealers to have a standard work like the one named, to which we can refer all who have a thirst for knowledge; whilst to those who desire to go a step further "'The Honey-bee" is grand, to learn what is, indeed, a liberal education in apiculture.

Second, have a bee-keeping friend to show how to carry out the various operations and to fully explain the reason why; having wisely decided on the "WBC" hive, how necessary it is to buy all additional hives of the same make and size; and to explain that shallow frames for extracting are easier for a novice to obtain than finished sections, etc.

Third, get a gentle strain of bees. This is very important and then when neighbours make remarks about the danger to children of being stung, etc, speak gently to them. Personally, I make it a rule, when the honey harvest is on, to fill a few small jars (small empty stone jars which have contained cream are especially suitable) and distribute them amongst the nearest neighbours and those whose gardens I have walked over in pursuit of swarms, and find it has a wonderfully soothing effect.

Fourth, obtain the help of the womenfolk of your own home. A penny per pound profit on each jar of honey sold helps to hide from view the speck-stains on clothes hung out to dry, which in the somewhat limited garden may occur in the spring, and also helps to brighten the sight and hearing when swarming may take place, and to nerve them to shake a swarm into a skep and then leave them in a shady position until the master (?) returns in the evening.

Fifth, do not go "Volunteering" all through the month of August. The writer, like every true Englishman, would if necessary die in defence of his hearth and home; but as a believer in the "brotherhood of man" and that "reign of peace" which has yet to dawn, I would suggest apportioning a week for "training" and three weeks for the bees.

In conclusion, bee-keeping in towns is profitable, as from an apiary of six hives, kept within seven minutes walk of the far-famed Nottingham Market Place, over 2 cwt. of honey has been taken in one season. As it is the "stickers" who help forward our lovable craft, I always tell inquirers that in bee-keeping there is some amount of work, a large amount of waiting, a fair amount of energy and perseverance, together with intelligent observation, needed; but these combined with a favourable season, and results like the above-mentioned may easily be attained in the suburban gardens of our highly favoured land.   Thos. N. Harrison, Yew Tree House, Carrington

BBJ 11 Feb 1909
The twenty-eighth AGM of the Derbyshire BKA was held on January 30th, at 2.30pm at Smith's Cafe, Derby. In addition to the musical entertainment, a lecture was given by Mr. Geo. Hayes, hon. Secretary NBKA, on the "Anatomy and Life-History of the Bee," illustrated by a fine set of lantern-slides, which was greatly appreciated, a hearty vote of thanks being accorded to Mr. Hayes for his services.
RH Coltman, Secretary.

BBJ 25 Feb 1909
The monthly meeting of the BBKA Council was held on the 18th inst., at Jermyn Street, Mr. TW Cowan in the chair. Letters were received from Dr. Elliot, Mr. AG Pugh, and a number of others expressing appreciation of the late Mr. WB Carr's services to the Association, and wishing to be associated with their colleagues in the resolution of regret at his decease.

BBJ 11 Mar 1909
Despite the terrible weather there was a considerable attendance of members at the Annual Meeting of the NBKA which was held at the Peoples' Hall on March 6th. Mr. Harry Wyles, of Cropwell Butler, presided, and among those present were Messrs. WS Ellis, WH Windle, W. Darrington, RJ Turner, G. Smithurst, GE Puttergill, J. Bickley. H. Vessey, JC Wadsworth, AH Hill, WC Moult, H. Dickman, WP Meadows, G. Hayes (secretary), Dr. Willoughby, Mrs. Hayes, Mrs. Turner and Miss Pugh.

The Annual Report recorded continued prosperity and an increased membership, the latter having advanced from 207 to 269 during the year. Reference was made to the deaths of Mr. P. Scattergood, who had been connected with the association since its inception in 1884; of Viscount St. Vincent, who, a practical bee-keeper and President of the association, had helped them in many ways; and of Mr. SW Marriott, a veteran bee-keeper and valued colleague.

The report touched upon the shows held in the district, and the lectures given under the auspices of the association in various villages. No fewer than 225 apiaries had been visited, and 698 stocks examined. Only 4% were found with "Foul Brood".

Arrangements had been made for the Annual Show to be held with the Mansfield Horticultural Society this year. The Financial Statement showed an income of £100 9s 7d, including £4 4s 10d brought forward. The expenses were £85 15s 2d, leaving a credit balance of £14 14s 5d. Speaking on the Report and Balance Sheet, the chairman expressed pleasure at the highly satisfactory position of the association and remarked that bee-keeping exercised a refining influence on the lives of those who were engaged in the industry. The Report and Balance Sheet were adopted.

The Duchess of Portland, who has been President of the association for the past two years, was re-elected; Mr. George Hayes was re-elected secretary and treasurer; Mr. W. Darrington was re-elected auditor. Messrs. Hayes and Pugh were re-elected representatives to meetings of the BBKA in London.

The committee was appointed as follows: Dr. TS Elliot, Messrs. W. Adams, TN Harrison, AG Pugh, GE Puttergill, W. Darrington, G. Smithurst, H. Dickman, G. White, FG Vessey, JC Wadsworth (Newark), G. Marshall (Norwell), J. Bickley (Chilwell), and WH Hoyte (Nottingham).

During the meeting two competitions were held, for the best single 1-lb. bottle of granulated honey of any year and the best bottle of liquid honey. Mr AG Pugh secured the premier award in both classes, whilst Mr. H. Wood and Mr. T. Marshall obtained the second prizes in each class respectively.

In the evening the medals, pendants, and certificates won at the Annual Show were distributed, and some interesting discussions and papers of interest to bee-keepers followed.

BBJ 18 Mar 1909
In a good deal that I have read on the fertilisation of blooms by bees and other insects, there seems to be some confusion of two sides of the question, viz. fruit-growing and seed-growing. Experiments are recorded where a limb or a whole tree is netted to exclude insects, the result being no fruit, and apparently in support of this theory clover is treated in the same way, but in this case the conclusion arrived at refers to seed. It is not suggested that clover heads do not grow. Although it seems to be well established that apples, pears, and especially stone fruits, unless fertilised or cross-fertilised, will not bear fruit, or if insufficiently fertilised the fruit will be imperfectly developed or deformed, I contend that it is a mistake to conclude from this that all trees and plants require similar fertilisation.

With cucumbers, for instance, it is quite the reverse. If the blossoms are allowed to become fertilised the fruit is clubbed, bitter, and unfit to eat, but will, on the other hand, contain seeds, while those unfertilised are long, straight specimens and contain no seed. In this country cucumber growers for the market keep their houses or frames closed in order to exclude bees, and also pick off all male flowers they may see. I do not know what variety they grow in the United States, or what they do with their cucumbers, but (I feel rather diffident in mentioning this) Mr. Root, quotes the editor of the Rural New Yorker: "In those great greenhouses near Boston, where early cucumbers are grown, it is always necessary to have one or two hives of bees inside to fertilise the flowers. No bees, no cucumbers, unless men go round with a brush and dust the pollen from one flower to another." It is mentioned further down the column that they grow them for the early market and get fancy prices, so it is not a question of seed-growing.

If seed could not be grown without the help of bees, of course "no bees, no cucumbers" would be correct.

Linoleum makes a capital covering for hives, and will last ten years or more. Waste pieces left after covering the floor cost little. One piece is perhaps best, but is not absolutely necessary, for if the pieces are cut straight, the joints fitted closely together, tacked down, and all round the edge (not turned over), the linoleum being kept right side up, the paint will fill up the crevices. This makes a perfectly flat and smooth roof, from which the water runs off readily. A flat or halfspan roof is best, and as a finish a strip of zinc along the front can be tacked over the linoleum, and another along the back tacked under the linoleum would be better. Old oilcloth is not satisfactory.

Another good covering could be made of tailors' old pattern-cards. These are made of stiff paper covered with linen. Three strips cover a hive top; over-lapped, turned over the edge, tacked all round, and painted, they make a waterproof roof. You can obtain them from your tailor when the last season's patterns are out of date.

W. Doleman, Keyworth.

[We can assure our correspondent that the description given in the book mentioned is quite correct, for we have ourselves seen such cucumber houses. The methods of growing are different in the two countries. In America seedless cucumbers are not esteemed as they are here. In fact, one rarely sees the English type of cucumber in that country, those grown there being much shorter, thicker, and always contain seeds. They do not become clubbed or bitter, and if gathered at the proper time, before the seeds become hard, are far more delicious than our seedless ones. Ed]

BBJ 18 Mar 1909

The Christmas Rose. With the pen of a ready writer "DMM" makes every bee-keeper take up the words of the poet and say, "'T'were glorious in the dawning of those days to be alive; but to be young was very heaven indeed" In referring to *Helleborus niger* (Christmas rose) I notice what appears to me to be a slight mistake. "DMM" says: "Propagate by division of the clumps. They grow freely in any soil; if planted in different aspects the bloom may be prolonged."

My experience has been that this plant is one of the most difficult to grow, that division of the clumps should only be made in September, and that only in deeply prepared soil (roots have been traced 3 ft. or more deep), planted in partial shade, and protected with glass during the winter months. Can the words of the song which says, "Youthful hearts will find for ever roses underneath the snow," be an actual reality.

Nearly all bee-keepers are gardeners, hence my writing you upon this point.

Thos. N. Harrison, Carrington

BBJ 25 Mar 1909

The AGM of BBKA members was held at Jermyn Street, on March 18th, Mr. TW Cowan, FGS, in the chair. There was also present AG Pugh. Apology for enforced absence was read from G. Hayes,

Conversazione.

At the conclusion of the meeting the members, after refreshments had been served, reassembled for the usual conversazione. A hearty vote of thanks to the Chairman, proposed by Mr. Pugh, brought the proceedings to a close.

BBJ 1 Apr 1909

Bees Robbing by Agreement. Will you be good enough to advise me on the following condition? I have four stocks, of which one, by far the strongest in numbers, is headed by a "British Golden" queen mated with a native drone.

During a few warm days in the middle of February I ascertained that all had sealed stores present, and I noticed that the yellow bees seemed very much interested in the entrance of the weakest black stock. I put a cloth soaked in carbolic acid solution (1 in 20) on the alighting-board, but the bees settled on it in swarms and drank it, apparently with gusto. True, I saw a considerable number later on who seemed to be in pain on the ground in front of their own hive and then the bad weather came and put a stop to it all.

During the present fine weather the same game seems to have started again. I have not seen any fighting but there are several yellow bees who make a business of watching for blacks on the alighting-board and seizing upon them and, after what looks like a very vigorous rub-down all over, allow them to enter the hive. I have not seen any blacks doing this but have seen the yellow bees treating their own race in the same way at their own hive.

When I examined the hive I found a few yellow bees apparently on good terms with the natives but they were in nothing like the same proportion as in front of the hive. Does this mean that robbing is going on and would it be advisable to change the positions of the hives? If so, should the queens be caged?

BEEKEEPING BETWEEN TWO QUEENS

The weak hive contained only enough bees to cover two frames with a small patch of brood and queen. Is this too weak to do any good? All my hives have a considerable amount of stores left and pea-flour is being carried into all of them.

Medico, Nottingham.

Reply. The yellow bees are no doubt robbing by agreement among the bees themselves. The robbed colony is evidently weak and without energy and instead of offering resistance, the bees of the hive fraternise with the intruders. As the stock is so weak your best plan is to unite it with another.

Minute Book – Quarterly Committee Meeting held in the Peoples' Hall at 3pm on April 3rd, 1909. Present: Mr RJ Turner in the chair, Messrs Windle, Hallam, Vessey, White, Doleman, Harrison, Darrington, Smithurst, Dickman, Bickley, Munro, Ellis, Wadsworth, Hoyte and secretary.

The minutes of the previous meeting were read and confirmed by the signature of the chairman.

Letters from Messrs Skelhorne and Mackender were read stating their inability to be present owing to illness.

A letter from Messrs Thos N Harrison stating that they were about to reprint their catalogue and they intended to give a page to an advertisement of the association and for which the committee tendered their thanks.

A letter from the County and Farmers' Insurance Association offering a prize of a cruet to holders of their policies, to the exhibitors gaining most points at the Annual Show. This was accepted.

A letter from the Imperial International Exhibition soliciting educational or other exhibits. This was ordered to be laid on the table.

The quarterly statement of accounts was next gone into and being considered satisfactory, it was ordered the same be accepted.

A letter from Mr W Herrod offering to give a silver cup, to be known as "The William Herrod Perpetual Challenge Cup" for competition at the Annual Shows. This was accepted and it was ordered the same be included in the show schedule.

Mr Munro rose to a point of order in connection with the Technical sub-committee and pointed out that the old committee should have died with the old year. It was explained that following the precedent of the last sixteen years, it was not considered to be finished until this meeting. However, it was agreed by all to be the proper course for it to die with the committee on the Annual Meeting day; and it was so arranged for the future. A report of the sub-committee was read and Mr Munro proposed and Mr Wadsworth seconded the same be confirmed.

"My dear friend,
Many thanks for your letter to hand. I am pleased the Committee have accepted my offer and in due course will send the cup on'
I wish it to benefit the Assocn. as much as possible in the years to come and make the entries at Shows buck up. I give it to the Assocn in perpetuity, not to be won outright no matter how many times one person wins it.
Kindest regards to all,
Yours very truly,
Will.

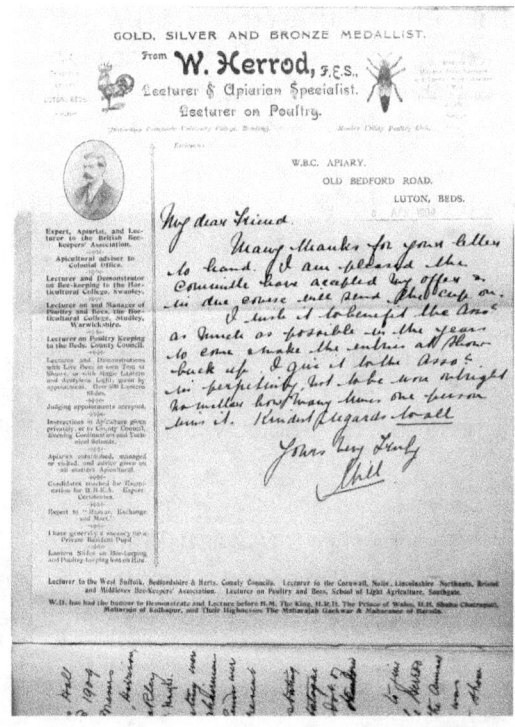

Resolved that the following gentlemen form the sub-committee for the remainder of this year - Messrs Turner, Munro, Darrington, Pugh, Windle, Puttergill, Harrison, Smithurst, Bickley and Dr Elliott.

The names of sixteen new members were submitted and it was resolved they be accepted.

Mr Turner was appointed judge for the Southwell and Mr Hayes for the East Bridgford Shows.

The question of again holding a Summer Meeting was brought forward by Mr Harrison and it was resolved that one be held and that details of arrangements be left to a sub-committee composed of the following gentlemen - Messr WS Ellis, Munro, Harrison and Pugh.

The advisability of holding examinations at the Annual Show was next considered and after considerable discussion pro and con, it was decided to go on the same lines as previously except that where the candidate is willing to pay the fees and other expenses; arrangements will be made by the secretary for the examinations to take place at the candidate's own apiary or one nearby.

It was also decided that each candidate provide his own stock of bees in a skep for this examination and he may, if he wishes, also provide his own stock in a bar frame hive; or may use the one that is at the show and being used for the lectures and further that a notice to this effect appear in the schedule.

Mr Pugh gave a short resumé of what was done at the BBKA Annual Meeting in March last.

BBJ 8 Apr 1909
GMH (Notts). The comb contains nothing but sealed honey and pollen and there is no evidence of "Foul Brood." The bees do not seem to have been strong enough to live through the winter, as those sent have no outward evidence of disease.

BBJ 15 Apr 1909
The Annual Meeting of the Lincolnshire BKA was held at the Louth Municipal Technical School on March 27th, and was well attended. Alderman HD. Simpson presided. The officers for the ensuing year were elected as follows with the addition of Mr. FHK Fisher (Ewerby formerly of NBKA)

BBJ 29 Apr 1909
Law Relating to Bees. From time to time you publish letters, evidently written by thoughtful, law-abiding citizens, describing their plan of a proposed apiary, and asking if they may keep bees. Believing that many dwellers in the suburbs of our great cities will yet join our ranks, may I through the columns of your widely-read journal suggest that you briefly review the laws of our land relating to bees, and so strengthen the hands of inquirers?

The simple position I take up is: My neighbour keeps pigeons; I bees. My neighbour and I both stand in the same position in the eye of the law, he having tamed pigeons, I bees, both of which are wild by nature, but, being tamed, they become our property by the law of the land.

Swarming time is near at hand. What is our position in following and claiming swarms? Your valued opinion on these and many other points would be welcomed by Thos. N. Harrison, Carrington, Nottingham.
[An article by a barrister on the subject of "Ownership of Swarms" appeared in the BBJ, September 17th, 1908.)

BBJ 20 May 1909
Good Natural Swarms for sale, 10s each.   Holleworth, Widmerpool

BBJ 27 May 1909
The BBJ as an Advertising Medium. As a buyer of bees, I can fully endorse Mr. HF Grimes' letter as to the value of the BBJ, as an advertising medium; but in my opinion he proves several other points, to which, with your kind permission, I will briefly refer.

First, Mr. Grimes has discovered a fact well known to our worthy Editor and to all appliance dealers - namely, that bees are a marketable commodity, and that if no demand exists in the particular locality where the apiary is situated, then an advertisement in the BBJ, will bring applications from north, south, east, and west.

Secondly, he has also found out what everyone in the trade knows, that bees are scarce this spring, and at present there are more buyers than sellers, and therefore, by the law of supply and demand, bees are dear and eagerly sought after. There are several reasons for scarcity
> the Arctic weather last spring which killed many stocks;
> the almost total absence of swarms during the swarming season of 1908;
> and the increased demand, owing to the growing popularity of beekeeping.

It is noteworthy that the great interest in bees, now shown in many ways, is not amongst those whom, living in the country, by the favourable position of their homes and by reason of the missionary work done by the parent and affiliated associations, we anticipated would take up bee-keeping. By this I refer to the cottager of this our highly favoured land; he, speaking generally, is not interested in apiculture. But the increased interest is shown by those who earn their living in our great cities, and who keep bees in their country or suburban homes.

My experience leads me to the conclusion that even amongst experienced beekeepers the preparation for packing of bees for transit by rail is not understood, and the plain directions in the "Guide Book," are not carried out.

During the last few weeks I have received skeps in the same position as when on their stands, all ventilation stopped, combs broken down, and with honey running out; also stocks in travelling-boxes received with so little attention to ventilation that the bees must have endured a veritable "Black Hole of Calcutta." Further, so many frames have broken down - that is, come away from top bars - that I would like to add a word of advice to those who send bees by rail. They should remember that railway porters are frightened of bees, and the present system of hurriedly rolling boxes out of guard-vans - not carefully lifting - makes it imperative that all frames have a fine pin driven through each joint, and when foundation is fixed frames should invariably be wired. I am not seeking a free advertisement; it is a pleasure to pay your reasonable charges, so I enclose my card and sign myself.   A Buyer of Bees, Notts.

BBJ 27 May 1909
Trade Catalogue Received. Thos. W. Harrison and Son (5 and 7 Cheapside, Nottingham) send us an advance copy of their fully-illustrated and up-to-date catalogue of forty pages of all that is required for the apiary in the way of bee-hives and appliances. It also contains a selected list of garden requisites, dairy utensils, and poultry appliances, as well as a few useful hints for beginners in apiculture.

BBJ 10 Jun 1909
For Sale, two "King's" Patent Bee-hives, new, 6 and 8 draw, to be sold under cost price, or exchange for good old violin or an English 'cello.
                        E. McNish, Linden Cottage, Abbotsford-street, Nottingham.

NEP 21 Jun 1909
The Summer Conference of the NBKA was held at Beeston on , interesting collection of diagrams, specimens and lantern and microscopic elides connected with the craft being shown in the Church Mission Hall. The secretary (Mr. George Hayes) gave a lecture on "The Metamorphosis of Insects" and Mr. Pugh read the report of a discussion by the Central Association on the feeding of bees.

NEP 29 Jun 1909
Ideal weather favoured the 46th Annual Show of the East Bridgford Horticultural Society, held this afternoon in a marquee opposite the old cricket field. The entries numbered nearly 60 which is an improvement on last year, and decidedly above the average. The quality of the exhibits, considering the unfavourable season, was all that could be desired, and especially strong were the strawberry and potato classes. The judges were Mr. A. Parr, gardener to Earl Manvers; Mr. J. Sears, gardener to Captain Hall; and Mr. G. Hayes, secretary of the NBKA.

BBJ 1 Jul 1909
3 "WBC" hives, new, complete, 10s each, or exchange Bees.

Green, Bigg House, Arnold

Minute Book – Quarterly Committee Meeting held in the Peoples' Hall on July 3rd, 1909. Present: Mr Wm Munro in the chair, Dr TS Elliott, Messrs AG Pugh, W Adams, RJ Turner, W Darrington, WH Windle, Jno Bickley, JC Wadsworth, Thos N Harrison, WH Moult, FG Vessey, G White and secretary.

The minutes of the previous meeting were read and confirmed by the signature of the chairman. Proposed by Mr Pugh and seconded by Mr Vessey.

After comments on the Summer Conference, Mr Pugh proposed a vote of thanks to the secretary and all those who had assisted to make the Conference so interesting.

Letters were read from Messrs Smithurst and Mountney stating their inability to be present at the meeting.

The quarterly account was then gone into and, as it was considered satisfactory, it was resolved the same be accepted. Mr Darrington proposed and Mr Pugh seconded.

A list of 32 new members was read and it was resolved the same be accepted. This, with sixteen at the previous meeting, made a total of 48 for the season. Proposed by Mr Turner and seconded by Mr White.

It was proposed to ask the following gentlemen, in the order named, to judge and lecture and examine at the Annual Show in Mansfield on August 2nd:
        Mr Herrod Mr Crawshaw Dr Sharpe Mr R Brown
Resolved that Mr RJ Turner be appointed as assistant judge at the Annual Show.

BBJ 1 Jul 1909
The seventieth Annual Exhibition of the Royal Agricultural Society of England opened on June 22nd in the grand old-fashioned city of Gloucester. Messrs. WF Reid, Addlestone, Surrey; AG Pugh, Beeston; and S. Jordan, Bishopston, Bristol, acted as judges.

Taking advantage of the impetus given by the "Royal" Show being held at Gloucester, the Gloucestershire bee-keepers have bestirred themselves and held a meeting on the 22nd ult., at the class-room of the Friends' Meeting House to inaugurate, if possible, a county association. They had secured a strong contingent of well-known bee-keepers to address the meeting, including Mr AG Pugh.

BBJ 1 Jul 1909
3 "WBC" hives, new, complete, 10s each, or exchange Bees.
                                         Green, Bigg House, Arnold

BBJ 8 Jul 1909

Disinfectants for Bee-Hives. It is probable that if properly used chlorine would be one of the very best disinfectants for hives and a good remedy for "Foul Brood." It is more diffusive and penetrating than carbolic acid, and would doubtless play havoc amongst "Foul Brood" bacilli. However, to give it in the form of chloride of lime is not the best way, because this substance has no standard of proportion between the lime and chlorine, so that a pound of calcium chloride ($CaCl_2$) bought at one shop might be stronger than that bought at another, as all lime is not fully saturated, and some may be "old stock" from which a good deal of the chlorine has evaporated.

For the bee-keeper the lime would be better done away with altogether; and if it was desired to use it in a fine spray then the chlorine should be absorbed by water and this diluted to a proper strength. But the best way to use it would be to fill a gas-bag with air to a known pressure, and then inject a given amount of chlorine gas into the confined air, and with this mixture the hive could be fumigated.

Greatly diluted with air, chlorine can be inhaled by human beings without injury: and why not by bees also? Chlorine is a very heavy gas and if used as chloride of lime for disinfecting purposes, and placed at the bottom of hive in rather cold weather when the bees were quiet, and not ventilating much, the chlorine given off would be so dense at the bottom of the hive after an hour or two that the bees descending the combs to drive it out would be stupefied before they could even begin to drive out the deadly gas. A more scientific method is necessary to ensure success.

A. Green, Notts.

BBJ 15 Jul 1909

Perhaps Mr. Green would add to his very helpful explanations by indicating a simple method of using the chlorine safely in a full stock. But why use any form of chlorine at all? Has it any advantage over formalin?   L.S Crawshaw, Norton, Malton.

BBJ 5 Aug 1909

In reply to Mr. Crawshaw, I am designing a smoker for the using of chlorine. Mr. Crawshaw asks, "Why chlorine and not formalin?" Without saying a word either for or against formalin, I will state why I think chlorine is suitable as a remedy for "Foul Brood".

First let us go to the root of the matter, and ask, "What is the nature, or rather the action, of a 'disinfectant' in stopping the progress of disease?" I presume Mr. Crawshaw accepts the theory that most diseases are the result of a vast number of minute collections of gases having regular forms and each in a vitalised state. Call them germs, microbes, bacilli, or what you like, they are living: and whether they have the power of reproduction or are only the result of continuous spontaneous generation under favourable conditions is of little importance in this case.

By what means does a disinfectant destroy these germs? Do they have formalin for, say, breakfast, dinner, tea, and supper, and then go to sleep, never to wake again? The typhoid microbe, which is one of the largest and most easily produced and formed, and enjoys much popularity, in consequence, is generated in filthy water. Water is known to contain hydrogen, and the filth would be certain to contain carbon, and as these gases easily unite it is probable that the typhoid germ will contain both.

If Mr. Crawshaw can find a gas having a stronger affinity for either the hydrogen or the carbon than they have for each other, we shall be able to destroy the unity, the life, and the germ itself, and it is because chlorine has a stronger affinity for hydrogen than any other known gas that I think it will prove useful in cases of "Foul Brood". I feel certain that "Foul Brood" is generated in a damp atmosphere, and probably what we call an unhealthy one, and therefore the bacilli would contain hydrogen, and if so chlorine would prove an irresistible "Dreadnought."

Upon the grounds above stated I rely upon chlorine as a certain remedy when rightly used.
A. Green, Notts.

Readers should realise that this discussion took place before the devastating use of chlorine gas in WW1.

BBJ 12 Aug 1909
NBKA held its Annual Show, in conjunction with the Horticultural Society, at Mansfield on Bank Holiday, August 2nd, the weather being on that day an improvement on the past few weeks. The entries were slightly below last year's, but, contrary to expectations, exhibits turned up well and the tent set apart for these was well filled.

Three years ago the Annual Show was held at the same place, when the late Wm. Broughton Carr, Esq., undertook the duties of judging, and I believe this was the last show, outside the "Royal," at which he officiated. I am glad to remember him saying, after the day's work was over, how pleased he was with every arrangement and how very much he had enjoyed the day. This year Mr. W. Herrod undertook the judging of exhibits, assisted by Mr. RJ Turner of Radcliffe-on-Trent. Mr. Herrod also gave lectures in the bee tent to crowded audiences and examined three candidates for third-class expert certificates, working very hard all day.

The awards were as follow:
Honey Trophy - 1st, WL Betts, Mansfield Woodhouse; 2nd, Uriah Wood, Arnold
Exhibit of Light-coloured Extracted Honey - 1st, AG Pugh, Beeston; 2nd, JT Wilson, Shirebrook; 3rd, WL Betts; 4th, Geo. Marshall, Norwell
Exhibit of Dark-coloured Extracted Honey - 1st, Dr. TS Elliot, Southwell; 2nd,

Geo. Marshall; 3rd, W. Lee, Southwell; 4th, Jno. Woods, Nettleworth
Six 1-lb Sections - 1st, W. Lee; 2nd, GH Pepper, Oxton; 3rd, WL Betts.
Six 1-lb Jars Granulated Honey - 1st, JT Wilson; 2nd, John Woods; 3rd, H. Merryweather, Southwell; hc, Geo. Marshall
Exhibit of Extracted Honey (Amateurs) - 1st, CF Fincham, Nottingham
Six 1-lb Sections (Amateurs) - 1st, John Woods
One Shallow Frame - 1st, John Woods; 2nd, WL Betts; 3rd, Geo. Marshall
Specimens of Bees - 1st, WL Betts; 2nd, Geo. Marshall
Beeswax - 1st, Geo. Marshall; 2nd, Uriah Wood; 3rd, AH Hill, Balderton

The William Herrod Perpetual Challenge Cup (offered for the highest number of points obtained by an exhibitor) - WL Betts, Mansfield Woodhouse.

William Herrod writes in 'Bee-Keeping Old and New' "In 1909 we visited Nottinghamshire to hold an examination for the third expert certificate of the BBKA. To our astonishment our old tutor, Mr R Mackender, was one of the candidates, so the unique position arose of the pupil becoming the examiner of the master. It was a compliment, the significance of which we fully appreciated." An image of Robert Mackender being examined by William Herrod appears below.

Surprisingly he makes no mention in his book of the awarding of the trophy he had given to NBKA especially as this was the first time it had been awarded.

W Lewin Betts was born in 1870 in Mansfield. In the 1891 census he is listed as an agricultural labourer and would have been designated then as a 'cottager'.

NEP 19 Aug 1909
Mr. J. Thomson, of Pym-street, Nottingham, writes as follows:
Of late years a great amount has been written and published concerning bee-stings as a cure for rheumatism. Apart from the thought which you suggest that few there be who would care to undergo the ordeal. I suppose it is like many more beliefs; someone who has suffered from rheumatism happened to get stung by bees on the

affected part and, thinking got relief, mentioned it and the thing turned into a belief that relief from rheumatic pains is had in this way.

The true explanation is that through fear the attention is drawn for the time being from the rheumatic pains and centred in that of the sting thereby affording to the victim temporary relief from his rheumatism. At least so he thinks.

Another way looking at it is this. We learn that it is a belief that one poison acts as an antidote to another and the effect of the action of the one is to neutralise the effect of the other. Now, the poison which the bee injects into the wound made by the lancets of its sting is formic acid and it might be thought that it has a neutralising effect on the uric acid said be preponderant in the systems of those who are sufferers from rheumatism, gout, and such kindred troubles. As a matter of fact, neither is the case.

The person who suffers most from bee-stings is the person who fears. Having kept bees for years, as much from a scientific as from a commercial standpoint, I have noticed a person extracting a sting turn as pale as death with downright fear. The person who has no control of himself in this way is the same as those we read of sometimes in newspapers as having died from the effects of bee stings whereas the cause of death was fear, pure and simple.

I have been stung repeatedly whilst manipulating bees, often receiving several stings at once, but I never suffered any great pain, only uncomfortable itching as the swelling and inflammation began to subside. It is self-possession and confidence which are the secret of manipulating bees. The experienced beekeeper can tell at once by the hum of his bees whether they mean 'business', and he likewise knows that though a bee is quick in its movements that there is a period which elapses before the bee can sting, for it has to get a foothold before it can get the necessary leverage to push home the lancets of the sting, and calmly but adroitly using a finger gives the bee a gentle push off and by thus doing he often avoids getting stung.

Bees sting only in defence. It is quite possible to take one of the old-fashioned straw skeps containing a swarm of bees, hold it above your bare head and give it gentle tap, dislodging all the bees, without receiving a solitary sting.

In regard to the great importance of bees to large growers of fruit and flowers, it is through insect agency that the fertilisation of flowers must take place. It takes five separate acts of fertilisation to form a perfectly shaped apple and this fertilisation is almost solely the work of bees. It will be seen how important it is that if we keep on planting orchards must supply more stocks of bees if we want fruit in proportion to the trees planted.

One drawback for many persons keeping bees, apart from the fear of being stung, is the dread of that greatest of all pestilence to beekeepers called "Foul Brood." but nothing short of gross carelessness or inexperience allows it to become so rampant.

BBJ 19 Aug 1909
Paper-Pulp honey-pots have for a considerable time been successfully used for cream, and, in the words of the one who, in my opinion, is the finest word-painter in the cataloguing of bee-appliances, "We offer them confidently to bee-keepers for honey."

From practical testing there is no doubt they do answer and, whilst being an additional article in which to sell honey, the lower price is a distinct advantage. The disadvantages as they appear to me are the honey cannot be seen and that the pulp pot when empty (unlike jars, tins, &c.) cannot be used for other household purposes.
Thos. N. Harrison, Carrington

Sheffield Independent 9 Sep 1909
The 26th Annual Show of Derbyshire BKA was held conjunction with the Agricultural Show and there was an interesting exhibit of hives and honey, and other appliances. The judge was Mr AG Pugh of Beeston, a member of NBKA.

BBJ 2 Sep 1909
The sixteenth exhibition of honey, under the auspices of North Norfolk BKA was held in the Park, Melton Constable, on August 2nd, and attracted an entry of about eighty-five exhibits, or about 320 lb of honey, as compared with 790 lb last year. The judge was Dr. TS Elliot, of Southwell and for the season the honey was very fair, but, as was to be expected, not up to last year's standard.

BBJ 16 Sep 1909
J. Jessop (Nottingham) The "greenish" colour is usual with lime honey of which yours is a sample. Some people like the strong flavour of limes and would not consider it "poor," though it cannot be classed as a first-class honey.

BBJ 16 Sep 1909
T. Gillett (Nottingham). Sample is spoilt by honey-dew, which is the principal cause of the bad colour. It also has a slightly bitter taste. No precautions or care will prevent bees gathering honey-dew in seasons like the present one.

BBJ 23 Sep 1909
A meeting of the BBKA Council was held at 105, Jermyn Street, London, SW. on September 16th. Present: Mr. Thomas W. Cowan (in the chair), Mr. Arthur G. Pugh. Reports upon examination of candidates for third-class experts' certificates held in Notts, were received.

BBJ 23 Sep 1909

The seventeenth International Exhibition of the Grocers and Allied Trades was opened on the 18th inst. and continues during the present week.

Three Shallow Frames Comb Honey - 2nd, J. Herrod, The Manse, Sutton-on-Trent

Twelve 1-lb Jars Light-coloured Extracted Honey - HE Pugh, Beech House, Beeston

Twelve 1-lb Jars Medium-coloured Extracted Honey - 1st, G. Marshall, Norwell; 3rd, J. Herrod

Twelve 1-lb Jars Dark Extracted Honey - 1st, T. Marshall, Ivy Cottage, Sutton-on-Trent

Twelve 1-lb Jars Heather Honey - 1st, G. Hunt, 66, Hawton Road, Newark; vhc, T. Marshall

Twelve 1-lb Jars Heather-blend Honey - 4th, G. Hunt; hc, AG Pugh, Beeston

Twelve 1-lb Jars Granulated Honey - 3rd, T. Marshall; 4th, JT Wilson, York Villas, Shirebrook

(HE Pugh is as listed – some relation of AG Pugh?)

BBJ 23 Sep 1909

Regarding bees being killed near copper-works the matter is a very simple one as the poisonous gases given off during the reduction of the copper pyrites would unite with the atmosphere sufficiently to cause the deadly gases to be carried some distance, this being regulated by the temperature, humidity, and stillness or otherwise of the air. Condensation would go on from the time the gases left the cupola and as the quantity of poisonous gases deposited near the works would be the greatest, the bees placed nearest the works would suffer most.

As these condensed gases fell to the earth they would be certain to come in contact with flowers, and especially the pollen, which would be gathered by the bees and given as food to their young which would die as a result. An atmosphere charged with these gases would not be very healthy for bees but as they survived the winter in it the cause must be sought in the poisoned pollen and honey.

In this country there is a law to prohibit the reckless diffusion of deadly gases from chemical works. At one time all vegetation in the vicinity of works where washing soda is made was destroyed by muriatic acid (impure hydrochloric acid), which, leaving the works as a gas combined later with the moisture in the air and fell in its hydrated form upon vegetation and by degrees destroyed it.   A. Green. Notts.

NEP 25 Sep 1909

This year's honey harvest has been, generally speaking, a disastrous failure, owing to the cold and wet weather which prevailed in the early part of the summer, and especially the lack of sun in July.

BBJ 30 Sep 1909
JPT (Trowell). Wintering on Shallow Frames.
1. It is very risky wintering a small lot of driven bees on shallow frames. Keep the bees well protected, and place a frame filled with candy on the outside of the cluster, putting the empty combs in the centre. With the four full combs they should be sufficiently provided for until the spring, when, if they have survived, you can slip a cake of candy under the quilt and renew it when consumed until they are able to collect enough for their wants.
2. They can be allowed to transfer themselves in the way recommended for bees in skeps.

Minute Book – Quarterly Committee Meeting held in the Peoples' Hall at 3pm on October 3rd, 1909. Present: Mr AG Pugh in the chair, Messrs Windle, Harrison, Skelhorn, Vessey, Smithurst, Puttergill, Bickley, Dickman, Doleman and secretary.

The minutes of the previous meeting were read and confirmed. Proposed by Mr Vessey and seconded by Mr Smithurst.

Letters of apology were read from several members stating their inability to attend.

The quarterly statement was next gone into and the same, being considered satisfactory, was passed. Mr Smithurst proposed and Mr Bickley seconded.

A report of shows held and a list of prizes awarded was rendered and received as satisfactory.

The names of two new members were submitted and accepted, making a total for the year of 50.

A report on the doings of the sub-committee at their previous meeting was given and the same was accepted as satisfactory.

It was suggested that a press correspondent or representative be appointed to secure better reports of our meetings, shows, etc. but the matter was kept in abeyance. Proposed by Mr Vessey and seconded by Mr Harrison.

BBJ 7 Oct 1909
Notts (Stapleford). Granulation of Honey. The white frothy appearance is due to the air not escaping and the shrinkage of the honey during granulation. It can be avoided by stirring the honey while it is granulating, thus causing the air-bubbles to rise to the

surface.

BBJ 14 Oct 1909
The monthly meeting of the BBKA Council was held on October 7th in the boardroom of the RSPCA, when Mr. TW Cowan presided. There were also present Mr. G. Hayes and Mr. AG Pugh. Reports upon examinations of candidates for third-class certificates held in Notts. were received.

Conversazione.
On the conclusion of the Council meeting, a short adjournment was made for light refreshments, the members assembling at 5.30pm for the usual Conversazione, over which Mr. TW Cowan presided. Amongst those present were Geo. Hayes and AG Pugh.

BBJ 14 Oct 1909
The thirty-fourth Annual Exhibition of the British Dairy Farmers' Association opened on October 5th at the Agricultural Hall with one of the poorest shows of honey ever staged, only eighty-seven entries being made, and of these thirteen exhibits failed to arrive.
Twelve 1-lb Jars (Dark) Extracted Honey (including Heather Mixture) - 1st, JT Wilson, Shirebrook

BBJ 21 Oct 1909
A special meeting of the BBKA Council was held at the offices of the BBJ, 8, Henrietta Street, Covent Garden, London, on October 13th. Present: Mr. WF Reid (vice-chairman), in the chair. A letter of regret for absence was received from Mr. AG Pugh.

Mr. Hayes said he would like to ask, with regard to putting bees in greenhouses and vineries, where they would have abundance of food, whether they gave off a swarm. His experience in such cases was that, although the fruit blossoms might be well fertilised, the majority of the bees were lost.

Mr. Ellison replied that he did not find that swarms were prevented when the bees were taken out again. He did not, however, think they ever swarmed when in the houses because peach blossom, as everyone knew, came early in the season, and when the skep was taken out again the weather was still cold. Very few bees were found dead about the greenhouse. He expected there would be a lot, and the first year there were a good many, but considerably fewer afterwards. He often watched the bees, which soon found out the arrangements of their new abode and never made any attempt to get out by the glass.

BEEKEEPING BETWEEN TWO QUEENS

Leicester Chronicle 4 Nov 1909
A very autumnal conference under the auspices of Leicestershire BKA was held at the Highcross Restaurant. Leicester. Amongst those present was Mr G Hayes (Secretary, Notts. BKA).

BBJ 11 Nov 1909
In the BBJ for October 21st we were agreeably surprised to see the appointment as secretary *pro tem*, of our old friend Mr. W. Herrod. We are wondering if this is to be permanent, and sincerely hope that is the case, as amongst all the prominent bee-keepers, we know of no one who would fill the post better.

As old exhibitors we are grateful for many kindnesses shown to us and consider that his management of honey competitions is all that can be desired.

His masterly and tactful handling of the "Royal" this year under great difficulties is still fresh in the minds of the exhibitors and officials closely connected therewith. Never has the department been so well managed.

R. Brown and Son, Flora Apiary, Somersham, Hunts.

[In addition to those we have published, we have received a letter commending the action of the Council from G. Hayes (Hon. Sec, NBKA).]

BBJ 24 Dec 1909
The monthly meeting of the BBKA Council was held on December 16th, in the boardroom of the RSPCA when Mr. TW Cowan presided. There was also present Mr. AG Pugh.

BBJ 6 Jan 1910
Mr. Smallwood says that in 300BCE bee-keeping was in absolute darkness, but he makes a mistake in going to Aristotle for this statement, as it is quite clear that this philosopher was not a bee-keeper himself, or he would have found out at least some of the errors into which Greeks had fallen regarding bees. Aristotle wrote on wax leaves or slabs the ideas of the Greeks and the Romans learnt what little they knew about bees from the Greek writers.

Absurd as it may seem now, it is also easy to see how the mistake was made regarding bees being the product of flesh maggots (gentles), as chilled brood thrown out by bees is something like the meat-fly at about fourteen days old, and once the notion got a start it would live as mistakes live in our own day. However, it is not to Greece or to Rome we have to go for ancient bee-keeping but to the seat of learning - Egypt.

It is clear that about 100BCE Egyptian bee-masters were not far behind Mr. Smallwood himself for they had observatory-hives and could also expand or contract their hives and we can do no more. They may not have had a standard frame and Bee Journal but they must have used either frames or boxes and as we are but mere babies compared with them in the designing and making of furniture, it is probable that their hives were far superior to our own and bee-keeping amongst the ancient Egyptians must have reached a high state of perfection.

I have searched for signs of apiculture in the world's greatest procession in honour of Ptolemy II. at his coronation but in vain. Aviculture was represented but bees, being considered dangerous, would no doubt be kept away. A. Green, Notts.

BBJ 13 Jan 1910
Ancient Egyptian Bee-Keeping. Will Mr. A. Green be so kind as to inform the readers of the BBJ, on what authority he makes this statement?

"It is clear that about 100BCE Egyptian bee-masters . . . had observatory-hives, and could also expand or contract their hives... They must have used either frames or boxes." I have devoted much time to this particular matter, without finding anything to make me believe that the ancient Egyptians were advanced beekeepers.

There is ample evidence that for various purposes honey and wax were used in Egypt from very ancient time and probably in large quantities; but much of this may have been produced by wild bees. The only instance known to me of a hive being depicted on an ancient monument is related by Sir Gardner Wilkinson in 'The Manners, and Customs of the Ancient Egyptians' as follows: " ... to the garden department belonged the care of the bees, which were kept in hives similar to our own (I remember to have seen them so represented in a tomb at Thebes)."

As this passage was written in 1837, it is almost certain that the author believed the vessel depicted to be a bee-hive in the nature of a straw skep. He gives no illustration to enable us to judge for ourselves. No one can doubt that by such a highly civilized people as the ancient Egyptians, bees were kept in a state of domesticity, nor will any bee-keeper fail to recognize the ruler of the hive in the conventional symbol of governance that occurs so frequently in their hieroglyphic inscriptions.

There is nothing in this, however, to warrant our crediting them with observatory hives, or anything more than elementary bee-keeping and it seems to me very unlikely that any hives were commonly employed but the roughly-made cylindrical hives of clay or clay and ashes, such as have been noticed by travellers, and described in various books and journals during the last fifty years as being still in use in Egypt and Palestine.

It is only in the hope of obtaining further information that I raise a doubt as to the correctness of Mr. Green's deductions. I shall be delighted to learn that another bee-keeper has been more fortunate than me in his researches.

HJO Walker (Lt.-Col), Budleigh Salterton

BBJ 20 Jan 1910

"Foul Brood" Legislation. Mr. G. Thomas says, "Before long bee-keepers will be urging another attempt at a "Foul Brood Bill", and this is given as a reason why the BBKA should be reconstructed and a very good reason, too, if the BBKA is going to annihilate this notion. Let us see what such a proposal amounts to.

Before you can make a man legally responsible for a thing the man must have power to prevent that for which he is to be held responsible. For instance, a man is responsible for being drunk because he can prevent it by drinking water instead of wine. But where is the man who can prevent "Foul Brood", when there is not a person who knows how it is caused?

Imagine a respectable bee-keeper, with a few hives as a scientific hobby and to get honey for his own table while away on his daily employment having the privacy of his home, his garden, his apiary invaded by a bombastic well-paid Government official, who ransacks every hive and every comb in search of something he himself does not understand. A comb is suspected, it has not been reported, a summons follows; the bee-keeper is so disgusted he does not appear. Next comes the policeman with a warrant, drags him like a felon from his children and throws him into a prison-cell to await his trial.

There is one thing belonging to modern bee-keeping which ought to be cried down more than another it is over-manipulation and yet to satisfy a "Foul Brood Bill" every frame would have to be hauled out of the hive once a fortnight or, perhaps, weekly and even then the bee-keeper's mind would never be at rest, for he would be uncertain whether he had missed a speck or not.

What a pity these people cannot find something better for bee-keeping than the fussy parade of "Foul Brood" before the public and House of Commons with a view to placing every bee-keeper under police supervision. I agree with Mr. Thomas; it is time for the BBKA to get into harness, not so much for constructive as for destructive work.

A. Green, Notts

[It is needless to say that we do not fear the dire results anticipated by our correspondent. Legislation has been adopted by most progressive countries where the bee-keeping industry is of commercial importance, and although it has been in force in many countries now for a good many years, we have yet to hear of a single case of trial and imprisonment. Nor have we heard of a single instance of an endeavour to repeal such laws owing to their pressing heavily on bee-keepers; on the contrary, they have enabled beekeeping to be carried on where formerly it was impossible to do so. Ed.]

BBJ 27 Jan 1910
The monthly meeting of the BBKA Council was held on January 20th, 1910, at 8, Henrietta Street when Mr. WF Reid (Vice-Chairman) presided. Amongst those present was Mr. G. Hayes (Notts). A letter expressing regret at inability to attend was received from Dr. Elliott

BBJ 27 Jan 1910
My acknowledgments to Mr. Green for the courteous references he makes to my last letter, and for his intimation as to where I may find traces of a generation advanced in bee-culture. But no one expects instruction from the tales retold of the era.
When Music, heavenly maid, was young,
While yet in early Greece she sung.
They ante-date by 200 years the reign of Ptolemy and are only a record of the craft as it was at that particular period. I thought it might interest to turn back the pages of history for 2,000 years in order to peruse the "Grecian bee-keepers' guide book" of those days. Personally, too, I have experienced much pleasure in renewing my school-day acquaintance with my old Greek Lexicon.

As to the fiction that the larvae bred in decaying flesh turn into bees, which Mr. Green alludes to, this is none of Aristotle's yarns. Wise men of those days knew not how to account for the generation of bees. The correct solution of the enigma was impossible to them because, in their pride, they had cut the ground from beneath their feet. Those large bees, they said, which we see the other bees follow, must be males. Who ever heard of a female capable of government? Wherefore they called them *"reges et duces,"* kings and leaders, and one complication led to another.

First, it is only too evident that the bee has a sting, and so has the "king," although seldom used. Now, as Aristotle quaintly words it "Weapons for the exercise of bravery are given to no females" - proof positive that the bees and kings were males. He might have added that females were quite able to take care of themselves.

Secondly, a further proof that both bees and drones were of the same sex was that, seeing the great number of both in one hive, the act of intercourse must have been frequently noticed, as in flies, if the contrary had been the case. Yet, arguing from the same premises, the drones, having no stings, might be females. But this, too, could not be, because the drones were never noticed to feed the young and attend to them, which is the province of females. They collected no honey and were idle. I quote again from Aristotle: "The drones are lazy and have no weapon by which they can struggle for food," and yet, having given the kings governing power, they add to it a parental power, for he goes on to say: "It is agreed among all bees to follow their king (for, unless it were so, they would lose all claim to be governed), and as they concede that kings as parents do no work, they chastise the drones as sons, for it is most just that sons should be chastised if they do no work."

Now, having got into this delightful quandary, and having almost established the fact that the home exists without a mother, how is the difficulty to be surmounted? It is patent that there is brood in the hive in due season. Well, one theory is that of the fishes. Repeating my previous remark that the fertilisation and generation of fishes had never been studied sufficiently by the ancients, it was seriously propounded that fishes generated within themselves their progeny, and they confirmed the possibility by reference to the vegetable kingdom, where male and female, they maintained, were to be found in the same plant. But Aristotle seems to have had his "doots," for he says if this is possible in bees, why not in other animals? They had also another theory that the brood was brought in from plants and flowers, especially the olives - "An abundance of olives is simultaneous with an abundance of swarms" are his words - but here again he is an agnostic. "If they are brought from there," he asks, "who put them there?" J. Smallwood, Hendon.

BBJ 27 Jan 1910
In reply to Colonel Walker, I may say that the evidence regarding Egyptian apiculture is somewhat fragmentary and no work on ancient Egypt deals largely with the subject. However, it seems clear that the Egyptians were large consumers of honey and that wax was much used. Egypt, not being a well-wooded country, and its woods chiefly hard, would not be favourable for wild bees on a large scale. To show our methods are not modern, Varro, in the last century before Christ, advised that hives should be made of wood, basket-work, &c, and that they should be contractible to suit the size of the swarm. He also recommends a pane of glass (*lapes speculares*), transparent stone, to enable the bee-keeper to see his bees at work.

Now, if our worthy Editor could have found Varro's original writing and put it in BBJ would anyone have detected that it was anything but a modern production. I ask. How far have we got beyond Varro? Practically we are beginning where he left off? How is it possible for a man to be writing about just the very things we are using ourselves, 2,000 years later, without bee-keeping at that time being in an advanced state if we ourselves are in that condition? Sallust recommends cork for hives. Why, if Sallust lived today remember names of books in which references have been made to ancient beekeeping but the one alluded to by Knight. However, Varro's advice shows that our present methods are 2,000 years old - I mean our best methods, for no Egyptian would be silly enough to use the sulphur pit. A. Green, Notts.
[The hive described by Varro was a round or square horizontal one in which the front and back walls could slide inside in such a way that they could be pushed forwards or backwards according to requirements. It was round if made of osiers, pottery, or

cork, and square when made of fennel stalks. Similar hives are used to this day and we have in this office a cork hive and one made of fennel stems which we brought with us from Africa some years ago. These, however, are not observatory-hives and could not be called such even if they had a pane of glass in them, as they do not carry out the principle of an observatory hive such as we understand it. With our movable-comb hives and observatory hives where both sides of every comb can be seen, we have advanced considerably since the time of Varro who wrote his "*Rerum Rusticum Libri*" in his eightieth year (37 BCE). The hive described by Pepys and which was given to Evelyn by Dr. Wilkins, of Wadham College, was called by him a "transparent" apiary. From the description it is evident that such hives were simply boxes with glass in the sides, just such as were used until Huber introduced his leaf observatory hive in 1789. There is no evidence in Varro's writings to show that he knew anything about our modern methods of bee-keeping, which have nothing in common with those of his time. Ed.]

BBJ 29 Jan 1910

The BBKA of the Future. The hearty thanks of all the associations are undoubtedly due to the gentlemen who have bestowed so much thought and care on the preparation of a project for uniting and strengthening the bee-keeping fraternity. But our gratitude - well deserved though it be - cannot counteract the misgivings with which many of us regard a scheme overloaded with unpractical suggestions. It does not look as if it could have emanated from anyone familiar with the working of local associations: it is clogged with "propolis" and "brace-comb." When these hindrances are cleared away we may be able to get to work.

But at the outset we are faced with the drawback that the secretary has no vote on the Council. As Mr. Samways remarks, "the hour and the man have arrived, and . . . the future success or failure of the BBKA is in the balance." We want a National Union of Bee-keepers (call it BBKA, Royal Apicultural Society, or what you like), with a strong general secretary in touch with all local associations and alive to their needs and it seems strange that the one who should know most about local peculiarities, prejudices, and requirements, and who, I dare say, would have the confidence of British beekeepers, should have no vote on the Council!

The position in the new Union of the proposed "Central Branch" appears somewhat anomalous. Why "Central"? Is there any reason for a special title suggesting, and probably intended to suggest, a special position? If Metropolitan members wish to unite let them form a "London County Branch," to be placed on exactly the same footing as any other county branch, while the financial business proposed to be placed in their hands is relegated, where it of right belongs, to the General Council and its financial officers.

BEEKEEPING BETWEEN TWO QUEENS

It seems to me most unpractical to suggest the formation of two separate and distinct corporations (viz., the "General Council" and the "Central Branch"), each having to conduct financial business with the local associations.

The supplementary recommendation bears strong evidence of its framers' lack of experience of local organisation. To anyone who knows the conditions under which new members are procured, and who realises the folly of inventing difficulties, the "standard form of application for membership" will supply "argument for a week, laughter for a month, and a good jest for ever."

No one, Sir, would wish to seem wanting in appreciation of any efforts undertaken for the good of the craft. Still, those who try to do expert work (such as projecting the architecture of a national society), without the necessary expert qualifications, cannot be surprised if their plans are criticised as faulty and inadequate.

The suggestion that the AGM be held at the same time and place as the Royal Agricultural Show cannot be too highly praised. Everything declares in favour of its adoption. The desirability of bee-keeping asserting its rightful place as an important agricultural pursuit, the practical advantage to be gained by bringing it prominently 'before the great yearly concourse of agriculturists', and the railway facilities granted in connection therewith – all proclaim the intensely practical nature of his proposal.

William Munro, Mapperley

[It is not usual for a salaried secretary to have a vote on the Council and we do not see clearly what advantage it would be if he had one. A secretary can be of more use to an association without a vote than with one. It is proposed that the BBKA should give up all its present members and transfer them to a Central Branch which would be the branch of the BBKA for London and for all other counties not having branches.

One duty of the Central Branch would be to secure as members all bee-keepers who do not reside within the area of any county association. The General Council would be composed of delegates from the county associations and would, as we understand it, form the Central Branch, and we do not think the scheme proposes "the formation of two separate and distinct corporations."

The supplementary recommendations are simply suggestions of the committee for the consideration of a General Council, and as it is proposed to be formed by representatives of the different branches they would no doubt on getting together adopt that which they considered workable and reject what was not so. Ed.]

BBJ 24 Feb 1910
The monthly meeting of the BBKA Council was held on February 17th at 11, Chandos

Street when Mr. TW Cowan presided. Letters expressing regret at inability to attend were received from Mr. AG Pugh and Dr. Elliott. The following new member was elected Mr. J. Herrod, Trentside Apiary, Sutton-on-Trent.

BBJ 3 March 1910
Felt Covering for Hive Roofs. I always find a small supply of any of the modern vulcanite felts useful in an apiary. For covering a skep to keep all snug and warm, if well weighted down with an earthenware pancheon on top, I know nothing to equal it; whilst for temporarily covering a leaky hive-roof when made secure from being blown off by means of a brick or two, it is very useful. After examining hive roofs covered with lead, zinc, vulcanite felt, linoleum, &c, in my opinion nothing can approach the paint and calico method, as described in " British Bee-keeper's Guide Book"; this is perfection, and combines lightness, efficiency, and cheapness in a marked degree.
Applicance Dealer, Carrington (Thos. N Harrison)

BBJ 3 Mar 1910
Thos. N. Harrison and Son (5 and 7, Cheapside, Nottingham). This firm sends us a revised illustrated catalogue of beehives and appliances, which also includes a selected list of garden requisites, as well as dairy and poultry appliances. They supply only the best quality appliances, and keep a large stock of all that the beekeeper may require. They have also acquired the business of a wire-worker, and are prepared to supply any description of wire-work. Descriptive catalogue of forty pages, containing hints for beginners in apiculture, post free on application.

NEP 5 Mar 1910
The Production of Honey. A satisfactory and encouraging record of work was submitted at the AGM of the NBKA, which was held at the Peoples' Hall this afternoon. Mr. JF Blackshaw (principal of the Agricultural College, Kingston), presided, and was supported by Captain JA Morrison, MP, Mr. WS Ellis, Mr. AG Pugh, and Mr. G. Hayes (secretary).

The Annual Report stated that the association started the year with a total membership of 233. Fifty new members were enrolled, but of these 33 had resigned or left from other causes. The financial statement was satisfactory in so far as the association began the year with a balance in hand of £14 14s. 5d. and finished with one of £l6 5s 6d.

The past season had not been one of the best for beekeeping and those who started beekeeping this year would, it was feared, fail to make their venture profitable. Beekeeping must not be judged by one, two, three years' experience; as generally one good year would make up for a number of bad seasons. Nine experts visited amongst them 214 apiaries, containing 809 stocks of bees, 715 of which were satisfactory and advice and assistance generally given. About 4% of the stock were found diseased, and were either destroyed or, if only slightly affected, remedial measures were taken to put them in a healthy state.

In the discussion on the report, Mr. Harrison commented on the great many beekeepers in the county who were not members of the association. One of the failings of the association, he thought, was that it did not advertise itself sufficiently. Mr. AG Pugh, on the other hand, spoke favourably of the progress which the association was making, the membership being steadily augmented.

In the subsequent discussion mention was made of prevalence of "Foul Brood" in the Hucknall district, and it was suggested that the experts should have attended to this. The report was adopted.

The officers were elected as follows: President, the Duchess of Portland; auditor, Mr. W. Harrington; secretary, Mr. G. Hayes. Mr. Hayes and Mr. Pugh were appointed delegates to the London meeting; and thanks were expressed to all the retiring officials. Replying to a vote of thanks, Mr. Blackshaw said that beekeeping was an important if small industry. Beside the profit arising from the production of honey, it had an indirect beneficial result in so far as it made a man think. Beekeeping was one of the things which would make smallholdings pay.

About ninety members and friends were at the afternoon meeting and partook of tea at 4.30pm. The evening meeting commenced at 6pm with the distribution of medals, certificates, etc, followed by a most instructive address on "The Production of Comb-honey" by Mr. W. Herrod, secretary of the BBKA.

The next item was to consider the proposed scheme for the reorganisation of the BBKA. The matter was well discussed in all its bearings, and eventually it was resolved "that the members of the NBKA, having considered and discussed the suggested scheme for re-organisation of the BBKA, desire to express their emphatic disapproval of the same, believing the present association possesses all the necessary machinery, rules, and regulations for carrying out any of the suggestions which have recently appeared in the BBJ, for its improvement." This was carried with only one dissentient. The usual prize drawing brought the meeting to a close.

Geo. Hayes, Secretary.

Captain (later Major) James Archibald Morrison (1873-1934) was MP for Nottingham East 1900-1912. In WW1 he was wounded on the Somme and was awarded the DSO in 1916.

NEP 4 Apr 1910
A well-attended meeting of the committee of the NBKA was held in Nottingham on Saturday, presided over by Mr. AG Pugh. The secretary reported that fifteen new

members had been added the roll. It was arranged to send the bee tent with a lecturer to the Newark Agricultural Society's Show and various other applications were considered. Ten experts were appointed to districts covering the whole county for visiting apiaries, to examine stocks and to give advice in apiculture to the owners. The Annual County Show of honey, bees, &c., was fixed to be held in July, in connection with the Southwell Horticultural Show.

BBJ 7 Apr 1910
Bees. Wanted, healthy stocks, on wired frames, no Hives.
                                            Harrison, Cheapside, Nottingham.

5 hives, frames, supers, extractor, good condition. Bargain, room wanted.
                                            Collier, Hendon Rise, Wells-road, Nottingham.

BBJ 21 Apr 1910
The AGM of members of BBKA was held at 11, Chandos Street, on April 14th, Mr. TW Cowan, FLS, in the chair. There was also present Dr. TS Elliot. Apology for enforced absence was read from G. Hayes. Without going into further detail, Mr Herrod moved that the report and Balance Sheet be received and adopted. The motion was seconded by Colonel Walker, supported by Mr. AG Pugh, and carried unanimously.

Mr. Watson proposed, and Mr. Salmon then rose to move his resolution "that all experts touring for affiliated associations shall hold BBKA certificates." He understood from conversations he had with Dr. Elliot he considered the idea a good one but did not see what power the Association had in the matter. Mr. Pugh thought it would be a difficult matter to carry out. He agreed that it would be very nice if it could be done.

Mr. Falkner stated that in Leicestershire they had copied the NBKA and had local experts and in doing this they found it was not always possible to get certificated men, though he had no doubt that those employed were quite capable of passing if they presented themselves. It would be a pity to debar these though he quite agreed that where possible qualified men should be employed.

After the Annual Meeting a meeting of the new Council was held for the purpose of electing officers and committees, amongst those present being AG Pugh and Dr. TS Elliot. The following was elected amongst others to the Exhibition Committee: AG Pugh.

At the conclusion of the BBKA Council meeting, a short adjournment was made for light refreshments, the members assembling at six o'clock for the Conversazione over which Mr. TW Cowan presided. Amongst those present were Dr. TS Elliot and AG Pugh.

Dr. Elliot thought that in those cases where honey-dew was not due to aphis it might be due to bacteria. It seemed certain that it was not always due to aphis and it can hardly be a beneficial secretion to the plant because it does so much damage to it. He thought that it had been noticed that after the cold north-east winds in May aphis was very abundant in the summer. Last year it was very bad, the elder-trees being especially covered with it, and many of them were killed in consequence.

BBJ 12 May 1910
The Death of King Edward VII.
With overwhelming suddenness and only a few hours' notice of the illness of King Edward, the Empire has been plunged in grief and now mourns the loss of a good Sovereign. Millions of English-speaking people awoke on morning to a sorrow so profound that it only has a parallel to that of the death of Queen Victoria nine years ago.

Not only do we all feel the loss, but deep and sincere regret will be felt throughout the world, where the dead monarch was so well known and loved and where he earned the title of "Peacemaker." It is by this grand and glorious title that he will be known to future generations a title which will go down to posterity unsullied by any unworthy act during his reign of only nine years, one of the shortest in our national history, but also one of the noblest, for there have been few monarchs who so thoroughly understood the people, who won their confidence and sympathy and whose popularity was so universal. His greatness and sagacity came out pronouncedly in a great crisis and it is a sad loss to the country that he should have been removed at the present moment when his wisdom was so much needed. No King in passing away has ever been more regretted by a sorrowing people, and he will always be regarded as a ruler who sought the welfare of his subjects and whose personal characteristics were those of a noble-hearted gentleman. All will feel deep sympathy with Queen Alexandra and the Royal family and give their loyal allegiance to King George.

King Edward VII died on 6th May 1910 at Buckingham Palace.

GEORGE V

King George V was born on June 3rd, 1865, the son of Edward VII. He was crowned King and Emperor of India in the year 1911 and changed his family name to "Windsor". Among other things, he is credited for starting the Royal Christmas Broadcast tradition. He was a renowned stamp collector. A popular monarch, many people thought he was a good King.

BBJ 19 May 1910
The AGM of the Leicestershire and Rutland BKA was held at the Highcross Coffee House, Leicester, on April 9th, Mr. EJ Underwood, Chairman of the association presiding over a good attendance, nearly a hundred members being present. During the evening Dr. TS Elliott of Southwell delivered an interesting address dealing with the scientific side of apiculture.

NEP 19 May 1910
The attendance at the second day's show of the Notts. Agricultural Society, held at Worksop, was undoubtedly sadly interfered with by the rain. From an early hour yesterday morning the rain descended in torrents and it was not until near midday that the weather cleared. Lectures were given on bee-keeping, practical apiculture, honey extracting and wax extracting. The manipulations with live bees in a specially-erected tent were also watched with interest by fairly large crowds.

BBJ 2 Jun 1910
The monthly meeting of the BBKA Council was held on May 18th, at 11, Chandos Street when Mr. TW Cowan presided. There were also present AG Pugh, Dr. Elliot and G. Hayes (Notts). The following new member was elected Mr. TN Harrison, 5 and 7, Cheapside, Nottingham.

That to obtain the views of bee-keepers throughout the country with regard to legislation respecting the diseases of bees and to consider the best means of promoting such legislation, the following committee was appointed with power to add to their number and that the first meeting of the committee to be held in the show-yard at Liverpool - General Sir Stanley Edwardes (Kent), Mr. JB Lamb (Middlesex), Dr.TS Elliot and Mr. G. Hayes, Mr. E. Walker (Surrey), Captain Sitwell and Mr. Kidd.

BEEKEEPING BETWEEN TWO QUEENS

Examinations for third-class certificates were sanctioned at Mansfield by Dr. P Sharp.

The special meeting to consider the reorganisation of the BBKA followed.

Reports had been received from various associations and Mr Cowan found that NBKA objected to the scheme. It was now open for discussion and he hoped all would give their views.

Mr. Pugh, in rising to move an amendment, said he considered the scheme a move in the wrong direction considering the improvement shown in the condition of the Association during the past few months, both in finance and, with a little alteration of rules, they had all the machinery necessary to make a strong association. It would be a fatal mistake to change the name. He therefore moved and Mr. Hayes seconded, "that taking into consideration the improved position of this Association, as shown by the recently-issued report and Balance Sheet, this meeting of representatives of the various associations, together with Council here assembled, desire to express their opinion that the present machinery, rules, and regulations (with a few modifications and additions) are quite sufficient to carry on the work of this Association without going into a reconstruction scheme as now suggested."

Mr. Hayes understood Mr. Garcke to say that people would not join both county associations and the BBKA. He maintained that the greater number of members of the BBKA also belonged to county associations. He appealed to the secretary to say if this was not the case. The Secretary: "That is so." Captain Sitwell quite agreed. Mr. Pugh's amendment was then put and defeated.

Captain Sitwell then moved as an amendment: "that the BBKA be reconstructed on the lines of a scientific society, to be recognised as the head centre of all apicultural interests by all affiliated associations; its position and functions towards such associations being scientific, advisory, and social." After discussion, in which Mr. Pugh took part, this was put and defeated.

Mr. Edwards then moved as an amendment: "that the constitution of the BBKA needs altering so that it may be thoroughly representative of bee-keeping interests and that a central body should be formed, to constitute a Federation of the County Associations and of the BBKA." Mr. Pugh having expressed his willingness to accept the amendment, it was put to the vote and carried. It was then put as a substantive motion and carried. The following delegate was selected to assist the special committee in their deliberations - AG Pugh.

BBJ 9 Jun 1910
Bee Shows to come - July 21st at Southwell. Annual Show of the NBKA, in connection

with the Horticultural Society's Show. Open class for Single 1-lb Jar. First prize, 20s. Schedules from Geo. Hayes, Mona Street, Beeston

BBJ 9 Jun 1910
W. Munro (Nottingham). Future of the BBKA. You will see that at the special meeting of members and delegates of county associations, held on May 18th, certain resolutions were carried as amendments to the scheme which completely change the constitution of the BBKA and if adopted will extinguish it as an association. The scheme as originally proposed has not been carried and the alterations embody the principles you advocate.

Should the members of the BBKA consent to the proposed changes, it will be for them to decide what they will do with the capital and other property of the Association. The financial part of the scheme has been referred back to the committee, which was strengthened by the addition of eight of the county representatives. When they present their report, no doubt another meeting with the delegates will be arranged.

William Munro was born in Rochester, Kent in 1887. By 1910 he was a metallurgist.

BBJ 16 Jun 1910
WD (Notts). Comb is infected with a form of sour brood.

New Superior "WBC" Hives, 10s 6d each; large Bee-shed, 40s.
                                                                             Green, Bigg House, Arnold

NEP 20 Jun 1910
The escapades of a Mansfield youngster, who endowed with remarkable precocity which unfortunately taking a most undesirable turn, are causing not a little trouble and anxiety to his parents and others concerned in his welfare. John Herbert Reginald Bowskill is a schoolboy, nine years of age, and his home is in Ratcliffe-gate. It is, however, only little over a fortnight ago since young Bowskill received four strokes with the birch rod and was placed under the probation officer for twelve months for breaking into shops and gardens and stealing money. He has also a penchant for visiting gardens and it was found he had been to one or more on the day mentioned. In one garden he came across several beehives and fancying honey, he, as a precautionary step, provided himself with a hat, veil and gloves, all of which he borrowed without leave and started to help himself from a hive. Naturally the tenants thereof objected but young Bowskill only admitted to the police being stung twice. Last Tuesday Ald. JE Alcock had the boy before him and remanded him to the Union Workhouse for eight days.

BBJ 23 Jun 1910
The monthly meeting of the BBKA Council was held on June 16th at 11, Chandos

Street when Mr. WF Reid presided. The following new member was elected - Mr. AH Margetson, Woodthorpe Avenue, Sherwood.

NEP 29 Jun 1910
For forty-seven years the East Bridgford Horticultural Society has held its Annual Show. There was exhibition of bees and honey, open to all members of the NBKA and East Bridgford Horticultural Society, the prize-winners in the section being D. Gower, Mrs. Cartwright, J. Higgs and W. Millington.

BBJ 7 Jul 1910
The Summer Conference of NBKA was held in the Peoples' Hall on July 25th, when about sixty members and friends from districts pretty well covering the whole county were present.

Up to the time appointed for tea the visitors spent an enjoyable hour examining a large collection of mounted nectar-producing plants, drawings of pollen grains of same, wire excluders, dummies and dividers, a simple slow-feeder, an apparatus for burning out odd diseased cells, etc.

After tea an excellent paper was read by Dr. TS Elliot, of Southwell, on "The Scientific Aspect of Foul Brood," which was profusely illustrated by diagrams, specimens of growths, &c, and lantern-slides. The lecturer pointed out how the organism was cultivated, measured and defined and gave instruction as to various means of keeping the disease subdued and at bay, concluding with the hope that ere long the Government would be induced to pass a Bill for the suppression of "Foul Brood" and other diseases of bees.

Mr. Darrington, in proposing a vote of thanks to Dr. Elliot, said how very instructive, lucid and interesting his paper had been. The secretary (Mr. G. Hayes) gave a description of twenty-five nectar producing plants and their pollens, illustrated by photo-micrographic slides of the latter and photo slides of the plants. Mr. Hayes' extremely interesting paper was listened to with interest and pleasure by all present, a most successful meeting being brought to a close at about 9pm.

BBJ 7 Jul 1910
Artificial Swarming. I have one hive, in which the bees are working in a rack of sections and two supers of shallow frames, there being ten frames well filled with brood in the brood box. The sections and upper rack of frames are nearly ready to come off and I am anxious to make some increase, although there is no sign of preparation for swarming. I propose to make an artificial swarm, giving the swarm the old stand and the supers and when the queen cells are ripe to divide the stock into two. I shall be very much obliged if you will tell me:

1. How late can I safely leave this for the stocks to build up before winter?
2. Should I give larvae in artificial queen cells, or can the bees be trusted to select larvae young enough when they are not swarming naturally?
3. Would it be advisable to make four nuclei and then unite as required to allow for mishaps in fertilisation?
4. What is the object of lugs $1\frac{1}{2}$ in. long on frames? It seems to me that lugs 1 in. long would very much simplify hive construction, particularly with hives made on the "WBC" principle.   Medico, Notts.

Reply.
1. It should be done at once.
2. If after being queenless twenty-four hours you enlarge a few of the cells containing eggs and destroy all other queen cells made by the bees, there will be no risk of failure.
3. No; this would be dividing up too much and would only be safe in experienced hands.
4. The $1\frac{1}{2}$ in. lugs found most convenient to enable the frames to be manipulated with comfort. A $1\frac{1}{2}$ in. top bar was tried some years ago but was discarded, as it did not give sufficient room to hold the frames.

BBJ 14 Jul 1910
The first meeting of the "Foul Brood" Legislation Committee was held at the Royal Show Ground, Liverpool, on June 23rd. Captain F. Sitwell (Northumberland) presided. Members present included George Hayes and Dr. TS Elliot (Notts).

BBJ 27 July 1910
When the hawthorn was in bloom this year, on three days especially it yielded nectar copiously. I got up into a hedge for the purpose of taking bees from the hawthorn and found they were very numerous and were, without doubt, gathering nectar from it, as could be seen both by their action and their distended abdomens.

Last year, although there was an abundance of blossom, I do not think it yielded any nectar, for though I frequently searched the bushes in the hope of finding hive-bees, I failed to discover any during the whole time it was in bloom, while very few wild bees visited the flowers. I have had honey which had both the scent and taste of "May," but from observation I am led to the conclusion that the climatic conditions must be very favourable for hawthorn to yield any appreciable amount of honey and these conditions prevailed on the three days mentioned.   Geo. Hayes, Beeston.

BBJ 28 Jul 1910
The NBKA Annual Show was held on the 21st inst. at Southwell, in connection with the Horticultural Society's show, under ideal weather conditions. This proved to be the largest show of honey, bees, and appliances ever held in the history of

the association, the number of entries and of exhibits staged being a record one. The trophies and collections of appliances made a good display and altogether the association may congratulate itself upon a distinctly successful exhibition.

Dr. P. Sharp, Brant Broughton, and Dr. Elliot, Southwell, were the judges, their awards being as follows:
Collection of Bee-appliances - 1st, Thos. W. Harrison and Sons, Nottingham; 2nd, W. Mountney, Southwell
Honey Trophy - 1st, WL Betts, Mansfield Woodhouse; 2nd, Uriah Wood, Arnold
Twelve 1-lb Jars Light Extracted Honey - 1st, Hy. Hill, Carlton-le-Moorland, Lincs.; 2nd, JT Duckmanton, Langwith; 3rd, WL Betts; 4th, Geo. Marshall, Norwell; vhc, GH Pepper, Oxton; hc, BC Craven, Southwell; c. W. Doleman, Keyworth
Twelve 1-lb Jars Dark Extracted Honey - 1st, WL Betts; 2nd, T. Gillott, Sherwood; 3rd, G. Marshall; 4th, W. Lee, Southwell; vhc, JR Almond, Gotham; hc, J. Breward, Staythorpe
Twelve 1-lb Sections - 1st, GH Pepper; 2nd, JT Woods, Nettleworth, Mansfield; 3rd, Geo. Marshall
Twelve 1-lb Jars Granulated Honey - 1st, Geo. Marshall; 2nd, H. Merryweather, Southwell; 3rd, JT Woods; hc, WL Betts
Three 1-lb Jars Extracted Honey and three 1-lb Sections (amateurs only) - 2nd, T. Gillott
Single Shallow-frame of Comb-honey - 1st, B. Mackender, Newark; 2nd, WL Betts; 3rd, G. Marshall; vhc, Uriah Wood; hc, H. Mackender
Observatory hive - 1st, G. Marshall; 2nd, R. Mackender; 3rd, EG Ive, Boughton; 4th, WL Betts
Beeswax - 1st, AH Hill, Balderton; 2nd, Geo. Marshall; 3rd, John Bee, Southwell; vhc, Uriah Wood
Extracted Honey (local) - 1st, J. Breward; 2nd, G. Noton, Southwell; 3rd, W. Lee

Mr. Geo. Marshall, of Norwell, was awarded the Herrod William Perpetual Challenge Cup for the highest number of points.　　　　　　　　　　　Geo. Hayes, Hon. Sec.

Uriah Wood was born in 1869 in Arnold. He was a chemist.

BBJ 4 Aug 1910
The twenty-seventh Annual Show of hives, bees, honey, wax, and appliances, under the auspices of the Derbyshire BKA, was held in connection with the Derbyshire Agricultural Society's show on July 13th and 14th at Osmaston Park, Derby. In spite of the unfavourable honey season in Derbyshire, some splendid exhibits were staged and the bee and honey section attracted a considerable number of visitors. Mr. Geo. Hayes, the secretary of NBKA, was the judge and he also conducted two examinations for third-class certificates of the BBKA.

BBJ 4 Aug 1910
TD (Mansfield)
1. There is no trace of carbolic in the honey you send.
2. You treated the combs properly.
3. To prevent any possible chance of tainting the honey it is best to use formaldehyde as a disinfectant.

BBJ 18 Aug 1910
Limnanthes douglasii, I consider is valuable for bees, for under suitable weather influences it secretes an abundance of nectar as anyone may see with the naked eye. If a beekeeper, on a suitable day, will take up a flower he may see at the base of each petal in the corolla that every nectary is overflowing. On a patch of two square yards I have seen forty or fifty bees at a time busy rifling the flowers of their contents and this is a very large number for the area, as I do not think one would ever find so many on the same area in a clover field. That a good patch in early spring is helpful to the bees is to my mind without doubt, but of course if the whole garden were full of it, it would be of no avail for surplus and I am very much afraid - although I have not proved it - that it would be unsuitable for a section or surplus comb.

Geo. Hayes, Beeston

BBJ 18 Aug 1910
JHB (Beeston). Dead Queen. The queen is quite dried up, but appears to be an old one and was probably a fertile one.

BBJ 18 Aug 1910
I would mention that seed of *L. Douglasii* should be sown in the autumn to be available for bees in early spring.

Geo. Hayes, Beeston

NEP 20 Aug 1910
The yield of honey locally is stated be considerably better than last year, though not what might reasonably expected. In June the bees were very busy, but the variable weather that has been experienced since, together with the chilly atmosphere, has seriously interfered with their work and several beekeepers complain of the small returns made. A gentleman in Cotgrave, however, has enjoyed remarkable success, whilst others in Arnold and Bulwell have also good reports to offer. Notts. is a good honey-yielding county but large quantity is wasted owing to a want of proper knowledge of bee culture.

BBJ 22 Aug 1910
The monthly meeting of the BBKA Council was held on September 15th at 11, Chandos Street when Mr. TW Cowan presided. There was also present AG Pugh. A letter expressing regret at inability to attend was received from Dr. TS Elliot. The

examiners' report on third-class examinations at Southwell was received,

BBJ 25 Aug 1910
NS (Worksop). Race of Bees. They are all pure British bees.

BBJ 25 Aug 1910
Wanted, Double Breech-loading Gun, 12 bore, left choke. Will give good value in stocks on frames, for really good weapon.   R Ive, Boughton

NEP 7 Sep 1910
Moorgreen Show. Except the section for hackney horses practically every department the Moorgreen Agricultural Show held yesterday showed a slight decline in entries. In the bee-keeping section, the prize winners were H. Meakin, F. Hopkin, W. Brooks and W. Darrington.

BBJ 22 Sep 1910
The eighteenth International Exhibition of the Grocery and Allied Trades was opened on 17th inst. and will continue till 24th inst.
Twelve 1-lb Sections Heather Honey - 2nd, T. Marshall, Ivy Cottage, Sutton-on-Trent
Three Shallow Frames Comb Honey - 3rd, Dr. TS Elliot, The Old Rectory, Southwell
Twelve 1-lb Jars Medium-coloured Extracted Honey - 2nd, J. Herrod, Sutton-on-Trent
Twelve 1-lb Jars Dark Extracted Honey - 1st, G. Marshall, Norwell
Twelve 1-lb Jars Heather Honey - 3rd, G. Hunt, Newark; vhc, T. Marshall
Twelve 1-lb Jars Heather-blend Honey - 1st, J. Woods, Nettleworth Manor, Mansfield; 2nd, T. Marshall; 3rd, J. Herrod; 4th, AG Pugh, Beeston; hc, Dr. TS Elliot
Twelve 1-lb Jars Granulated Honey - 1st, T. Marshall; vhc, G. Marshall

NEP 3 Oct 1910
The quarterly meeting of the NBKA was held on 2nd October and the statement of accounts showed the association to be in a flourishing condition. Forty new members have been enrolled so far this year. Demonstrative lectures have been given at four centres; 228 apiaries have been visited and 729 stocks examined. It was found that only 3.6%, of stocks examined were affected with disease, which is a very low rate. The season, although not a good one, has been considerably better than the last and the quality of the honey obtained has been generally good. Beekeeping is developing greatly in the county but a real good season would serve as a great impetus to the industry.

BBJ 6 Oct 1910
Sutton-in-Ashfield. A very nice clover honey smoothly granulated. Quite good enough for showing.

BBJ 13 Oct 1910
The monthly meeting of the BBKA Council was held on October 6th at 23, Bedford Street. Mr. TW Cowan presided, and there were also present AG Pugh and G Hayes (Notts),

The Conversazione was held at the "Eustace Miles" Restaurant at 5pm., when about sixty members and friends assembled. Refreshments were provided in the reserved balcony, after partaking of which the company adjourned to the Green Salon, where Mr. TW Cowan presided over the Conversazione. Just one hundred were present, amongst whom was Dr. TS Elliott

BBJ 13 Oct 1910
The thirty-fifth Annual Exhibition of the British Dairy Farmers' Association opened on October 4th at the Agricultural Hall and continued until the 7th inst. Mr. G. Hayes, Beeston staged a very interesting exhibit of lantern-slides from micro-photographs of pollen from various flowers, which was deservedly awarded a first prize. A great deal of labour must have been expended on their preparation.
Interesting and Instructive Exhibits of a Practical or Scientific Nature - 1st, Geo. Hayes, Mona Street, Beeston

BBJ 20 Oct 1910
JR Baxter (Notts). Insect Nomenclature. The name of the fly is *Eristales tenax*

BBJ 10 Nov 1910
Nectar-Producing Plants and their Pollen. Geo. Hayes, Beeston, Notts.

Through the kindness of the Editor I am able to give the result of my three years' study of this subject for the benefit of the readers of the Bee Journal. Probably many have felt the need of some work of reference for helping them to diagnose the source of the various honeys, for I know of none that treats of the subject especially as it concerns beekeepers. Up to now I have been working more particularly on British nectar-producing plants, but hope to continue my investigations with those of countries from which we import most of the foreign honey, so that we may have, as far as possible, a complete record for the diagnosis of any honey put before us.

This was a series of essays on this subject published by the BBJ over several months. Each one highlighted one species. Subsequently George Hayes combined these notes into a book – a copy of which can be found in the NBKA Library which was founded by the donation of his personal library and is now named after him. This book is required study for anyone attempting the BBKA examination in microscopy.

W Herrod wrote, "The study of pollen grains under the microscope is both interesting and fascinating and those who wish to spend pleasant hours in this pursuit should purchase and read the book".

BBJ 17 Nov 1910
An Autumn Conference of members and friends of Leicestershire and Rutland BKA, numbering upwards of sixty, was held at the Highcross Restaurant, Leicester on October 29th, Mr. GO Nicholson, Market Harborough, occupying the chair. Tea was followed by the distribution of prizes won at the Abbey Park Show and Mr. Geo. Hayes (lecturer to the Notts County Council and Midland Dairy Institute) gave an interesting and instructive lecture on "Nectar-Producing Plants and their Pollen," which was illustrated by photo-micrographic slides.

BBJ 8 Dec 1910
Nectar-Producing Plants and their Pollen - White Clover.

BBJ 8 Dec 1910
Legislation and Infected Honey. A number of your correspondents seem to be much perturbed that honey should be included among the articles liable to be destroyed in an infected apiary in the proposed Bee Diseases Bill. Some years ago I went over to a friend's apiary twelve miles distant to fetch some bees that were in frame-hives. My friend was giving up bee-keeping and I was to have the bees. The first hive I looked into was reeking with "Foul Brood" and as the other hives were in the same condition, instead of taking the bees home I suffocated the lot. Had I been given a free hand I would have burnt the hives, &c, as well.

The combs contained a fair amount of honey, which the owner insisted on being extracted before they were burnt, as he knew that honey from "Foul Brood" stocks was harmless to human beings if they cared to eat it. I extracted about 60 lbs from the combs, but the stench was so bad that I had to turn my head away from the extractor when turning the handle and the honey itself had a distinctly 'Foul Broody' smell. Would any of your readers advise that honey such as this should not be destroyed when endeavouring to clear an apiary of "Foul Brood"?

It is well known that honey from a hive infected with "Foul Brood" will, if given to a healthy stock, cause that stock also to become diseased and it is quite possible - I was going to say probable - that in extracting from diseased combs a little of the infected honey may be spilt or smeared in some place where bees may have access to it.

We often hear the saying that it is possible to drive a coach-and-four through any Act of Parliament and I venture to think that omitting honey from the list of articles that may be ordered to be burnt would be leaving an already open coach-road and would, in many instances, seriously handicap an inspector who was dealing with an infected apiary. It would not matter how little honey or how much disease there was in the combs, the owner would be able to claim the right to have the honey extracted and the inspector would be powerless. Of course, an inspector would have to use his discretion with respect to honey in supers, but he should make it certain that infected honey was not to be used for bee-food, either in syrup or honey-candy. If the honey was of poor quality and not very saleable, the best course would be to burn it.

J. Herrod, Sutton-on-Trent.

BBJ 22 Dec 1910
The monthly meeting of the BBKA Council was held on December 15th at Cavendish Square, London. Mr. WF Reid presided and also present was Mr AG Pugh. A letter expressing regret at inability to attend was received from G. Hayes.

BBJ 12 Jan 1911
Nectar-Producing Plants and their Pollen - Borage (contd.)

BBJ 19 Jan 1911
Nectar-Producing Plants and their Pollen - Apple

BBJ 2 Mar 1911
"Foul Brood" Legislation. Bee-keepers Record, April 1910, Notts: 729 stocks, 3.6% diseased - "a very low rate."

BBJ 9 Mar 1911
Nectar-Producing Plants and their Pollen - Hawthorn
NEP 15 Mar 1911
Best quality Bees and Hives for Sale, cheap.   Wood, Expert Beekeeper, Arnold

BBJ 16 Mar 1911
The AGM of NBKA was held in the Peoples' Hall on March 4th. Captain JA Morrison, MP presiding. There was a large attendance of members. Mr. George Hayes, hon. secretary and treasurer, presented the report and Balance Sheet.

In his report he stated that of late years they had not been able to speak of the honey harvest in very eulogistic terms, and this applies to the last season, which, although somewhat better than the few preceding years, left much to be desired. However, there was every opportunity for putting bees into winter quarters, well fitted to come out this spring in the best condition. One bright aspect of the coming season was that owing to the late scarcity, there would be very little honey on the market, and it would be an easy matter to dispose of what they might get, while a better price should be obtained.

The Association commenced the year with 269 members, and 41 others joined during the season. There were, however, numbers of bee-keepers in the county, many of whom would, most probably, join the Association if the advantages were put before them. There was plenty of scope here for those members who were willing to assist the industry and the Association.

He had pleasure in again being able to put before the members a satisfactory Balance Sheet for, although there was a slight loss on the year's working, there was still a small balance to the good, and they would see that the funds had been well administered.

The Annual County Show was held in connection with the Southwell Horticultural Society on July 21st and from almost every point of view it may be considered a very successful one. The entries were more numerous than at any previous show and, what was still more pleasing, most of them were staged. The following members had entered and passed the BBKA examination for expert certificates (third-class) - WH Windle, West Bridgford, John C. Mellars, Norton Cuckney, Thos. N. Harrison, Carrington.

The report was adopted and ordered to be printed and circulated in the usual way.

Mr. Pugh proposed and Dr. Elliott seconded, that the thanks of the Association be accorded to all the retiring ofIficers and that her Grace the Duchess of Portland be re-elected President for the ensuing year; the resolution being carried with applause. Mr. G. Hayes was then re-elected secretary and treasurer; Mr. W. Darrington hon. auditor; and the Committee was re-elected *en bloc*.
A vote of thanks was accorded to Capt. Morrison for presiding who, in replying, complimented the Association on the great amount of good work they were doing and wished them every success. He hoped that every member might have a good return in the coming season. Of all the minor industries he considered bee-keeping the best, and could recommend it as a good thing for all smallholders to adopt.

Tea was then partaken of by about 120 members and friends, after which the meeting was resumed, with Wm. S. Ellis in the chair. The medals, certificates and prizes won at the Annual Show were distributed, and the delegates to the BBKA made their reports on the business done at the meetings.

Mr. Puttergill (Beeston) then read a short paper on "The Return of Exhibits from Shows," and suggested a new form of label to be used for this purpose so that each exhibit could be easily traced. The label was passed round for examination and it was finally resolved to refer the matter to the Committee.

Mr. Doleman (Keyworth) then brought forward the question of "How are our judges to be trained?" stating how he saw great difficulty in getting to know the qualities, etc., of different honeys other than those produced in their own county and suggested a little co-operation of members desirous of advancing in this knowledge. This brought forth a lively and interesting discussion.

A vote of thanks was accorded these two gentlemen for their papers. The usual prize drawing brought to a close the meeting, which was considered by all present to have been the most successful on record.   Geo. Haves, Hon. Sec.

BBJ 16 Mar 1911
Nectar-Producing Plants and their Pollen - Primrose

BBJ 23 Mar 1911
The AGM of members of BBKA was held at the "Gardenia" Restaurant, 6, Catherine-street, Strand on March 16th. Mr. TW Cowan presided. There were over one hundred members present from all parts of the country.

Mr. Reid rose and asked the chairman, if he was not out of order, that he might say a few words. He thought it was only right at that juncture to propose a vote of thanks to their secretary. He knew he was a paid official, and the proposition might be a little out of order, but he was quite certain Mr. Herrod had done a lot of hard work outside his ordinary duties.

Mr. Pugh heartily supported the motion. He knew the secretary's heart was in his work and to show that the County Associations had his sympathy, he would like to say that he (Mr Herrod) visited the Annual Meeting of the NBKA representing the BBKA and this without an invitation although it meant a journey of over 200 miles.

The election of Council for 1911 was then proceeded with and AG Pugh was one of those elected.

At the conclusion of the Annual Meeting the members and friends assembled for the usual conversazione and among those present were G Hayes and AG Pugh.

BBJ 6 April 1911
Several Stocks of Bees, 1910 Queens, on Bar Frames, for Sale, guaranteed healthy; also several Nuclei.   Rector, Elston. Newark.

Charles Hubert Whitfield was born in 1869 in Warwickshire. He was Rector of Elston from 1904 to 1922.

BBJ 13 Apr 1911

BEEKEEPING BETWEEN TWO QUEENS

Wanted, Simmins' Conqueror Hives, warranted free from disease. Gillman, Stapleford

BBJ 27 Apr 1911
What Offers for Plate Glass Trophy Stand, up to date, taken first prizes at the County Show?
Wood, Expert, Arnold

BBJ 27 Apr 1911
The monthly meeting of the BBKA Council was held on April 20th at 23, Bedford-street. Mr. TW Cowan presided. The Staffordshire Association applied for the appointment of Mr. AG Pugh as judge and examiner at their show on July 19th and 20th; this was granted.

BBJ 27 Apr 1911
Honey Wanted. Price and sample to SW Catton, Rose Cottage, Old Lenton, Nottingham,

BBJ 4 May 1911
White Fantail Pigeons. Several pairs of last year's birds for sale, 3s a pair.
Schoolgirl, Elston Rectory, Newark

BBJ 18 May 1911
Four Good Strong Stocks, English, on 8 combs, 25s each; guaranteed free from any disease.
Ive, Boughton

Derbyshire Times 23 May 1911
An inquest was held at Pinxton, yesterday, on Kate Hill who died from the effects of carbolic acid poisoning. The deceased appeared with a bottle containing carbolic acid in her hand. She drank some of the contents of the bottle. Her husband had used the carbolic acid in the course of his bee-keeping business and he purchased this and other chemicals for disinfecting the hives when he wanted to drive bees into other positions. Since his time it had only been used once. The poison had been the house for upwards of 20 years, locked up in a cupboard. The jury returned a verdict of "Suicide while of unsound mind."

BBJ 1 Jun 1911
Nectar-Producing Plants and their Pollen – Blackberry or bramble

BBJ 15 Jun 1911
Shows to come. July 20th at Southwell. Annual Show of NBKA, in connection with the Horticultural Society's Show. Open class for Single 1-lb Jar - First prize 20s. Schedules from Geo. Hayes, Mona-street, Beeston

BBJ 22 Jun 1911

Nectar-Producing Plants and their Pollen – Horse chestnut

BBJ 29 Jun 1911
The seventy-second Annual Show of the Royal Agricultural Society of England, held this week at Norwich has, we hear, been a very successful one so far as entries are concerned. Amongst the judges was AG Pugh (Notts).
Twelve 1-lb Jars of Extracted Light-coloured Honey - 2nd, Dr. TS Elliott, Southwell
Twelve 1-lb Jars of Granulated Honey - 3rd, J. Woods, Nettleworth Manor, near Mansfield

BBJ 29 Jun 1911
JA (Nottingham). "Wells" system". It is not worth while giving particulars as it has been tried and found to be a failure.

BBJ 3 Jul 1911
Nectar-Producing Plants and their Pollen - Lime

BBJ 13 Jul 1911
Fertile Queens, Native, post free, 2s 9d each.   JC Mellors, Norton Cuckney

BBJ 27 Jul 1911
OKW (Basford). No honeydew in the sample sent: it is a good sample of Notts honey (which we know so well) from mixed sources.

East Bridgford Parish magazine August 1911
The Annual Show was acknowledged to be one of the best for many years.

Bees and Honey
Class I   Best 6 1-lb sections – Mr D Gower
Class II   Best 6 1-lb bottles run – 1st, Mrs Robinson; 2nd, Mr J Higgs; 3rd, Mr D Gower; hc, Mr Doleman
Class III   Hive of bees – Mr J Fletcher

BBJ 3 Aug 1911
Nectar-Producing Plants and their Pollen - Hawthorn

BBJ 7 Sep 1911
The Warwickshire BKA held their Annual Honey Fair on August 30th and 31st, in connection with the Warwickshire Agricultural Society's Show at Coventry. A good display of honey was staged, but the whole effect of the late strike had not yet passed away, and, owing to the quantity of goods at the station, the Railway Company were unable to get many of the exhibits to the show in time, beekeepers having to suffer

with the rest. A number of the entries for honey were also cancelled owing to the difficulty of getting the exhibits through to the show ground. Mr. J. Herrod of Sutton-on-Trent, acted as judge.

The National Railway strike of 1911 was the first national strike of railway workers in Britain. The strike lasted only two days but its effects were long lasting.

BBJ 21 Sep 1911
The Nineteenth International Exhibition of the Grocery and Allied Trades was opened on last, and will continue until, the 23rd inst. The good honey season has enabled beekeepers and traders to make a splendid show of honey.
Twelve 1-lb Heather sections (ten entries) - 3rd, T. Marshall, Sutton-on-Trent
Three shallow frames (eight entries) – 2nd, Herrod, Sutton-on-Trent; 3rd, G Hunt, Newark
Twelve 1-lb Jars Light-coloured Extracted Honey (seventy-five entries) - c, Dr TS Elliott, Southwell
Twelve 1-lb Jars Medium Extracted Honey (forty seven entries) - 2nd, T. Manfield, Newark; vhc, Dr. TS Elliot; c. G. Marshall, Norwell
Twelve 1-lb Jars Dark Extracted Honey (thirteen entries) - 2nd, J. Herrod; 3rd, G. Marshall.
Twelve 1-lb Jars Heather-blend Honey (seventeen entries) - 1st, T. Marshall; 3rd, AG Pugh, Beeston, vhc, T. Walker; vhc, Dr. TS Elliot; hc, AG Pugh
Beeswax (judged for quality of wax only) (sixteen entries) - 4th, T. Marshall
Extracted Honey in 1-lb Jars - Certificate, AG Pugh

BBJ 28 Sep 1911
The monthly meeting of the BBKA Council was held on September 21st at 23, Bedford-street, Strand, when Mr. CLM Eales presided. There were also present Messrs. AG Pugh and Dr. TS Elliot (Affiliated Association delegates), G. Hayes (Notts). The report of the Treasury Grant Committee was presented and after a long discussion, it was proposed by Mr. A. Richards, seconded by Mr. OR Frankenstein and carried that the report of the Treasury Grant Committee be referred back to that committee for further specific recommendations, which committee shall have the assistance of AG Pugh amongst others.

BBJ 5 Oct 1911
Nectar-Producing Plants and their Pollen. - Coltsfoot

BBJ 12 Oct 1911
The monthly meeting of the BBKA Council was held on October 5th at 23. Bedford-street, Strand when Mr. TW Cowan presided. There was also present AG Pugh. Letters expressing regret at inability to attend were read from G. Hayes and Dr. TS

Elliot.

BBJ 12 Oct 1911
The Thirty-sixth Annual Exhibition of the British Dairy Farmers' Association opened on Oct. 3rd at the Agricultural Hall, London, and closed on the 6th inst. The Honey Show was an improvement on last year.
Twelve 1-lb Jars of Dark Extracted Honey (including heather blends) - r, Dr. TS Elliott, Southwell.

BBJ 11 Nov 1911
The Annual Show of the Ayrshire Agricultural Society was held on October 19th and 20th, at Kilmarnock, and proved in many ways one of the most successful exhibitions ever held by this association, the show being in point of size a record one for Ayrshire.
Six 1-lb Jars of Granulated Honey (10 entries) - rd, J. Woods, Mansfield.

BBJ 11 Nov 1911
Would you (please?) mind answering the following questions in your next issue of "BBJ"?
1. Could I insure against injury to cattle by a barbed wire fence, of course for an extra premium?
2. Will honey keep well in tins if kept airtight? Will it be contaminated in any way by the metal if allowed to stand for months?
3. Is there any danger of bees balling the queen if the hive is opened on a fine day in March or the beginning of April?
4. Is it useless introducing a fertile queen (or virgin) to a hive with a laying worker in possession? Would the bees prefer the fertile worker and does the season make a difference?
5. In uniting two stocks, which queen is sacrificed, the incoming one or the one in possession? Of course, if both were equally valuable we should take one away.
6. We are Derbyshire people, but live on the borders of Notts. Would it matter if we joined the Notts Association, as that county is the handiest for us?
7. I may say we appreciate the Journal very much, and although bee-keepers of some years standing we have picked up several ideas from it.
WAH Alfreton.

Reply.
1. The BBKA do not issue such a policy.
2. Honey will keep for any length of time without contamination in tins if hermetically sealed.
3. There is always a certain amount of danger of balling on the first examination after a winter's rest.
4. The only way it can be done is to cage the queen on a comb of unsealed

brood from another hive.
5. Always select the youngest and most prolific queen irrespective of which colony she is in and cage her.
6. The Notts Association, like most of the others, take members from two miles over the border and as the Derbyshire Association is not affiliated to the BBKA its members are not eligible for the benefits therefrom. We should strongly recommend you to join the NBKA.

BBJ 4 Jan 1912
"Nectar-Producing plants and their Pollen - Willowherb

BBJ 18 Jan 1912
Brand new Brice Observatory Hive, maker 'Lee and Son', 18s.   Ive, Boughton

Nectar-Producing plants and their Pollen - Dandelion

BBJ 29 Feb 1912
Wanted, healthy 1911 fertile queen.   H. Bowmar, 9, Nuncar Gate, Kirkby

Nectar-Producing plants and their Pollen - Wallflower

BBJ 7 Mar 1912
For Sale. WBC Hives and best quality Bees.   Uriah Wood, Expert, Arnold

BBJ 21 Mar 1912
Two WBC hives, secondhand, shallow frames, fitted wire foundation, excluder, rapid feeders, smoker, good order, perfectly healthy; what offers?
Murray, Shearing Hill, Gedling

BBJ 28 Mar 1912
The AGM of members of BBKA was held on March 21st, 1912. Mr. TW Cowan presided. Owing to the dislocation of the train service it was not expected that there would be a large attendance, but in spite of this there was a very large number of members present, though many others sent letters expressing regret at their inability to attend on account of the restricted train service.

Mr. Sladen proposed the re-election of the Council for 1912 as printed on the agenda, with the addition of the two new names; this was seconded by Mr. Bocock and carried unanimously. The names included AG Pugh.

The National Coal strike began at the end of February in Alfreton and spread nationwide. It ended on 6th April. The strike caused considerable disruption to train schedules.

BBJ 28 Mar 1912

May I congratulate the committee on the text of the Bill, as given in your issue of March 14th? The only suggestion I would make is, that in Clause v., Sec. 1 (b) the words "or appliances for bees" should be inserted after "colonies," and in (c) "or products of" should be inserted after "hive."

The committee have done well not to exempt those who call themselves big bee-keepers; such class legislation is totally opposed to present-day ideas, or to an Englishman's notion of fair play. The number of stocks a man has is no guide or guarantee as to his knowledge or capabilities of dealing with disease. I know one so-called bee-keeper who, when I last visited him, had over sixty stocks of bees and did not even know how or when the queens were mated. I was only once allowed to examine any of his stocks and half-a-dozen taken at random were all found to be more or less diseased; at the same time he was doing a fair trade in selling bees, combs, and appliances in the neighbourhood. In several instances I saw the disastrous effects of purchases made from him. I could give other instances of big bee-keepers dealing in diseased bees and combs, &c.

With regard to skeps, in some parts of the country it would no doubt be a hardship and very unwise to exterminate them but possibly that could be left to the local authority to deal with. In a county like Lancashire, for instance, where I have had the pleasure of doing the expert work for the last four years, there have been less than forty skeps, among from 300 to 350 members. It would be very little hardship there to make the driving of all skeps, etc, at the end of the season compulsory but the case would be very different in some of the southern counties, where one could find more skeps in a week's tour than there are in the whole of Lancashire.

To say that skeps or skeppists do not spread disease is simply nonsense, as I know to my cost. This district was quite free from disease until it was brought in by a skeppist from a village a few miles away about eight years ago. It was not at all difficult to get the skep destroyed, but the mischief was done and it has been a very different matter to get rid of the effects of that one diseased skep. The fact is that skeps are difficult to examine. I once examined thirteen in one apiary by cutting cut a comb from each, when every one was found to be diseased. I have also seen skeps and boxes, in which bees have died from disease, turned up for other bees to clean out.

To my mind your correspondent makes out a very good case for the Bee Diseases Prevention Act. It appears that if he knew his bees were diseased he would still move them up to the moors, probably among other stocks, only this Act will prevent him doing so. "Truly a nice state of affairs!"

This is just one of the instances in which the Act will do good. I have heard beekeepers complain of those who will take infected stocks on to the moors just spread disease to those around. I should advise "J. and B." to keep his bees as far as he possibly can from other hives on the moors, or anywhere else; then should that "prying expert," who has been given a hint, spot the disease, other healthy stocks may be allowed to be moved by their owners. He also seems to overlook the fact that the "prying expert" who has "reasonable grounds for supposing that disease exists," has to give "reasonable notice" of his intention to examine hives.

I hope the Act will come into force at the earliest possible moment. It has been postponed and delayed far too long already.   J. Herrod, Sutton-on-Trent.

BBJ 28 Mar 1912
Nottingham Castle. The Secretary of the Ontario BKA, Toronto, would inform you regarding beekeeping in Canada. There are also a number of small branch associations in various parts of the country. The Canadian Bee Journal, a monthly periodical, is published at Brantford, Ontario. It can be had from "BBJ" Office, at 5s 6d per annum.

BBJ 4 Apr 1912
The monthly meeting of the BBKA Council was held immediately after the Annual Meeting. Mr. TW Cowan presided and there was also present AG Pugh. A letter expressing regret at inability to attend was read from Dr. TS Elliott. The following officers were elected:
    Exhibition Committee included AG Pugh.
    The names of delegates included Mr. G. Hayes (Notts)

BBJ 11 Apr 1912
BBKA Conversazionne. Mr. Pugh said there was a good deal of misconception as to what constituted "heather blend." Some seemed to think it should be principally clover with a dash of heather, while others held it should be heather with a dash of clover. He had won many medals with bell heather honey mixed with clover, but he found that clover honey with a dash of heather was most suited to the public taste. The bell heather honey came from the combs but with the true heather honey it was quite impossible to remove it with the extractor. What was considered to be the correct thing on the show bench? Were the prizes to be awarded to the pure heather honey or to the sample that suited the judges' taste the best? In Derbyshire they were wintering entirely on heather honey and the bees were doing well.

Mr. Pugh considered the taste for pure heather honey was an acquired one. There is no doubt that a heather man will hear of nothing else. As to the difference of opinion between certain bee-keepers respecting different moors he thinks it is easily accounted for by the rainfall. The high moors are drier on the eastern side than on the western side, which will account for the difference.

NBKA holds the record of nearly a clean bill of health - only 2% of diseased stocks in the county and no "Foul Brood".

BBJ 25 Apr 1912
In Derbyshire, under a system of examination by the association experts, we have reduced "Foul Brood" cases among members' stocks to even a less percentage than that reported as 2% for Notts, and this, although our touring experts constantly complain that here and there are apiaries whose owners are not members of the association and who constantly refuse them admission, whilst there are the best of reasons for believing these apiaries to be centres of infection and re-infection for the whole district.

BBJ 2 May 1912
Exhibitors Lee's Observatory Hive, new, 18s, an absolute bargain.   Ive, Boughton

Nectar-Producing plants and their Pollen - Gooseberry

NEP 14 May 1912
Whilst many of the larger agricultural organisations find it a difficult matter to make both ends meet, the Newark Society continues to flourish. In accordance with a time-honoured custom the Annual Show opened to-day in conjunction with the May Fair and the weather being beautifully fine there was a large influx of visitors to the town. The exhibition once more took place on the Sconce Hill grounds and in many respects gave promise of being one of the best ever held. There was a record entry. Lectures on beekeeping and demonstrations with live bees were given by representatives of NBKA.

NEP 28 May 1912
To-day and to-morrow Mansfield entertains the Notts. Agricultural Society for the fourth time and if the town assists the show as well as it did on the occasion of the last visit 13 years ago, no one will be better pleased than the secretary, Mr. WH Bradwell, for he states that then the receipts on one day amounted to a record which has stood up till to-day. Fortunately, the weather was beautifully fine and the result was large attendance of visitors. NBKA gave lectures and the manipulations with live bees were watched by large crowds.

BBJ 30 May 1912
GFB Mansfield. I am hoping my bees will not swarm this year, but in the event of their doing so, if I take the queen away from the swarm, can I return it to the parent hive, and how?

BEEKEEPING BETWEEN TWO QUEENS

Reply. You can return the swarm to the parent hive after having removed all the queen-cells but one. The swarm should be returned to the hive in the evening in the same way that a swarm is hived into a new home.

Nectar-Producing plants and their Pollen - Currant

BBJ 6 Jun 1912
Three strong healthy Stocks in strong hives, one Observatory Hive with Swarm, and all accessories, for sale. Doell, Laburnum Grove, Beeston

BBJ 6 Jun 1912
For Sale, or exchange for Bees, etc., six good black Airedale and Spaniel Puppies, will make good workers or guards, etc. price 12s 6d each.
 Apply, WA Allfree, Talbot Inn, Mansfield.

Nectar-Producing plants and their Pollen - Willowherb

BBJ 13 Jun 1912
Nectar-Producing plants and their Pollen – Plum and Cherry

East Bridgford Parish Magazine 25th June 1912
In spite of the early date of the show (June 25th) and peculiar season, the Horticultural Show was considered a good one. Unfortunately the rain which fell at intervals the whole of the day adversely affected the attendance.

Special Prizes	Honey in comb - 1st, Cartwight; 2nd, Bond; 3rd, Robinson
	Honey extracted - 1st, Dolman; 2nd, Bond; 3rd, Colville
	Living bees in Unicomb hive - Gower

BBJ 27 Jun 1912
August 23rd and 24th, at Nottingham. Grand Exhibition of Appliances, Honey, Beeswax, collections of objects of interest and instruction. Demonstrations, etc. to be held in the Mechanics' Hall Nottingham. Open classes for appliances, extracted honey, sections, fitting-up frames, fitting up sections, judging competition, etc. Schedules ready July 15th, from G. Hayes, Mona-street, Beeston. Entries close August 12th.

NEP 8 Jul 1912
Hundreds of passers-by on Alfreton-road. Nottingham, yesterday and again this morning, had their attention drawn to swarm of bees, which had settled on the shop front of Mr. G. Newmarch, at No. 26. Where the swarm came from it is impossible to say, but the probable explanation of this desertion from the hive is that the housing

accommodation had become insufficient - in which event the queen bee and the elderly section of her subjects have an accommodating habit in relinquishing their tenure of office in favour of the more juvenile portion of the community. Mr. Newmarch had been acquainted with the nature of the visitation when he visited the shop this morning and proceeded to take down the shutters warily. The task was half completed before he discovered the swarm, upon which he decided that discretion was the better part of valour.

An experienced beekeeper in Mr. C. Shaw. Lowdham, was called to the scene. Mr. Shaw proceeded to remove the remaining shutters and with a carbolic cloth, gently induced the swarm to walk into an inverted hat box, after which the lid, suitably perforated, was clapped on. Mr. Shaw explained to a Post representative that the swarm weighed about $2^1/_2$ lbs. As a good swarm of about 30,000 would weigh about 7 lbs, he calculated that the number caught would about 10,000. They were a common type of bee, were not of much value and would be no use this year for collecting honey. He understood that if a swarm left the hive and was not followed, the ownership lapsed, but if satisfied as to the ownership of these, he would return them. He proposed to hive them at Lowdham for 12 months.

BBJ 11 Jul 1912
The monthly meeting of the BBKA Council was held at 23, Bedford-street, Strand, on June 20th. Mr. WF Reid presided. A letter, expressing regret at inability to attend, was read from AG Pugh. It was resolved that the following be appointed on the Development Fund Committee - AG Pugh. It was resolved that the Committee be asked "to draw up a scheme to be laid before the Council for expending the Government grant in regard to organising bee-keeping in Counties and to continue the necessary work of the experimental apiary, and that they further be authorized to expend a sum not exceeding £150."

BBJ 11 Jul 1912
The monthly meeting of the BBKA Council was held in the Show Ground at the Royal Agricultural Society's Show at Doncaster on July 4th. Mr. AG Pugh presided.

Mr. Pugh said how pleased they were to welcome the delegate of the Yorkshire Association and he hoped that it was the first of many meetings Mr. Richardson would attend, as the Council were anxious to get into touch with all the Associations. London was a long distance for the Nottingham delegates to travel to meetings and he hoped the new departure, which was to be a permanent one, of holding a Council meeting at the Royal Show each year, would be taken full advantage of by the delegates residing in the immediate vicinity of the centre in which the Show was being held in that year. No doubt the small attendance was due to the fact that delegates had not quite realised that a Council meeting was being held out of

BEEKEEPING BETWEEN TWO QUEENS

London and as this came to be more generally recognised the attendance would improve each year.

Mr. Richardson replied, saying how he had enjoyed the meeting and the opportunity of seeing the excellent way in which the affairs of the Association were managed. He felt quite at home, had been heartily welcomed and he intended attending all the meetings he possibly could.

BBJ 11 Jul 1912
The seventy-third show of the Royal Agricultural Society of England was held at Doncaster from July 2nd to July 6th. It was expected that a record in attendance would be created, as had been done in the case of entries, but disappointment came in the form of the removal of all cattle during Monday night by order of the Board of Agriculture, on account of the outbreak of Foot-and-Mouth Disease. This, of course, made a great many people give up the idea of visiting the Show. Also on the Tuesday rain poured in torrents during the whole of the day, with the result that there was not a single visitor in the Hives and Honey Section and other departments shared a like fate.

Lectures and demonstrations were given each day to crowded audiences. WE APPEND A LIST OF THE awards made by the judges, Mr. WF Reid (London), Mr. G. Hayes (Notts) and Mr. JH Hadfield (Lincolnshire) NBKA members.
Class 519. Twelve 1-lb Sections - 3rd, G. Marshall, Norwell, Newark
Class 521.Twelve 1-lb Jars of Extracted Medium or Dark-coloured Honey - vhc, G. Marshall
Class 522. Twelve 1-lb. Jars of Granulated Honey - 1st, J. Woods, Nettleworth Manor, Mansfield
Class 529. Six Jars of Heather-mixture Extracted Honey - 1st, GH and TS Elliott Southwell

BBJ 18 Jul 1912
Bee Veils, black or white, 8d. each, post free; also net for demonstrating tent.
Harrison, Rockville, Stapleford

BBJ 25 Jul 1912
Nectar-Producing plants and their Pollen - Charlock

BBJ 22 Aug 1912
The Royal Lancashire Show at Preston, though much affected by the weather in point of general attendance, was quite a success from the bee-keepers' outlook. Owing to the outbreak of "Foot-and-Mouth" disease and the absence of cattle from the show, the committee arranged for extra attractions and among them the Lancashire expert, Mr. J. Herrod, was engaged to lecture and demonstrate each day of the show on

bees and bee-keeping and despite the mud and dirt consequent upon heavy rains, he held on each occasion an intensely interested large audience for over an hour.
Twelve 1-lb Jars 1912 Medium Honey (Open) - 1st, T. Manfield, Newark-on-Trent
Twelve 1-lb Jars Granulated Honey (Open) - c, J. Woods, Church Warsop

NEP 22 Aug 1912
Don't Miss This! A New and Unique Exhibition of Bee-keeping. Honey Bees at Work. Manipulation of Live Bees at Intervals. (Without the Danger to the Public), the whole arranged by the NBKA. President: Her Grace the Duchess of Portland. To be held at the Mechanics' Hall, Nottm. Friday and Saturday, Aug.23rd and 24th, 1912. An excellent string band under the Conductorship of Mr RW Liddle, (organist of Southwell Cathedral). Prices of Admission and Saturday, from Opening until 6pm 1s, after 6 6d. , 10 to close. 6d. Children Half-price.

NEP 23 Aug 1912
An important departure has been made this year by the NBKA. Hitherto the organisation has always held its exhibition in conjunction with some other association, but this afternoon the association launched its own venture in the Nottingham Mechanics' Lecture Hall. The results quite justified the move for there were 190 exhibits comprising a thousand bottles of honey, 235 sections, 15 hives of bees, and 20 lbs of wax. In addition there were many exhibits of an educational character, such as the life history of a bee, the constituents of honey, queen-rearing, wax production, etc, while bees were manipulated by experts and competitions were arranged. The Mayor (Edwin Mellor) and Sheriff (HB Halford) were present at the opening ceremony, which was performed by the Mayoress and in the course of the speech-making, it was mentioned by Mr. AG Pugh that in every rural village a ton of honey was wasted annually because there were no bees collect it.

Henry (Harry) Baker Halford, (1867-1935), estate agent and surveyor with premises on St. Peters' Gate lived at the White House, Edwalton.

BBJ 29 Aug 1912
GEH (Sherwood). The section was smashed to pulp through insufficient packing. So far as we can tell from the remains, the honey is from the limes.

BBJ 12 Sep 1912
The NBKA held their Annual Show at the Mechanics' Hall on August 23rd last. A large number of the general public as well as many members of the Association visited the show, which was opened by the Mayoress of Nottingham; the Mayor and Sheriff also being present at the opening ceremony. There were some 190 exhibits staged, the following exhibitors being successful in securing awards:
Collection of Appliances - 1st, TW Harrison and Son, Nottingham

Beginner's Outfit - 2nd, R. Mackender and Son, Newark
Interesting Exhibit - 1st, GH and TS Elliott, Southwell
Honey Trophy - 1st, GH and TS Elliott; 2nd, D. Marshall, Carrington; 3rd, WL Betts, Mansfield Woodhouse
Twelve 1-lb Jars Light Extracted Honey - 1st, J. North, Sutton-in-Ashfield; 2nd, W. Doleman, Keyworth; 3rd, JT Duckmanton, Langwith; 4th, JB Curtis, Carlton-on-Trent
Twelve 1-lb Jars Dark Honey - 1st, G. Marshall, Norwell; 2nd, RH Mackender; 3rd, T. Gillett
Twelve 1-lb Jars Heather-blend Honey - 1st, GH and TS Elliott; 2nd, AG Pugh, Beeston
Twelve 1-lb. Sections - 1st, GE Puttergill, Beeston; 2nd, G. Marshall; 3rd, F. Gillett.
Twelve 1-lb Jars Granulated Honey - 1st, J. Woods, Nettleworth; 2nd, JT Duckmanton; 3rd, G. Marshall
Pair of Shallow Frames - 1st, J. North; 2nd, G. Marshall; 3rd, GH and TS Elliott; 4th, JT Wilson, Shirebrook
Amateur - 1st, CE Smith
Observatory Hive - 1st, G. Marshall; 2nd, D. Marshall; 3rd, WL Betts; 4th, Mrs. Copping, Beeston
Honey Cake - 1st, AG Pugh; 2nd, G. Smithurst, Watnall
Mead - 1st, J. Woods
Honey Vinegar - 1st, W. Doleman
Beeswax - 1st, G. Marshall; 2nd, GE Puttergill; 3rd, J. Woods
Wiring and Fitting-up Frame - 1st, H. Mackender; 2nd, R. Mackender; 3rd, D. Marshall
Judging Honey - 1st, W. Darrington; 2nd, H. Mackender; 3rd, J, Wilson
Folding and Fitting Sections - 1st, H. Mackender; 2nd, D. Marshall; 3rd, W. Doleman

The William Herrod Perpetual Challenge Cup for the member who obtained the highest number of points in their exhibits and this was won by Mr. G. Marshall, of Norwell, with 32 points.

BBJ 12 Sep 1912
Nectar-Producing plants and their Pollen - Melliott

BBJ 26 Sep 1912
The monthly meeting of the BBKA Council was held at 23, Bedford-street, Strand, on September 19th. Mr. WF Reid presided and there was also present AG Pugh.

BBJ 26 Sep 1912
The Twentieth International Exhibition of the Grocery and Allied Trades at the Agricultural Hall was opened on September 21st, and will continue to the 28th inst.
Twelve 1-lb Sections (twenty-two entries) - 5th, T. Marshall, Sutton-on-Trent
Twelve Heather Sections (eight entries) - 3rd, J. Herrod, Sutton-on-Trent

Twelve Jars Light Extracted Honey (fifty-six entries) - 1st, and BBKA certificate, T. Marshall; vhc, J. Herrod; hc, J. North, Sutton-in-Ashfield
Twelve Jars Medium Extracted Honey (thirty-three entries) - 2nd, T. Marshall
Twelve Jars Dark Honey (fourteen entries) - 2nd, T. Marshall
Twelve Jars Heather Honey (ten entries) - 1st, J. Herrod
Twelve Jars Heather Blend Honey (twelve entries) - 2nd, GH and TS Elliott, Southwell; 3rd, G. Hunt, Newark
Twelve Jars Granulated Honey (twenty four entries) - 1st, J. Herrod
Beeswax (Three 1-lb cakes) (twenty-two entries) - 1st, J. Herrod

BBJ 10 Oct 1912
The Thirty-seventh Annual Exhibition of the British Dairy Farmers' Association opened on October 8th at the Agricultural Hall, London and will close on the 11th. The honey and bee appliances make an attractive display.
Twelve 1-lb Jars of Medium-coloured Extracted Honey - 3rd, T. Manfield, Hillside Lodge, Newark
Twelve 1-lb Jars of Dark Extracted Honey (including heather-blends) - 1st, GH and TS Elliot, Southwell

BBJ 17 Oct 1912
The monthly meeting of the BBKA Council was held at the Zoological Gardens, Regent's Park on October 10th. Mr. TW Cowan presided and there were also present AG Pugh and G. Hayes. A letter expressing regret at inability to attend was read from Dr. TS Elliot. A report on the Third Class Examinations, held in Nottingham, was presented and it was resolved to grant certificates.

BBJ 17 Oct 1912
Nectar-Producing plants and their Pollen. - Lucerne

BBJ 17 Oct 1913
Could you kindly give the publishers' names and price of the following books referred to by TW Cowan in his book "The Honey Bee," numbered as follows:
(15) Lord H. Brougham, "Observations, Demonstrations, and Experiments upon the Structure of the Cells of Bees":
(52) JD Haviland, " The Social Instincts of Bees: Their Origin and Natural Selection";
(101) Sir J. Lubbock, "Ants, Bees, Wasps";
(102) "The Senses, Instincts and Intelligence of Animals"
(167) GR Waterhouse, "On the Formation of the Cells of Bees and Wasps";
(171) J. Wyman, "Notes on the Cells of the Bee"? Interested, Pinxton.
Reply. (No. 15) This has long since been out of print and can only be obtained from a second-hand bookseller. It might be consulted in the British Museum.
(No. 62) Out of print, but can be seen in the library of the BBKA

(Nos. 101 and 102) Published by Kegan Paul and Co., London; price 10s
(No. 167) Published in "Transactions of the Entomological Society of London," 1864, Vol. II., 3rd Series, Part II. Apply to Secretary, 11, Chandos-street, Cavendish Square, London, W.
(No. 171) "Proceedings of the American Academy of Sciences and Arts," Vol. VII, January 9, 1866. Published in Cambridge, Mass. Probably out of print, but this and all such pamphlets are occasionally to be picked up through second-hand booksellers. There are several such on the Continent who make a speciality of these pamphlets. In this country they may be had sometimes from W. Wesley and Son. 25 Sussex Street, Strand, London, who would send a catalogue on application.

Sir John Lubbock (1834-1913) "Ants, Bees, and Wasps" was published in 1882. Based on painstaking research and a thorough acquaintance with previous investigations and written in a clear and attractive style with an abundance of interesting anecdotes and curious information, this is a book which appeals both to the scientist and to the general reader. The author had kept numerous colonies of ants under continuous observation and made some important experiments. An electronic copy of this book can be found in the NBKA Library.

Caption to the above cartoon:
How both the banking busy bee
Improve his shining hours?
By studying on Bank Holidays
Strange insects and wild flowers

BBJ 24 Oct 1912
For many years past it has been the practice of the BBKA to hold a conversazione on the Thursday in "Dairy Show" week thus enabling country members to take advantage of the cheap railway tickets and attend the meeting. This precedent was again followed this year and the large number of members and friends who attended on October 10th (there being over 150 present) indicates very clearly that the Association continues to advance in popularity. Amongst those present was G Hayes

BBJ 14 Nov 1912
AW (Notts). The symptoms indicate "Isle of Wight" disease, from which we are of opinion your bees are suffering.

BBJ 28 Nov 1912
The monthly meeting of the BBKA Council was held at 23, Bedford-street, Strand, on November 21st, 1912. Mr. WF Reid presided for a portion of the meeting and upon his

leaving, Mr. AG Pugh was voted to the chair. The report of the Exhibition Committee was presented and the schedule for "Royal" Show at Bristol was passed, with the inclusion of classes for members of Somersetshire Association only. Amongst the reserve judges appointed was Mr. AG Pugh.

BBJ 12 Dec 1912
OES (Notts.). The sample is nicely blended and should stand a good chance in its class at a local show.

BBJ 19 Dec 1912
Nectar-Producing plants and their Pollen - Raspberry

BBJ 26 Dec 1912
The monthly meeting of the BBKA Council was held at 23, Bedford-street, Strand, London, WC. on December 19th, 1912. Mr. WF Reid presided for a portion of the meeting and upon his leaving on account of another important engagement, Mr. JB Lamb was voted to the chair. There was also present AG Pugh.

BBJ 23 Jan 1913
What is the best and quickest remedy for a sting in the mouth or throat, or any part likely to cause suffocation through swelling of air passages? Some recommend hot water, as it causes better circulation and the spread of the poison over the whole system.   Enquirer, Mansfield.
Reply. The person stung should be given a strong dose of *sal volatile*, also a purgative. The actual part stung should be fomented with very hot water to which a little vinegar has been added.

BBJ 13 Feb 1913
WYZ (Notts). Stock Found Dead in well-stored Hive.
1. The bees have died from "Isle of Wight" disease.
2. Do not use the food elsewhere; burn it.
3. Ordinary British bees.
4. The sample of honey is good and certainly worth showing. You have not complied with our rules in omitting to send full name and address. Please do this in future, or we cannot answer your queries.

BBJ 27 Feb 1913
Enquirer. (Mansfield)
1. Both lots have "Isle of Wight" disease.
2. One lot are blacks, the other hybrid Italians.
3. Yes, it is possible.

BBJ 6 Mar 1913

BEEKEEPING BETWEEN TWO QUEENS

In May of last year I purchased a prime swarm on ten frames. Early in July I was surprised to find a swarm on my garden hedge, but as there are several colonies of wild bees in the vicinity I concluded it had come from one of these, and hived it. Both lots appeared quite well through the summer.

I fed the original hive a little before packing down for winter, and have looked inside several times since, the last occasion being about a fortnight ago. On opening it last week, I was amazed to find the bees gone. There were about a dozen dead bees on the floor and the frames contained a fair amount of stores. They had no objectionable smell, but contained a good deal of the bees' cleansings. Did the swarm issue from my own hive, and have they again "joined forces?" WSH, Notts.
Reply. It is quite probable that it was your own bees which swarmed. The virgin was probably lost in mating, and, of course, the stock dwindled.

BBJ 13 Mar 1913
Some years ago I asked an old Worcestershire bee-keeper what was his belief on the subject of "Telling the Bees," and he replied that there is a great amount of commonsense in the old superstition; because the people who took enough interest in the bees to tell them of their master's death generally looked after them in other ways, and naturally the bees benefited. On the contrary, if no one took the trouble to "tell the bees" no one took any other trouble with them, with the result that sooner or later they died out.

This simple explanation (which I have never seen in print) enabled me to understand why this particular custom was so widely and implicitly believed in, and why it is dying so hard. Beeston

BBJ 13 Mar 1913
I understood Mr. Herrod to say at his lecture at Nottingham on Saturday last that the best queen-cells were those of good shape and deeply indented, while those cells containing drones, reared as queens, were perfectly smooth. I can bear him out as to cells with deep indentations producing good queens, but how do you account for the following: That two cells of equal size, shape, etc., produce two queens of equal merit. One cell is reared on a new comb, and is but slightly indented, while the other, reared on an old comb, is deeply indented? WAH Alfreton.
Reply: The condition you mention is due to the age of the comb. In a new comb the indentations will not be so deep, but the queens will be of equal merit.

Lincolnshire Echo 17 Mar 1913
Schoolmasters and people the Sleaford district will hear with deep regret of the death in his 57th year, at Ewerby, of Mr. FHK Fisher, which occurred Sunday afternoon. For upwards of 20 years Mr. Fisher has been headmaster of the village school. He was an

exceptionally energetic man, for apart from his scholastic duties he held several other positions, among those being Postmaster and organist. He an expert apiarist, being a prominent member of the Beekeepers' Association and a well-known lecturer. He was a great lover also his garden.

BBJ 20 Mar 1913
The AGM of the NBKA was held in the Peoples' Hall, Nottingham on March 1st. which was attended by a large number of members, the Mayor of Nottingham presiding.

After the reading of the minutes of the previous meeting the Balance Sheet was considered and passed as a most satisfactory one, although it showed a deficit. The committee's report referred to the large amount of work that had been done. Technical Instruction had been given at various shows, and the lecturer's report that large audiences were the rule, and that many enquiries were made at, and between the lectures; that 218 apiaries had been visited, the total number of stocks examined being 889, of which twenty-six were found to be diseased. A case of "Isle of Wight" disease had been reported, and all members were urged to do everything in their power to prevent its spreading. Exhibitions of honeybees, etc., have been held in connection with a good number of horticultural shows, in various parts of the country.

It was unanimously decided to hold the exhibition of honey, etc, at Nottingham again this year on the same lines as the last, which was so successful. The following officers were elected:
President, Her Grace the Duchess of Portland; committee, Dr. Elliott, Messrs. EJ Turner, W Adams, TN Harrison, GE Puttergill, AG Pugh, G Smethurst, GE Skelhorne, FG Vessev, G. White, and SC Hartston and MH Fox; hon. auditor, Mr. J Bickley Chilwell; hon. secretary, Mr. G Hayes; representatives to the BBKA, Messrs. Hayes and Pugh.

A vote of thanks to the Mayor concluded the business for the afternoon, and the company, then numbering about 140, sat down to tea. The meeting was resumed at 6pm, when a lecture was given by Mr. W. Herrod on "Queen-rearing and Introduction," to a still larger audience. The lecture was most lucid, instructive and interesting, and was very greatly appreciated by both the old and young in the craft, and at its conclusion a hearty vote of thanks was passed to Mr. Herrod, and also to the BBKA Council for sending the lecturer down.

The distribution of prizes, certificates, etc., and other matters, followed, the meeting concluding with a prize drawing for all who had paid their subscriptions.

G. Hayes, hon. sec.

BBJ 3 Apr 1913
The AGM of BBKA members was held in the Lecture Hall of the Zoological Society

of London, Regent's Park, London, on March 27th, 1913. Mr. TW Cowan presided.

The attendance was excellent, the facilities afforded by the meeting being held during the holidays was taken advantage of by many living at a distance, members from Sheffield, Doncaster, Nottingham, and Ipswich being present, while a large number of letters expressing regret at inability to attend were received.

Mr. Hayes proposed, and Mr. Bocock seconded, a very hearty vote of thanks to the retiring Council and officers. This was carried unanimously.

The monthly meeting of the BBKA council was held immediately after the Annual Meeting. Mr. TW Cowan presided, and among those present was AG Pugh who was elected as a member of the Exhibition Committee.

BBJ 3 Apr 1913
It is with extreme regret that I have to record the death of Mr. FHK Fisher (of Ewerby), a prominent Lincolnshire bee-keeper, who died on March 16th, after a short illness, at the age of fifty seven. Before coming to live in Lincolnshire, Mr. Fisher acted as Hon. Secretary of the NBKA, and for his services to that Association he was made an honorary Life Member. He took a prominent part in the work of the Lincolnshire Association. He was an active member of the committee and did useful work as lecturer, expert, and judge.   JH Hadfield, Hon. Sec, Lincs. BKA
[The junior Editor, as a personal friend of the late Mr. F. Fisher, extends his heartfelt sympathy to the family and friends of the deceased gentleman. The news came as a shock, for this sad loss removes a dear and valued friend. One by one the links with the past are broken, as the "Great Architect of the Universe" calls his labourers to eternal rest. During his busy life, Mr. Fisher has done much for bee-keeping, and there are many living to-day who had their first lesson and their interest first aroused through his genial talks about bees and bee-keeping.]

BBJ 24 Apr 1913
The monthly meeting of the BBKA Council was held at 23, Bedford-street, Strand, London on April 17th. 1913. Mr. TW Cowan presided for a portion of the time, and upon his leaving Mr. JB Lamb was voted to the chair. Letters expressing regret at inability to attend were read from Messrs. AG Pugh and G Hayes,

Nomination of representatives from affiliated associations were received and accepted (Nottinghamshire) Mr. G. Hayes. Application for third-class examinations was received from the NBKA. A letter of thanks for lectures were read from the NBKA.

BBJ 24 Apr 1913

It has been suggested that for the purpose of popularising honey we should have at our exhibition a tasting of all or at least some means of giving to the visitors just a taste of honey. I have been requested to enquire if those of your readers who have had any experience in this line - as I daresay some may have - will give us an outline of what they consider the best manner of doing this, and also what to avoid.

Geo. Hayes, Secretary, NBKA

NEP 25 Apr 1913

This time of the year is critical for the young brood, and great care should be taken to stop all draughts of cold air from entering the hive. All hives should be thoroughly examined externally, and cracks stopped up. The beekeeper should also see that the floor boards are quite sound and fit well to the bottom of the hive, that the plinths and ledges round the outside of the hive have not become loose, and where they have should make them quite secure.

Carefully looking after these apparently trivial details may be the means of saving many pounds and using bees to substantial profit. The worst enemy the modern beekeeper has to contend with is "Foul Brood". Nothing will kill the brood sooner than a draught of cold air. Therefore, it is very important that the hives should be made quite air and water tight. Once the young brood when in larva state and before it is sealed gets chilled it dies and liable to contaminate the whole of the stocks in the apiary. Do not take the bar frames out of the hive to examine them for mere curiosity. That involves great risks. Should it necessary to examine them, select a hot, still day.   I am. Sir. etc. Chas. F. Brearley. 13, Annarth-terrace. Kinglake-street.

NEP 30 Apr 1913

The close connection between beekeeping and fruit-growing was demonstrated by Mr. Jno. Thomson lecturing in Nottingham last night. With the increase which there will be in the number of smallholdings and allotments in the near future, he said that we should become a greater fruit-growing nation than in the past and he explained that fruit growing could not be successfully carried on without beekeeping, the two being inseparable.

Bee-Keepers' Record May 1913

At our meeting on the 5th inst., it was decided by the NBKA Committee to send the bee tent with lecturers to the following shows:

Newark May 14th and 16th
Notts. Agricultural Show Nottingham June 4th and 5th
Welbeck August 4th

The lecturers will be glad to meet both old and new friends at these rendezvous. It as also arranged to hold Honey Shows in connection with the Horticultural Societies at Arnold, Beeston, Southwell, Farnsfield, and East Bridgford, the schedules for which will be sent out later.

A committee was appointed to arrange for the next large exhibition, which is to be held in Nottingham in September.

It is with deep regret that we heard of the death of our late secretary, Mr FHK Fisher, and we extend our sincere sympathy to those he has left, in their bereavement.

The season has not yet opened out with us. Although it is dry, cold north-east winds prevail, and bees cannot get abroad very much. The blossoms of the pears and plums are ready for bursting forth as soon as genial weather prevails, and the same condition appears to be present within each hive. Geo. Hayes. Hon Sec. NBKA

BBJ 8 May 1913
GFB (Mansfield). You must not confine the bees to the hive or you will kill them. As they have not flown much this season very little harm will be done if they are moved straight to the new location. They should be moved without delay, the work being done at night.

BBJ 26 Jun 1913
The monthly meeting of the BBKA Council was held at 23, Bedford-street, Strand, London on June 19th, 1913. Mr. TW Cowan presided. There was also present Dr. TS Elliot. A letter expressing regret at inability to attend was read from AG Pugh.

BBJ 7 Jul 1913
BPE (Nottingham). Earwigs in Hives. To get rid of these, use powdered naphthaline under the lugs of the frames.

BBJ 7 Jul 1913
Shows to Come. September 12th and 13th, at Nottingham. Grand Exhibition of Appliances, Honey. Beeswax, collections of objects of interest and instruction. Demonstrations, etc. to be held in the Mechanics' Hall, Nottingham. Open classes, with liberal prizes for appliances, extracted honey, sections, fitting-up frames, fitting-up sections, judging competition, etc. Schedules from G. Hayes, Mona-street, Beeston.

BBJ 10 July 1913
The "Royal Show" at Bristol. An ideal place and glorious weather, the natural result being a highly successful show this year. Our interest, of course, centres where a bold advertisement tells us are to be found the "Hives and Honey." Other departments all seem crowded, but it seems that on this occasion we have a larger attendance of visitors interested in bee-keeping than usual, the inquiries are more numerous and intelligent; those who ask seemed to have previously acquainted themselves a little

with our industry. Evidently we are progressing. The man in the street wants to know more of bees and how they get their honey.
Class 540 Twelve 1-lb Jars of Granulated Honey - 1st, J. Woods, Nettleworth Manor, Mansfield; 3rd, T. Marshall, Ivy Cottage, Sutton-on-Trent
Class 547 Six Jars of heather mixture Extracted Honey - 2nd, CE Smith, Grayfield Place, Sutton-in-Ashfield

BBJ 24 July 1913
Sections Wanted. One gross, well filled clover honey; immediately. Send lowest price to Elliot, Old Rectory, Southwell, Notts.

BBJ 31 July 1913
The Annual Show of honey, hives, appliances, etc., in connection with the Lincolnshire Agricultural Society, was held at Lincoln on July 17th and 18th. This department was under the management of the Lincs. BKA. There was a good entry and the honey exhibits were of excellent quality. The judges were Dr. Percy Sharp, Mr. J Emerson, and Mr. George Hayes (Notts.).

NEP 31 Jul 1913
This is the season of the year when wasps are most numerous. They do an enormous amount of damage to the ripening fruit, and also kill thousands of bees. I used to give one farthing for each wasp captured in the months of April and May when I lived in the country and kept bees. I think it be very good plan for the fruit growers and beekeepers to combine and do the same for, as is well known, every wasp captured in those two months means the destruction or prevention of a nest. I have tried many ways of taking wasp nests, including potassium-cyanide, but I find turpentine the simplest and most satisfactory as it effectually destroys all the wasps and the same time does not injure the grubs.

Provide yourself with ordinary wine bottle with a neck about four inches long and put two tablespoonfuls of turpentine into it, then place the bottle neck right into the hole where the wasps come out. Take care that the contents do not run out of the bottle; this can be prevented by slanting it a little. The best time it is in the evening when all the wasps are inside. If there are more holes than one which the wasps use cover them up with soil. Also take care to place soil round the bottle neck to prevent the fumes from the turpentine escaping. The next morning you may take a spade and dig out the nest with safety, for all the wasps will be dead. You can then sweep them off the combs, which then may be used or sold. Cover all up again, and you will not see any more wasps there.   I am, sir, CF Brearley. 13, Annarth-terrace, Kinglake-street

BBJ 7 Aug 1913
JC Death (Notts.) We are sorry to say it is "Isle of Wight" disease.

BEEKEEPING BETWEEN TWO QUEENS

BBJ 14 Aug 1913
GW (Mansfield). The sample has no aroma, and absolutely no flavour, therefore, without analysing for pollen grains we are unable to tell its source.

BBJ 28 Aug 1913
Sale or Exchange; Euphonium B-flat, Higham, Manchester, splendid instrument.
                                                                        Bandmaster, Southwell

Bee-keepers' Record August 1913
The season in Nottinghamshire, so far as my knowledge extends, has been a fairly good one for honey, especially considering the backward state of the majority of stocks in the spring. In those parts of the county which were favoured with showers in June and July, the bees have done exceptionally well, but immediately round about Nottingham in a radius of eight to ten miles we have only had four very slight showers since May 18th and in consequence the clover was soon over, and the lime blooms of short duration. The honey gathered is mostly light and of good consistency; but I found during the last few days of July the bees began to gather honeydew.

The bee-tent was at Welbeck Agricultural Show on August 4th and the lectures throughout the day were well attended, good audiences being present at all the demonstrations, which left the visitors very little time to see the other exhibits and attractions of the show. At one of the lectures we were favoured with the presence of HRH Prince Arthur of Connaught, His Grace the Duke of Portland and Lord Kitchener, and they all appeared very interested.

Prince Arthur was appointed in 1911 as Governor General of Canada. He was succeeded by the Duke of Devonshire in 1916.

Show held at East Bridgford on July 1st. Judge Mr HJ Turner, who made the following awards:
Six 1-lb sections of comb honey in Frames – Equal 1st Mrs Cartwright and D Gower, 3rd RB Plowright
Six 1-lb Jars of Extracted Honey – 1st W Doleman, 2nd J North, 3rd J Bond
Specimen of Bees of any race in an Observatory Hive – 1st J Higgs, 2nd D Gower

This was also recorded in the East Bridgford parish magazine for August but not in such detail. It was the 50th annual Horticultural Show held in the village.

Southwell Show. Judge Mr HJ Turner, whose awards were as follows:
Six jars of Granulated Honey – 1st AG Pugh; 2nd JT Duckmanton; 3rd WG Lucas
Six 1-lb Sections – 1st D Marshall; 2nd D Marshall; 3rd Dr TS Elliot
Six 1-lb Jars of Extracted Honey – 1st WG Lucas; 2nd AG Pugh; 3rd Dr TS Elliot

Specimen of Bees of any race in an Observatory Hive – 1st W Mountney; 2nd G Marshall
Six 1-lb Jars of Extracted Honey – 1st CW Chappell; 2nd Spray; 3rd WH Mellors
Single 1-lb Jar – WG Lucas

Show at Arnold on July 26th and 28th. Judge Mr G Hayes. Awards were as follows:
Six 1-lb Sections – 1st J North; 2nd D Marshall; 3rd Dr TS Elliot
Six 1-lb Jars of Extracted – 1st D Marshall; 2nd Dr TS Elliot; 3rd A Atherley

Honey Show at Beeston, August 4th. Judge Dr TS Elliot. Awards:
Six 1-lb Sections – 1st D Marshall; 2nd J North; 3rd AG Pugh
Six 1-lb Jars of Extracted Honey – 1st AG Pugh; 2nd D Marshall; 3rd WG Lucas
Specimen of Bees of any race in an Observatory Hive – 1st SR Dawes; 2nd AG Pugh; 3rd D Marshall
Six 1lb Jars of Extracted Honey – 1st A Riley; 2nd AG Pugh

Derby Telegraph 28 Aug 1913
Derbyshire Agricultural Show.
Open classes
For the best 12 1-lb jars of run honey – 4th AG Pugh, Notts.

Don't Miss This! ! ! The Second Beekeeping Exhibition, A most interesting and instructive collection for all connected with the craft. Instructive Competitions. Honey Bees at Work. Manipulation of Live Bees at intervals (without danger to the public), the whole arranged by the Notts. Beekeepers' Association. President; Her Grace the Duchess of Portland. To be held in the Mechanics New Lecture Hall, Nottingham, Friday and Saturday, September 12th and 13th, 1913. An Excellent String Band. Prices of Admission: Friday and Saturday, from Opening until 6 pm, 1/-: after 6pm 6d.

NEP 12 Sep 1913
A high standard of excellence characterised the second exhibition held by the NBKA in the Nottingham Mechanics' Lecture Hall, this afternoon, when the Mayor (Councillor Thomas Ward) performed the opening ceremony. There were a number of competitions, for which Messrs. W. Herrod of London and AG Pugh, of Beeston, members of the BBKA, were the judges. The awards were;
Appliances – 1st, Messrs. TW Harrison and Son
Beginner's Outfit - Messrs TW Harrison and Son
Trophies – 1st, G. and TS Elliot, Southwell; 2nd, G Marshall, Norwell; 3rd, D Marshall, Carrington
Light Extracted Honey – 1st, WH Mellors, Norton; 2nd, W. Lee, Southwell; 3rd, G. Marshall; 4th, D. Marshall

Dark Extracted Honey -1st, D. Marshall; 2nd, H. Merryweather, Southwell; 3rd, GF Stubbs
Heather Honey - 1st, J. North, Sutton-in-Ashfield; 2nd, WG Lucas, Southwell
Section of Comb Honey - 1st, J. North; 2nd, D Marshall; 3rd, G. Marshall; 4th, D. Maher, Cropwell
Granulated - 1st, GH and TS Elliot; 2nd, J. Woods, Warsop; 3rd, JT Duckmanton
Shallow Frames - 1st, Mrs. Fidler, Hucknall; 2nd, J Woods, Warsop; 3rd, J. North
Amateur - 1st, J. Parkins, Sutton; 2nd, Mrs Fidler
Bees – 1st, D Marshall; 2nd, GF Stubbs; 3rd, G Marshall; 4th, Mrs Copping, Beeston
Cake – 1st, Mrs Riley, Beeston; 2nd, J North; 3rd, Mrs Copping
Wax – 1st, W Darrington, Eastwood; 2nd. WG Lucas; 3rd, Geo F Stubbs

Beekeepers' Record September 1913
Our second Annual Exhibition is to be held at Nottingham on September 12th and 13th, and bids fair to excel the previous one in many ways. We hope our members will do everything in their power to make it a success.

I am wanting an old 'Neighbours Cottager's' and a 'Buncefield' hive to complete an exhibit I am preparing. If anyone has these and would either give or loan me one, I shall be very glad. They should, of course, be without bees.

Geo. Hayes, Hon Sec.

BBJ 25 Sep 1913
The monthly meeting of the BBKA Council was held at 23, Bedford-street, Strand, London on September 18th, 1913. Mr. WF Reid presided. Among those present was AG Pugh. A report on the Preliminary examination held at Beeston was read.

BBJ 25 Sep 1913
The twenty-first Annual Exhibition of the Grocers and Allied Trades was opened at the Royal Agricultural Hall, Islington on Saturdaty, September 20th, and will close on the 27th inst.
Twelve 1-lb Jars Heather Blend (twelve entries) - 1st, AG Pugh, Beeston; 3rd, T. Marshall
Twelve 1-lb Jars Light-coloured Extracted Honey (sixty-one entries) – 1st and Certificate of Merit, J. Herrod, Sutton-on-Trent
Twelve 1-lb Jars Medium-coloured Extracted Honey (thirty-two entries) - 2nd, T. Marshall, Sutton-on-Trent
Twelve 1-lb Jars Dark Extracted Honey (eleven entries) - 1st, T. Marshall
Twelve 1-lb Jars Granulated Honey (twenty-one entries) - 1st, J. Herrod;
Extracted Honey in 1-lb Jars - Certificate of merit to JT Duckmanton, Langwith

BBJ 25 Sep 1913
WM (Mansfield) The honey is mainly from limes, but has an admixture of honeydew,

which causes the dark appearance and has not improved the flavour.

BBJ 9 Oct 1913
The Annual Show of the Cumberland and Westmorland BKA was held in conjunction with the Northern Counties Fruit Congress and Show in the Market Hall, Kendal, on September 34th and 20th.
One 1-lb Jar Extracted Honey - 2nd, Arthur G. Pugh, Beeston

BBJ 16 Oct 1913
The autumn conversazione of the BBKA, to be held at the Lecture Hall, Zoological Gardens, in Dairy Show week, on October 23rd, at 5pm promises to be a very successful and interesting one. Though Mr. TW Cowan has, owing to an important engagement, been obliged to defer giving the conclusion of his lecture on "Bee-keeping in Other Countries" until another occasion, Mr. AG Pugh has kindly consented to fill his place and read a paper on "Judging," a subject which is arousing a good deal of interest just now.

Mr. Chairman, Ladies and Gentlemen,
The subject of "Judging Honey," selected by the Council for our consideration this evening is, I think we all agree, a most important one. It has, however, the disadvantage, from my point of view as speaker, of having recently been so fully dealt with by Mr. Wm. Herrod in his book, entitled "Producing, Preparing, Exhibiting, and Judging Bee Produce," with which most of my hearers will be acquainted. There is, therefore, little possibility of many new ideas being propounded by me, especially when it is borne in mind that Mr. Herrod and I have worked together for so many years. Our views and opinions naturally run largely in the same direction.

An electronic version of William Herrod's book can be found in the NBKA library.

It is rather a strange coincidence that so many communications have appeared in the "BBJ" bearing upon this subject since it was selected as a topic for discussion at this meeting, so there is evidently a desire abroad that the question of judges and judging, as applied to our craft, should be kept well to the front.

In dealing with this subject, it seems to fall naturally under three heads: First, the judge; second, the article to be judged; third, the method of judging, etc.

Firstly, then, we have to consider the judge. What special equalities is it desirable that he should possess? Primarily, it is necessary that he should have experience. We are all familiar with the complaint that at small horticultural and similar shows judges of flowers, butter, etc., are frequently asked to undertake the duty of judging the honey exhibits and whilst being, no doubt, excellent fellows in their respective spheres, they sometimes have no idea as to what is required to obtain the highest points in the honey classes.

At small local shows this is a real difficulty, and can be best met by the local Bee-Keepers' Association (which is, presumably, granting a portion of the prize-money) making it a *sine qua non* that prizes will only be awarded upon the recommendation of a judge approved by the Association. Having agreed, then, that an experienced judge is necessary, it is desirable to consider how such experience can be acquired. Personally, I think this is the weak point in present-day conditions.

Sufficient opportunities are not always available for training the requisite number of judges. It seems to be generally conceded that a first or second class expert must be a competent person: but whilst this is often quite true, it is not always the case. The fact must be borne in mind that the examination for expert certificates makes little, if any, test of a person's competency in this direction.

At a recent county exhibition a honey judging class was inaugurated, prizes being offered to the competitors who placed six different samples of honey in their respective order of merit, and named the probable source from which each had been gathered. This competition proved very popular, and it was noticeable that some who were anxious to gain a little knowledge - not being sufficiently experienced to hope to get a prize – willingly paid their entrance fees to obtain the opportunity of becoming acquainted with the respective flavour, etc., of honey in competition.

The more noticeable feature, however, was the diversity of opinion expressed as to the merits of the honey and as to the sources from which it had emanated. The results of this competition proved to my mind that a competition on these lines, say with eight or ten well-defined samples of honey, the reputed source of which had previously been proved by microscopical examination to be held at a "Royal" or other important show, would be most helpful; and it is even desirable that a judging certificate should be granted to successful competitors in an examination on these lines.

The Grocers' Exhibition gives us a good lead in this direction by awarding valuable prizes in the competitions for tea and coffee blending and sampling. It is sometimes rather annoying to those who exhibit exceptionally fine extracted honey to find that the number of critics and would-be judges have taken a somewhat heavy toll of the honey by taking free samples, and secretaries and stewards are blamed for allowing this to take place: but as a great sinner in this respect I feel very lenient to those who in turn take their revenge out on my exhibits, and personally I do not think a competitor who has won a prize should be too hard upon such a practice, at any rate until the suggested honey-judging contests become a *fait accompli*, because at the present time practically the only way in which anyone can ascertain what kind of honey Judge So-and-so considers worthy of a first prize is by means of this surreptitious sampling.

A good way to get some practice in the art of judging is to be appointed assistant judge, and many of our present-day judges look back with pleasure to the time they were understudies to some of our departed veterans such as Messrs. Broughton Carr, Weston, Hooker, and others.

Where a show has a reasonable number of honey classes it is often advisable to appoint two judges, for, whilst the junior is gaining experience, the senior is often glad of a colleague's opinion when exhibits are nearly equal on all points, and the system of dual judges commends itself to all concerned.

Of course, it is most desirable that a person should be in a good state of health, with all his faculties well developed, at the time he undertakes the duty of judging honey and in addition to the five accredited senses, a sixth, the due sense of proportion, is a most valuable one, as different localities and seasons have their own peculiar effect upon the honey, and acute discrimination is often required. Fads, fancies, and idiosyncrasies are all to be avoided and a just and impartial decision, without fear or favour, should be unhesitatingly given when it has been carefully, patiently and conscientiously arrived at. Finally, I am of opinion that the best all-round judge will usually be found in a really good prize-winning exhibitor.

Having dealt with the personality, training and experience of the judge as fully as my limited time permits, I will now proceed to the consideration of the second point: 'The article to be judged.' Now, although at a fairly large show the judge will have to adjudicate such other classes as bee-keepers' appliances, bees in observatory hives, beeswax, honey vinegar, honey cake, mead, etc. our title and my instructions confine us solely to honey. If, so we will leave these interesting items to a more convenient time.

It has already been suggested that nothing but actual practice and experience will enable a person to fully appreciate and gauge the subtle, unique, delicate flavour and aroma that is so characteristic of good British honey. Such characteristics must, however, be present in all honey that has a chance of winning a prize at a large show and it will be well to deal with these and other points when considering honey in the various forms in which it will be submitted to the judge's inspection, ie. in the following five forms, in which it usually appears on the schedule: (1) Run or extracted honey; (2) Granulated honey; (3) Section honey; (4) Honey in frames; (5) Honey trophies.

Our first form, then, is run or extracted honey in its liquid form and here we usually have the keenest competition. The fact that honey is a food product naturally places flavour as the most important point in its consideration, for no matter how good or attractive it may be in other respects, failure in this, possibly prohibiting its use as food, makes it valueless. For instance, in a large single-bottle class, recently judged by myself, the exhibit that would otherwise have easily been first had to be relegated to obscurity owing to the fact that it had been so strongly impregnated with carbolic acid (no doubt caused by some carelessness of the exhibitor) that the flavour was obliterated and it was rendered useless for food.

Fortunately for judges and competitors alike, the result of years of experience brought out the standard colour scheme with which, I daresay, most of those present are familiar. It is, however, to be regretted that some exhibitors are still so careless that one frequently finds honey that would secure a prize in its proper class shown in a wrong one and consequently disqualified. The purchase and use of a set of colour glasses at a cost of a few pence would be a good investment for such competitors and prevent much unnecessary annoyance and expense.

I have sometimes thought that a still further sub-division of the huge light honey classes at the Royal, Grocers', and Dairy shows would be an advantage, as there are usually numerous exhibits at these shows of that special product usually termed "Water White" honey. This colour, or rather want of colour, has sometimes caused a little dissension amongst both exhibitors and judges and is somewhat difficult to deal with and it should be given a class to itself.

In those districts where heather honey is in evidence, upon no consideration should such honey be allowed to compete in the dark honey classes. It is far better to divide the prize-money so that each variety has its own class and stands upon its own merits. Again, heather honey and heather blends should not compete against each other, each having its own characteristics.

The class for heather blend honey is one upon which a variety of opinions are held. Those who prefer the heather flavour naturally liking a blend in which the strong flavour predominates, whilst others have an inclination towards a good clover honey, with just that "dash" of heather which to some palates produces the flavour which surpasses all others. Following close upon flavour and colour comes the question of density or consistency.

This is a most important point because thin honey is generally in an unripe condition and will frequently set up fermentation in a short time. It has poor keeping qualities and may soon become useless. I have often wondered whether it would not be possible to obtain some kind of instrument which would register the actual specific gravity upon being immersed in the honey. If such an instrument could be obtained at a reasonable price, and was easy to use, it would be of great assistance when so many samples are so nearly alike. Care must always be taken in making a just allowance for the necessarily high temperature often experienced in the show tent during the heat of summer, as this has a great effect upon the exhibits.

Aroma, although not so important as the other points already mentioned, adds a great charm to a first-class honey. It is, however, very elusive and often disappears to a great extent if the caps of the bottles are removed for any length of time.

In judging the general appearance, or "get up" as it is usually termed, of an exhibit in this class it is always considered a great defect and would almost invariably debar prize-winning for the bottles in an exhibit to vary in size or filling. It must, however, not be overlooked that the colour of the glass is sometimes of a distinctly objectionable green colour. Of course, it is easy to say the exhibitor should not buy or show such bottles, but sometimes he has little choice and as it is the honey and not the bottles, that is under consideration, perhaps it is as well for the judge to exercise leniency in such cases.

As we are taught "Cleanliness is next to godliness" an exhibit showing any impurity in, or on, the bottle, wad or cap must be passed over and, unless the honey is perfectly bright and clear, it should never be found amongst the prize-winners. Even the slightest sign of granulation is most objectionable in the liquid honey classes. Any extracted honey having froth or scum upon its surface, or containing particles of wax. etc., shows careless or hurried preparation and, no matter how good in other respects, will be passed over by a good judge. The particles may, however, sometimes be so small that good eyesight and good light are necessary to discover them.

In regard to granulated honey, it is worthy of mention that good, even granulation being always considered a proof of the purity of honey, this is a class that should commend itself to our Show Committees, as the mind of the public is not always clear upon the point. In judging this class, the points enumerated in respect to liquid honey will apply. The grain of granulation is an important point. A rough, coarse grain is not to be commended, whilst the very smooth or "greasy" grain is equally objectionable. The whole exhibit should, naturally, be uniform in colour, whilst this should not be too chalky. The air spaces often seen in this class, although most difficult to avoid, detract from its appearance and must be taken into account and care must be taken to see that no signs of fermentation are present on the surface of the honey.

Our third form of honey exhibits is that known as "sections," and here both the exhibitor's and judge's abilities will be put to the test, it being generally admitted that the production of "A1" sections brings out the bee-keeper's abilities to their greatest extent. Whilst the judge must take care, by careful inspection and keen observation, that those abilities have not taken the wrong direction in faking or trying to improve upon nature in an illegitimate manner by covering over any defect in sealing, etc. or in feeding syrup to the bees to enable them to keep up high pressure in comb building.

To prevent anything of the latter kind passing undetected, the judge should always insist upon sampling the contents of those sections to which it is proposed to award prizes. Some persons resent breaking open the glazing, etc. for this purpose, saying it is not necessary but my own experience at a large show, where sections which had been awarded a first prize were afterwards found quite unfit for food, proves to me the absolute necessity for this precaution. Uniformity throughout the whole exhibit, together with a happy medium of sealing - that is to say, not too heavy to hide formation of cells, or too thin to properly preserve the contents - is desirable. The honey should not touch the capping and thus giving it a weeping or watery appearance, and when held up to the light the comb should be sufficiently transparent to show that it does not contain an objectionable thick mid-rib.

The colour of the capping is a contentious matter in this class, for whilst a nice clear white is looked upon as the ideal, the pleasant light yellow of sainfoin honey sections has a very attractive appearance. Here, again, I would recommend, whenever possible, especially at large shows, that there be a special class for sainfoin sections. Clean, well-filled sections, without "popholes", empty or unsealed cells, with a good attachment all round are the ideals to be aimed at and, in the get-up of sections for exhibition, the careful and expert beekeeper has a chance of showing his abilities, for whilst the appliance dealers have produced some very useful section holders, in both cardboard and enamelled tin, I think all will agree that the sections glazed by the exhibitor, with a neat lace edging, such as it has been our privilege to see at some of our shows, is a very effective proof of their taste and ability and should receive due consideration in the award of prizes.

Unfortunately, whilst the lace edging of section cases improves the general appearance, it can also be used to cover up defective attachment to the sides, etc. and it has been found necessary to limit the width allowed, and a section showing less than $3\frac{1}{2}$ in. square of the comb, clear of the edging, should now never be allowed to take a prize. Although this is sometimes looked upon as a vexatious restriction, all concerned must be careful to see that it is adhered to strictly. In a general way drone cells seem to give the best appearance to comb honey but this should have little, if any, effect upon the awards.

Our fourth form of honey in competition is that of comb honey, shown in shallow or other frames. The idea in this class is to show that which may be considered the best means of producing good commercial honey in bulk and, as such, combs are not intended to come to the table as in the case of sections. It is necessary to judge it from a different standpoint and what is of the chief importance is that an even and full surface is presented so that the uncapping knife may remove all the capping from one side at a sweep of the blade.

The honey should, of course, be well ripened and in a good condition for extracting but the actual quality, except that it must be of a good commercial grade, is not of so much importance as in the other classes. The exhibits with the least number of pop-holes and best attachment to the frames will usually take a high position in this class.

Our final form of exhibition honey, that which consists of honey trophies, is no doubt the most attractive and at the same time most difficult to stage. The conditions which the schedule imposes upon the class must be carefully noted as some show committees permit and encourage the use of flowers, ferns, etc. to embellish the trophies whilst others strictly prohibit any extraneous decoration.

A trophy should be symmetrical and well balanced and not in any way bulky or top heavy. Its general appearance should be graceful from all points to which it is supposed to be exposed to public view. Monotony must be avoided. An exhibit composed of all one-sized bottles, even if they contained ideal honey, would be inadmissible.

A reasonable variety of the highest class of bee products is most desirable. Thus a fair quantity of comb honey in really good sections, and some little variety of colour in the liquid honey, together with samples of granulated honey, all in a variety of shapes and sizes of bottles, go to make a really attractive display and one upon which the eye loves to linger.

Where the schedule allows wax, or other accessories, these must be considered in making the award and the general excellence of the exhibit must not escape attention. It is desirable that some indication should be given in the schedules as to the maximum and minimum weight of honey to be staged on a trophy, otherwise it is necessary to take into consideration the amount, a large exhibit being more difficult to produce and stage than a much smaller one.

Having dealt with our first two heads, we will now proceed to consider the third, "The Method of Judging". The show secretary should take care that a schedule, giving full particulars of classes and rules applicable to the same, has been sent to the judge selected, in due course, so that he may have an opportunity of noting any special classes or conditions. Most judges will also appreciate a reminder, say, a couple of days before the show, enclosing an admission ticket, and giving the exact time his presence will be required.

If, as is not unusual, the judge has some distance to travel, a little consideration upon this point and consultation of the current time-table will be helpful. For a person who is habitually punctual it is anything but pleasant to be put to considerable inconvenience and perhaps the necessity of rising before the usual hour, to get to a show at the time suggested, say, 10.30am., and upon arrival to find everything in a state of chaos and the exhibits not ready for his inspection until, say, 11.10am; whereas, if his arrival had been timed for eleven o'clock a quicker and more convenient train or route might have been used and the original starting time have been a couple of hours later.

It is objectionable and unpleasant for a judge to be on the ground and practically compelled to see the exhibitors staging their exhibits, building their trophies, etc. An impartial judge prefers those arrangements which preclude the possibility of identifying owners of exhibits.

Presuming the judge has arrived and everything is in readiness for his inspection, the secretary will provide him with his 'badge of office' (which some judges preserve as souvenirs of duties often performed solely for the love of the work) and a judging book in which each class is shown on separate pages, giving the number of entries in each and having a perforated duplicate which may be torn off for each class as the work progresses so that the secretary can proceed with marking the prize cards, etc. A bowl, or bucket of water, and a towel should also be provided to enable the hands and clothes to be kept in a presentable condition. A glass of water, or soda water, and a biscuit are also often found useful in cleansing the mouth and palate when so many different lots of honey have to be sampled.

Personally I consider the disqualification of liquid honey owing to its being shown in a wrong colour class, or sections, through being over-laced, should be considered the business of the secretary and stewards, such gentlemen being, of course, non-exhibitors and as such disqualifications are founded on fact and not on anyone's opinion, this arrangement will save the judges' time, and disqualification cards can, unless it is deemed undesirable, be placed upon each of these exhibits stating why disqualified. I always enquire whether or not this question of disqualification is to be part of the judge's work, and if I am informed that these points have already been dealt with, take it that all the exhibits, unless otherwise indicated, are in order.

These points being settled, the judge will proceed to his duties accompanied by the steward, who will render assistance by taking off and replacing the caps, glazing, etc., as no other person should be allowed to be moving about, distracting the judge's attention from his duties, etc.

The question of judging by points has frequently been discussed, and whilst this method is necessary, it is obvious that in a fairly large show a hard-and-fast rule of submitting each exhibit to such a test is quite impracticable. I recommend judging by rejection and selection and a judge of experience coming to an average class will, upon little more than a casual glance be able to reject forty to sixty per cent of exhibits as having not a shadow of a chance of being amongst the prize-winners and upon a little closer examination, say, for density, etc. a further percentage will be discarded.

The remainder will now consist of those from which the winners will be selected and will be proportionately greater in accordance with the general excellence of he class, and points will be necessary to allocate each to its respective place. My table of points for run or liquid honey is given here.

Flavour	30
Colour	20
Granulation	25
Aroma	10
Uniformity and Get-up	15
TOTAL	**100**

Each of these items has already been dealt with. It may, however, be pointed out that it is helpful to a judge, for obvious reasons, if the exhibits are arranged in such a manner that he can view them from the rear as well as the front. In making the final selection, the chosen bottles should be brought alongside each other in a strong light.

In testing liquid honey for density, the old system of inverting the bottle is not as satisfactory as stirring the honey with the glass honey taster and here again experience counts strongly, as a practical person can judge very accurately by the quantity which adheres to the glass when withdrawn from the honey as to its relative density.

When dealing with granulated honey the points for prize winning will be different and the question of colour more a matter for the judge's discretion, as all present are aware that when honey granulates it frequently changes a great deal in this respect. A nice, even cream colour, or light yellow should receive full marks. My points in the granulated honey classes are given opposite.

Flavour	30
Colour	20
Granulation	25
Aroma	10
Uniformity and Get-up	15
TOTAL	**100**

In judging sections, the view from the rear is even more desirable than in the case previously mentioned, because a casual glance under such circumstances will enable a judgment to be passed upon their uniformity and imperfections in sealing, etc, are easily seen without one having to turn each section round. And again, any signs of granulation or patchiness, which are great defects in this class, are more easily detected.

Prize sections, as before remarked, must be sampled by the judge, mutilation of the comb, however, is not advocated: that admirable adjunct to judging, the "Reid "honey-taster", enabling one to do the needful without damaging the sections for further show, or sale, purposes. My marks for final selection of prize winning sections are as follows:

Flavour	30
Colour and cleanliness of comb surface	20
Uniformity and cleanliness when held up to light	20
Quality of capping, including attachment to wood	15
Glazing and general get-up of exhibit	15
TOTAL	100

Honey in shallow frames, as already mentioned, the quality of the honey is not so much under consideration, so long as the judge satisfies himself that it is genuine and of good grade, the question of quality is not of so much importance, neither does the colour of the capping matter so much as in the case of section honey. A table of marks in this class would be:

Complete sealing and evenness of surface	50
Absence of pollen, and suitability for extracting	25
Cleanness and general appearance	25
TOTAL	100

Finally, with regard to honey trophies, little need be added to what has already been said respecting this class. If the remarks applicable to the various parts of the exhibit are complied with, the experienced judge will have little difficulty in awarding the prizes in a just manner. Whilst variety is very essential, if an exhibitor has not a sufficient sample of each kind of produce of a show quality, the entire absence of lesser quantity of that particular object is better than the lowering of the standard of the whole exhibit by the presence of an inferior article.

The practice of having compound classes (except in the trophy class), that is to say, classes in which so many bottles of liquid honey and so many sections are staged and judged together is not to be commended as it often happens that one portion of the exhibit is so different in quality to the other, that an ordinary all-round, fairly good exhibit may gain a prize, and really good stuff have to be passed over, or, owing to the excellence of one part being so high some inferior produce may appear in a prize lot and give wrong impressions to the general public.

If any attempt at fraud or dishonest practices is noticed, the person making such discovery should at once report the same to the show authorities and not wait to see whether or not the judge will find it out. When the judging is completed and all concerned have done their best, we may find that some discontented exhibitor would like to discuss, and even revise, some of the decisions. Some good tempered judge, if time permits, may gently, but firmly, take such a competitor in hand and point out that until our show committees have funds sufficient to give every exhibit a prize, there must be losers, and as there are, inevitably, under present conditions, more blanks than prizes, it behoves him to become *a cheerful loser*, to try and try again, ever remembering that one's produce nearly always looks so much better than it really is until placed alongside its competitors on the show-bench.

The numerous candidates now taking the examinations for the BBKA expert's certificate is, I think, a happy augury, because while so many are willing to study and work for this coveted honour, there should also be many aspirants for the post of judge and if this lecture proves helpful in stimulating any such, it will not have been delivered in vain.

BBJ 30 Oct 1913
The monthly meeting of the BBKA Council was held at the Lecture Hall, Zoological Gardens, London, NW. on October 23rd, 1913. Mr. WF Reid presided. There was also present AG Pugh, The judges for the Royal Show, recommended by the Exhibitions Committee were accepted and appointed as follows:
    Messrs. WF Reid, AG Pugh, Rev. TJ Evans, Rev. GEH Pratt.

Beekeepers' Record November 1913
A very interesting exhibition of bees and their life and produce, and of the art and methods of utilising the services of the useful little insects in collecting honey, was opened at the Nottingham Mechanics' Lecture Hall on September 12th. The exhibition is the second which has been organised by the NBKA. Its primary object is to popularise Notts honey, the trade which constitutes an industry the extent of which few people are aware. There were 123 entries compared with 166 last year. The falling off is due to the dryness of the season which, in many districts locally, has been bad for the industry, for showers are necessary to cause the flowers to secrete nectar abundantly. The exhibition of sections was rather poor but there was a splendid show of extracted honey and of wax.

The NBKA and its courteous secretary, Mr. George Hayes, are to be cordially congratulated upon their interesting and comprehensive exhibition. A new feature this year, to induce a trial of honey by those who have not learnt its deliciousness, was the sale of penny sample jars.

The exhibition was formally opened by the Mayor who made a characteristic speech

in performing the ceremony. The bee he described as a perfect Socialist, because in gathering its honey it respected no man's flowers and it took it back for the benefit of the other bees. That was Socialism in the dream. It was a perfect Socialist because it knew what to do with the work-shys and drones. Those that produced nothing were cleared out. They were of no use, and were not allowed to sap the vitality of the community.
G Hayes, Hon Sec

Frederick Ball (1861-1915) was an architect based in Nottingham. He was Sheriff from 1906–07, and Mayor from 1913-14.

BBJ 6 Nov 1913
The bee-keeping industry was well represented on October 23rd and 24th, at Kilmarnock, at the Ayrshire Agriculture Association's Dairy Show.
Six 1-lb Jars of Granulated Honey (10 entries) - c. J. Woods, Mansfield

BBJ 6 Nov 1913
Sweet stuff. (Notts). We are afraid you make the candy a little too hard. You will have to take it off the hive and scrape the hard surface away to enable the bees to eat it.

BBJ 13 Nov 1913
I wonder how many bee-keepers have any idea of the work of an expert. I would like to reassure those who are afraid of compulsory inspection under a "Bee Diseases Act", and whose idea is that if we had such an Act inspectors would be sent round to inspect every apiary and stock at least once a year. In my opinion, such a proceeding would be found so costly as to be impracticable. One man would only be able to work a comparatively small area as every stock would need to be thoroughly examined. This would take on average at least ten minutes for each one. I find that when touring I can visit an average of five apiaries a day and then do not examine every stock. I do not think a Government inspector could do much more if every apiary was visited. Not only would he have to examine every stock but there would be the difficulty of locating the apiaries, as there would be no "list of members." I leave your readers to judge how long it would take a man to do a county or district.

Touring as expert is not by any means easy work, especially when done on a push-cycle, as a certain amount of gear has to be carried. A motor-cycle and side-car - or, better still, a cycle-car - would be a great help. Unfortunately, the remuneration of an expert will not allow such luxuries. The first cost of even a second-hand motorcycle would probably exceed the total amount received for one tour. There would then be Excise and driver's licences, also running expenses, which are about 1d per mile for a cycle and $1\frac{1}{2}$d per week for a cycle and side-car. When the tour was finished the machine would quite likely be more or less of a "white elephant."

In most associations the expert is paid a fixed sum per visit and in some cases the work is to a great extent a "labour of love." Unfortunately, nearly all county associations are hampered in their work by lack of funds and the remuneration of secretaries and experts, whose work is the most important to the well-being of the association, is not nearly as good as it should be.

I have a load that makes its presence felt, especially uphill, and towards the end of a day's work. My cycle is not of the motor variety, or suitable for a racing track, and every extra ounce of weight tells. One or two items that are necessary are omitted - a mackintosh cape and leggings, a sleeping suit, and a change of underline *(sic)*, etc. Although I go through my outfit each spring and discard everything not absolutely necessary, I still find there is enough to require an 18in. bag on the front carrier and a fair-sized basket on the back.

When disease, or a suspicion of disease, is found the owner should not only be informed of the fact but, if possible, shown the symptoms of disease so that he may he able to detect it himself in the future. I do not agree to closing other hives when examining a diseased stock. The returning bees, not being able to get into their own hive, would fly round for a yard or two and quite likely alight on the comb or hive under inspection whereas if their own hive is left open they will fly in and out attending to their own business.

Regarding the work that should be done by an expert when working for an association, that should be left to his own discretion. He should not be expected to undertake long operations. Most associations have a notice in their Annual Report to the effect that the expert is not expected to do the work but to give advice. This prevents the expert being imposed upon. Very often, however, it takes less time and is more useful to the bee-keeper for the expert to do certain work than to explain how it should be done. I find also that as a rule members are quite willing and pleased to pay for the extra work done. I often ask the owner to do the manipulating. My only objection to this is that it is very difficult to see the small unsealed larvae clearly enough to detect symptoms of disease when another person is handling the frames.

Cards for notifying disease were supplied by the Devon Association. I have brought them to the notice of several other associations, who have adopted them. As to the advice given in cases of disease, no hard and fast rule can be laid down, especially in cases of brood diseases – much depends on circumstances. Before giving advice as to treatment I first inquire if there is any disease known to be in the locality. If there is, and the owner of the diseased stocks will do nothing, it is very little use other bee-keepers adopting the B treatment *(sic)* as his stocks are sure to be re-infected as I know only too well.

Any stocks that are bad should be destroyed and the others treated with Apicure and naphthalene and, when fed, medicated syrup should be used. The disease may thus be kept at bay, or even cured. The difficulty often is to get bee-keepers to keep up the supply of disinfectant as it evaporates.

Should the locality be free from disease and an odd stock be found infected in the apiary under inspection, I generally advise its destruction. This is safer than to risk infecting other stocks while attempting to cure the one.

Printed matter is too heavy to carry in any quantity and an expert has not the time to make elaborate notes of every visit. There are other members waiting to see him, possibly staying home from work for the purpose and it is a great disappointment to them if the expert is hindered and unable to get there. It is well to keep a diary to enter up anything worth noting - the weather, etc. - but this can be done in the evening.

Many times I have known beekeepers astonished at the expert disinfecting himself and appliances; they had no idea the disease was so infectious and it causes them to think on their own methods. Gauntlets I never use on tour. I prefer taking off my coat and pulling up the sleeves of my sweater.

I am afraid I am taking up too much space, or would give a few experiences I have had when "on tour," but these must be left to a future number. One piece of advice I would give budding "touring experts": Leave your politics and religious creeds at home. Bee-keepers are of all shades of opinion on these matters and I have known an expert who entered into a political controversy with a member he was visiting being ordered off the premises. J. Herrod, Trentside Apiary, Sutton-on-Trent.

BBJ 11 Dec 1913
To obtain ripe honey, and the requisite density for extracting and showing in bottles, one leaves the honey on the hive as long as possible. The combs that have been on longest are naturally the most travel stained and presumably the ripest and best. As the cappings are cut off this travel-stain is immaterial. Why should a judge refuse to look at a section that is travel-stained? It ought to be richer and riper than one freshly sealed; therefore, why is it ignored absolutely if flavour comes first in the estimation of all good judges? He maintains that some of his ripest and best sections, under existing methods of judging, are unshowable. Is he right?

If some of our best judges would kindly give their views on this point, which I have never before seen raised, it would be both interesting and instructive to all bee-keepers. I for one should greatly value the opinions of Mr. AG Pugh and Mr. G. Hayes.

With kindly greeting for Christmas and good wishes for the New Year to all bee-keepers. F. Sitwell.

BBJ 18 Dec 1913

In reply to Captain Sitwell's letter above, I may say that at some shows where (under protest) I have judged, I have not been allowed to open sections to taste their flavour and test their consistency. I contend that these points should be taken into consideration because in some cases the difference is too marked to ignore.

I consider, however, the points given for sealing and general appearance should come slightly in front of those for flavour and consistency; so a travel-stained section would require having an exceptionally better quality honey to take its place before a well-filled clean white sealed sample of pure honey.

Comb honey, especially to the lay mind, is an altogether different article to extracted honey as it first appeals to it through sight. We have therefore to deal not only with the palate but with the eye as well and we all know that sight has an influence on the palate.

As an instance I may ask, of the two following, which would you most prefer? A good high-class meal served on a dirty tablecloth with uncleaned cutlery and silver or a slightly inferior but still wholesome meal on a clean white cloth with other things perfectly clean and attractive? Again, why is so much trouble taken to garnish our tables and food but to attract and influence the palate to the acceptance and enjoyment of that food?

A clean white comb which in this case has to be eaten with honey is attractive and will compensate to a fair degree for some loss in the quality of the honey. So, on the contrary the least sign of discolourment, which is known or appears to be foreign to the substance, would be repulsive and would nauseate the palate so that, although the honey was of good quality, it would not compensate for the effect of the discolourment. Moreover, people purchasing sections for consumption do so by sight and not by taste.

Wishing the Editors and all readers of our Journal a Happy Christmas and a most Prosperous New Year.   Geo. Hayes, Notts.

BBJ 25 Dec 1913
The monthly meeting of the BBKA Council was held at 23 Bedford-street, Strand, London, WC, on December 18th, 1913. Mr. WF Reid presided. Also amongst those present was AG Pugh.

BBJ 15 Jan 1914
Nectar-Producing Plants and their Pollen - Butterburr

WHW (Notts) Kent as a Bee-County. There is no Bee-keepers' Association in 'South Kent'. The districts you name are good for bee-keeping and so far as we know at present are free from "Isle of Wight" disease.

Bee-keepers' Record February 1914.
The monthly meeting of the BBKA Council was held at 23, Bedford-street, Strand on January 15th, 1914. Mr TW Cowan presided. A letter expressing regret at inability to attend was read from AG Pugh.

Certificates for success in the BBKA Intermediate Examination were given to FS Elliott (TS?) and W Doleman of NBKA.

BBJ 5 Feb 1914
Nectar-Producing Plants and their Pollen - Broom

Bee-Keepers' Record March 1914
I have only been a reader of your paper for a year and though a beginner in bee-keeping I have never yet written to ask a question, as I generally find the solution to my own difficulties in your query column in reply to some other correspondent.

I have often thought, however, that I should like to send you a line giving some of my experiences which may be of interest to some of your readers. I am a groom-gardener, motor-driver and general man by occupation and in September, 1911, my employer had sent to him a new hive full of bees.

My grandparents kept bees when I was a boy – about six skeps – and I was always interested in them but when I was left to see to the new acquisitions I was rather in a difficulty. However, after carefully covering myself up, I set to work to get them in order, following the instructions given in the 'Guide Book,' a copy of which I had.

I had a great desire to have some bees of my own and, having permission to keep them, I got a joiner to make me two hives after my own instructions. In March, 1912, I purchased bees to stock them from a local bee-keeper who hived them for me. Though it was a very wet season, I secured 100lbs of honey and divided one stock. This attempt at artificial increase was not successful as the oldest queen died. The bees raised another, but there were no drones, so I united the two lots again.

This was in March, 1913 and I bought a fine stock of black bees for 30s, but the weather being cold and dry, I only got 45lbs of surplus from my new purchase and not so much from the others.

In June a relative gave me some bees in an old wooden hive. It was literally falling to bits and had an old doormat on top with moss growing two inches thick on it. The bees had to get in and out as best they could through the long grass that grew up in front of the entrance.

I had to drive twenty-two miles to the place where the bees were and reached there about seven o'clock and started to pick up my hive. I put some wire on the entrance and wrapped two horse rugs round it to hold it together and, though the floor nearly fell out, I managed to get it home about nine o'clock.

It looked a deplorable object in the garden next morning but I scraped away the moss and cleaned it up and found I had a very strong stock of bees. I put the old hive over one fitted with ten frames of foundation and in a few weeks, on lifting it off, found six frames drawn out and brood in most of them.

Thinking the queen was in the lower hive I put on an excluder and replaced the old hive. However, three weeks later, on examination, I found the queen still above so I bought a new queen which I introduced successfully to the bees in the lower hive. I then repaired and painted the old hive, leaving the original stock still in it. As I divided my first two stocks I now have six, all healthy and busy flying today.

A good deal has been written and said about finding a market for honey. I can sell all mine and here is a tip which may be of use to other bee-keepers. If you cannot *sell* any to the local doctor *give* him a jar; one doctor here has sent me quite a lot of customers, among his patients, who are advised to give it to children or use it in cases of cold, sore throat, etc, and I get 1s per lb jar.

Hoping a report from this corner of the globe will interest other bee-keepers, and wishing them a successful season. G Ward, Langley Mills (*sic*)

BBJ 5 Mar 1914
WC (Long Eaton). Transferring from Skeps to Frame-hives - You should buy the "British Bee-keepers' Guide Book," in which you will find the method of doing this fully described. You should also join the NBKA. The Hon. Sec, G. Hayes, Mona-street. Beeston will send particulars if applied to.

Bees and their Keepers (source lost but suspect the NEP "Beekeepers' Corner")
It is bad news for Notts. beekeepers that what is known as the "Isle of Wight" disease - a very infectious malady - has appeared in this county. In the South of England the loss of stock in the past year or two from this cause has been very great, and apparently no effectual means of preventing it have yet been discovered. It was suggested at the meeting of NBKA that the use of artificial food might be a cause of the disease, but this does not seem to be the case because wild bees suffer equally

with the domesticated species. Moreover, we are assured that no beekeeper of repute would deprive the bees of sufficient natural food to carry them through the winter. The keeping of hives in a sanitary condition is not doubt likely to protect the bees and we can only hope that the epidemic will not prove as severe in the Midlands as it did in the South. Still, should it do so, there would be no reason for despair, for despite the decimation of the hives the honey harvest last year in the southern counties was a record one – the surviving bees seeming to have considered themselves bound in honour to supply the work of their dead comrades.

BBJ 19 Mar 1914
Nectar-Producing Plants and their Pollen - Ragwort

BBJ 19 Mar 1914
Novice (Notts). Disinfecting Hive.
1 and 2. We prefer to use carbolic acid, as recommended in "British Bee-keepers' Guide Book."
3. If the bees have enough sealed stores to last until April you need not give candy again, but start to feed with syrup then.

BBJ 19 Mar 1914
The AGM of NBKA was held in the Peoples' Hall, Nottingham, on March 7th, at 3pm, with the Deputy Mayor (Thomas Ward) presiding over a large assembly of members and friends.

The minutes of the previous meeting were read and confirmed. The annual Balance Sheet was next submitted, and was unanimously adopted, great satisfaction with same having been expressed.

The whole of the officers who served for 1913 were most heartily thanked for their services and re-elected for 1914.

A discussion arose at this point with regard to microsporial loses, or "Isle of Wight" disease which became interesting in several ways and no doubt information would be found by those who sought it. It was pointed out how very serious the disease was and that as it had shown itself in several places in the county it was necessary to be on the alert and for every bee-keeper to do all he could to prevent it's spreading. Members were advised not to purchase, during this season at least, any queens, stocks, swarms, or driven bees from outside the county but to get all required from those within it.

Mr. Pugh proposed a hearty vote of thanks to the Deputy Mayor for presiding over the meeting. Mr. Ellis seconded, and it was carried with acclamation. Mr. Ward, in reply, said how pleased he was to be able once more to be with them because he saw that they were doing all they could as an Association to push forward the industry of bee-keeping, and anyone could see on carefully scanning the Report and the Balance Sheet what a large amount of work had been done for the money expended.

The meeting was then adjourned for tea, to which about 120 members and friends sat down. After tea the prizes and medals were distributed to those winners who were present.

The Chairman then called upon Mr. Pugh to give his lecture on "Judging Honey." This proved to be, as was expected, a very interesting address and called forth considerable discussion which must have been very helpful to those privileged to hear it.

It was resolved: "that a resolution be sent from this meeting urging the Government to press forward the "Bee Disease Bill", as it was felt to be most urgently needed."

A very successful meeting was brought to a close with the usual prize drawing.

Geo. Hayes Hon. Sec.

BBJ 26 Mar 1914
BBKA Meeting. The following nomination of representative from affiliated Associations was received and accepted (Nottinghamshire) Mr. G. Hayes,

BBJ 30 Apr 114
Leicestershire and Rutland BKA. There was a large attendance at the Annual Meeting which was held at the High Cross Restaurant, Leicester on April 4th. Following the meeting an excellent tea was partaken of, after which the usual prize drawings took place. A very interesting and instructive lecture on 'Honey: Its Granulation and Density' was given by Mr Geo. Hayes, Nottingham, who was accorded a hearty vote of thanks. J Waterfield, Hon. Sec.

Bee-Keepers' Record, May 1914
The monthly meeting of the BBKA Council was held at 23, Bedford Street, Strand on April 16th. Mr WF Reid presided. Amongst those present was Mr AG Pugh.

Mr Pugh brought forward the matter of arranging a judging competition. It was resolved that a class be arranged at the consersazione to be held on October 22nd, 1914.

BBJ 7 May 1914
MEB (Southwell) The bees were too dry for us to determine cause of death.

BBJ 14 May 1914
AD (Notts). There is no sign of disease in the bees sent, they have overladen themselves apparently and died from exhaustion.

NG 15 May 1914
There is always an attractive programme on the two days of the Newark Agricultural Show. On both days members of the NBKA, assisted by a grant from the County Council, gave lectures on beekeeping as well as demonstrations.

The second is usually the popular day but as this year it fell on a Friday, it was hardly to be expected that there would be so many people present as yesterday, when a record was established and £168 was taken at the gates. The weather remained delightfully fine.

BBJ 28 May 1914
MB (Southwell). We regret to say there is every sign of "Isle of Wight" disease in the bees sent.

BBJ 28 May 1914
During the winter of 1912-13 I lost the whole of my bees through "Isle of Wight" disease. They all died before many bees commenced flying and as I destroyed dead bees, etc. and disinfected hives, there were no outbreaks of disease among neighbouring beekeepers as far as I am aware, although I have made numerous enquiries.

On May 1st this year I bought a stock of bees and placed them in one of my old hives, in a different garden altogether. I scorched the hive thoroughly and afterwards washed the whole with strong carbolic solution. The first week that I had bees was very wet and cold but it has improved a little during the last few days. Although there are plenty of wallflowers and apple trees in bloom the bees have not worked on them very much.
1. Can you suggest a reason for this?
2. Could you tell me the name of the flower enclosed? One rather cold, wet day bees were very busy working on it. This district seems very satisfactory for pollen and I am afraid the combs will soon become so filled with pollen that there will not be sufficient room for queen to lay eggs.
3. What is the remedy?
4. Is "white carbon" useful in keeping off attacks of "Foul Brood"?
5. Is there any trace of disease in the bees sent? I picked them up from the front of the hive where they had been accumulating during the last few days.
6. What kind of bees are they?

I trust your paper will continue to be as helpful to me as it has been in the past.

XYZ, Notts.
Reply.
1. We should say the paucity of bees is the reason you do not notice them working.
2. The flower is *Sedum hispanecum*, a native of Southern Europe.
3. Remove the combs and replace with foundation.
4. No, it will asphyxiate the bees. Use naphthalene.
5. We find no trace of disease in the bees sent.
6. They are of the English variety crossed with Carniolan.

BBJ 4 Jun 1914
Nectar-Producing Plants and their Pollen by Geo. Hayes, Beeston - Sainfoin

July 23rd, at Southwell. The Notts. Annual County Show, in connection with the Southwell Horticultural Society. Open class for single jar extracted honey. Schedules now ready from G.Hayes, Mona-street, Beeston, Notts.

BBJ 9 Jul 1914
At the "Royal Show" at Shrewsbury the "Hives and Honey" stand occupied an exceptionally good position this year. From the front stretched a wide avenue right across the ground past the "Horse Ring" to the stand of the "Education Department". The exhibits were excellent, as usual, and in a goodly number, very few entrants being unable to send their exhibits.

The lot of the judges was not at all enviable; not only were the exhibits all so good as to make it a difficult matter to decide which should take premier honours, but the weather was intensely hot. With such well-known exhibitors of "trophies" as Messrs. Pearman and others, that class was a strong one. The class for mead secured a large number of entries this year, as did the wax classes also.

The judging was undertaken amongst others by AG Pugh. No Notts members were awarded prizes.

BBJ 6 Aug 1914
The Annual Show of NBKA was held in connection with the Southwell Horticultural and Gardening Society on "Lowes Wong" in the Cathedral city on July 23rd, the Honey Show being accommodated in a separate tent which, rather unfortunately, proved too small for the effective display of the numerous exhibits. The weather was threatening all day but the rain kept off until towards the close. The honey tent was well patronised throughout the day and appeared to be one of the most attractive spots on the show-ground, judging by the number of people always in it and from the many remarks overheard concerning it. The bee tent was also there and

demonstrations were given to very interested and enquiring audiences. There was also a "Training Class for Judging Honey" at this show and a good number of beekeepers availed themselves of it.

The judges were Dr. Percy Sharp, Swallowbeck, Lincoln, and Dr Thos. S. Elliot, of Southwell and the following is a list of their awards to local beekeepers:
Winner of the Herrod Perpeutal Challenge Cup - James North, Sutton-in-Ashfield
Single Jar of Extracted Honey - 2nd, J. North
Most Attractive Display of Honey - 1st, D. Marshall, Nottingham; 2nd, G. Marshall, Norwell
Six Sections of Comb Honey to approximate 6-lbs produced in 1914 - 1st, J. North; 2nd, D. Marshall, 3rd, G. Marshall; h.c, W. Doleman, Keyworth
Twelve Jars of Light Colour Extracted Honey to approximate 12-lbs produced in 1914 - 1st, J. North; 2nd, JT Wilson, Shirebrook; 3rd, JT Duckmanton, Langwith; 4th, WH Mellors, Norton; hc, JC Mellors; c, W. Lee.
Twelve Jars of Dark Colour Extracted Honey to approximate 12-lbs produced in 1914 - 1st, J. North; 2nd, G. Marshall; 3rd, AG Pugh, Beeston; 4th, H. Merryweather; hc, W. Lee; c, Mrs. Waller
Six Jars of Granulated Honey to approximate 6-lbs produced in any year - 1st, AG Pugh; 2nd, JT Wilson; 3rd, W. Mountney, Southwell; hc, G. Marshall; c, JT Duckmanton.
Shallow Frame of Honey suitable for Extracting, produced in 1914 - 1st, JT Wilson; 2nd, J. North; 3rd, G. Marshall; hc, D. Marshall
Three Jars of Extracted Honey produced in 1914 (for members not having previously taken a prize) - 1st, RB Hutchinson, Bilsthorpe; 2nd, GH Worth, Long Eaton
Specimen of Bees of any race, with Queen on best and most complete Single Comb in Observatory Hive - 1st, H. Merryweather; 2nd, JT Duckmanton; 3rd, G. Marshall; 4th, J. North; c, W.Lucas; c, W. Mountney.
Sample of Beeswax in six pieces to approximate 2-oz each - 1st, G. Marshall; 2nd, John Bee, Southwell; 3rd, J. North.   Geo. Hayes.

BBJ 20 Aug 1914
New Beginner (Holme Pierrepont) Take the comb out at midday when all the old bees are out at work, put it into the observatory then brush the bees from two more combs in as well. It is very late to stock an observatory hive.

Bee-Keepers' Record, October 1914
The monthly meeting of the BBKA Council was held at 23, Bedford Street, Strand on September 17th 1914. Mr WF Reid presided. Amongst those present was Mr AG Pugh. A report on preliminary examinations held in Nottingham was presented.

BBJ 17 Sep 1914

Nectar-Producing Plants and their Pollen - Bean

BBJ 15 Oct 1914
Nectar-Producing Plants and their Pollen – Red Clover

BBJ 3 Dec 1914
Nectar-Producing Plants and their Pollen - Crocus

BBJ 10 Dec 1914
Notice - Owing to the dislike of one of my ancestors to a long signature he dropped a portion of his name. For several generations this caused no inconvenience, as they did very little business. Owing to the continued increase in my business interests, the dual situation of one signature for ordinary use and another for legal matters has become impossible therefore, on and after January 1st, 1915, I shall assume my full name of W. Herrod-Hempsall. I shall be grateful if all correspondents will kindly note this and address me, also make out all cheques or documents, and insert in any list my name as above.   W. Herrod-Hempsall, hitherto commonly known as W. Herrod.

The Bee-Keepers' Record, referring to a photograph of a group of prominent beekeepers, says: "Mr. Dadant's well-known features are easily spotted." We are sorry, but a little cold cream will sometimes do wonders.

BBJ 28 Jan 1915
BJ (Notts.).The honey is mainly from clover and the flavour is very good. Honey varies considerably in the time of granulation, possibly yours was warmed. If so, that would retard the process. Keeping it in the light in a cold, dry place will help it. The sample you sent is granulating. It will not be likely to ferment if kept in a dry place. When properly ripened before extracting, and free from pollen, honey will keep indefinitely with the above proviso.
				J. Allsop (Arnold) See answer to BJ (Notts.) above.
BBJ 18 Feb 1915
Obituary Notices. Mrs. Reid and Mrs. Pugh.
We are sorry to have to announce the death after prolonged suffering of Mrs. Reid, the wife of Mr. WF Reid, vice chairman of the BBKA. Also of Mrs. Pugh, the wife of Mr. AG Pugh, of Beeston, one of the council of the BBKA, who passed away last Friday after a long illness. We are sure our readers will join with us in tendering our sincerest sympathy to Mr. Reid and Mr. Pugh in their bereavement.

BBJ 25 Feb 1915
Nectar-Producing Plants and their Pollen - Mallow

BBJ 18 Mar 1915

BEEKEEPING BETWEEN TWO QUEENS

The NBKA has for many years carried out a useful work in the county, and is now a flourishing and active organisation with over 300 members

At the Annual Meeting held at the Peoples' Hall on Saturday afternoon, there was a large attendance, the Mayor (Ald. JH Gregg) presiding. He congratulated the association on its improved position, and hoped that success would continue to attend their efforts. It was a matter for regret, he said, that the "Isle of Wight" disease continued to assert itself, but he trusted that ere long something would be found to stamp out the scourge.

The secretary (Mr. G. Hayes) noted that the yield of honey for the county during the past year had been a very fair one: the quality was exceptionally good and free from the honeydew that was feared would be gathered with it. During the year the association had enrolled thirty-three new members, making a total of 323. At the beginning of 1914 there was a debt of over £3 but this had been wiped out and there was now a small balance in hand. The association had again been favoured with a grant from the County Council, and by this means had been enabled to send a bee tent and lecturer to several shows. Owing to the "Isle of Wight" disease there was a slight increase of unhealthy stocks. The disease had manifested itself in several localities and the committee urged extreme carefulness in the matter of trafficking in bees.

In view of the fact that the "Royal Show" was to be held at Nottingham, it was hoped that members would do everything they could in the way of exhibiting, so that the honey section might be a strong one. The annual bee-keepers' show would, it was stated, be held in conjunction with the "Royal Show" at Wollaton Park.

The benefits of insurance for their bees was impressed upon the members so that they would have a full sense of security on that point.

Her Grace the Duchess of Portland was thanked for her services as President and was unanimously re-elected for the current year.

The Committee elected were: Messrs. Adams, Harrison, Darrington, Fox, White, Vessey, Turner, Pugh, Riley, and Dr. Elliot. Mr. Bickley was re-elected Auditor, and Mr. Hayes Secretary and Treasurer. Representatives to BBKA: Messrs Hayes and Pugh. The meeting then adjourned for tea, to which about 100 members and friends sat down.

After tea, the Secretary lectured on the "Isle of Wight" disease, giving illustrations with photo-micrographic and other slides, whilst Mr. D. Wilson, of Belper, dealt with "The Advance of Apiculture during the Last Century." Each of these lectures called forth a good deal of discussion, which kept the members well engrossed and extended the meeting to a much longer period than usual.

Mr. Pugh, who presided at the evening meeting, was thanked for his able services.

John Henry Gregg, born 1857, was a retired builder. He was Sherriff in 1913/14 And Mayor of Nottingham 1914/15.

Don Wilson was the co-subject of the author's book "Bee-keeping on Two Fronts" published to commemorate the centenary of the First World War. A copy of this book can be found in the NBKA library.

BBJ 18 Mar 1915
With the coming of March the old feelings of interest for our bees come along. Last August, after taking off all the supers, I looked in my stocks, eight in number, and found all very strong. I keep four at home and four at my place of work in the garden. No. 4 hive at home was very strong. I went for ten days' holiday. When I came back the strong colony in No. 4 had dwindled so I wondered what had gone wrong. There is a bakehouse about 300 yards away and the next day the baker, who is a friend of mine, sent for me. He told me what a job he had had with my bees in his bakehouse.

There were four men at work there but only one got stung and he said that in taking up some buns he trapped one. (I think that speaks well for my bees.) In the mornings they had to clear out very many dead bees. There was a large bowl with sugar in that they put on the buns. I took it outside; it was one mass of bees and the sacks of sugar were covered. Fortunately, a few cold days set in and that brought the trouble to an end.

I sent for 4 lbs. of driven bees as I did not want to lose the stock. Three days before they arrived I was looking at the entrance to No. 4 hive and instead of them flying off as usual they were creeping about looking snuff coloured. I was very upset.

The next day they were worse so I looked in my "Guide Book" and found they answered the description of "May Pest" so I treated them with flowers of sulphur. When the driven bees arrived I did not know what to do. I thought the strong bees might carry away the sick ones so I decided to try my hand and run them in, this being my first attempt at uniting. I opened the hive and separated the frames half an inch, blew in flour till all the bees were covered with it. Then I propped up the hive front, opened the box of driven bees, dusted them all over with flour and threw them on the alighting board. It was some time before they would run in, the bees from inside meeting them. I had a fine time for half an hour; a lot went in and a lot flew out.

BEEKEEPING BETWEEN TWO QUEENS

Next morning when the sun shone on my garden there were thousands of white spots crawling about. A few came out for several days. I dug them in each day until only one or two were seen crawling. I looked in the hive and found two combs of eggs so I concluded I had got the young queen. As they had plenty of food I packed them up snugly to take their chance; they have been out often when it has been warm. I have seen just one or two down.

Looking in the entrance the other day one "got me" on the nose end in fine style. Now you may think I ought to have burnt them, but as there are no other bees near them except my own I thought I would try my luck I had a peep the other day, and they have food and a good number of bees so I am hoping they may pull through all right.

I live in a village ten miles from Nottingham and about the same from Derby. Potteries, foundries and mills are all around, so that it is not a great honey district but, honey or no honey, I am fond of my bees. Last year I had five hives at work, which had been divided from three the year before and I took 115 lbs. of most lovely honey. I sold out in five weeks at 1s per lb, and have sold 70 lbs. for another bee-keeper friend of mine. I had seven hives in all last year, but two had old queens and just when the flow came on out went the old queens.

I learnt my lesson; no more three year old queens. I divided one very strong stock at the end of the season, making eight stocks, all with young queens, so I am hoping for a good season this year. A farmer friend has promised to let me stand two hives on his farm, two miles away. There is much clover there; I think I shall take them.

I have never missed reading your most valuable little paper for more than three years now: it is so interesting. Pleased to read bits about the war and bees and bee-keepers. My mistress is keeping nineteen Belgians [refugees from the War], and I have a great deal to do for them, and have two of them with us at my home.

With best wishes from A Lover of the Bee

PS. If I ask you one question, it is do you think that the stuff the bees got at the bakehouse would upset them, as the "Bee Guide" says improper food would do them harm? In a day or two, when I hope to be better, I will give them a cake of candy. [It is quite likely the mixture of what would probably be beet sugar. flour and possibly yeast germs would be detrimental to the bees. Eds.]

BBJ 25 Mar 1915
BBKA Meeting. The following nomination of representative from affiliated Associations was received and accepted - Notts, Mr. G. Hayes.

Canaries (Yorkshires), one cock, two hens, long, clear birds, set for breeding, value £1 10s exchange for bees. D. Bowler, Mansfield Woodhouse

The Yorkshire Canary - also known as the "Gentleman of the Fancy" - was developed in the mid-1800s in Bradford, England, and was first shown in Yorkshire, hence its name.

BBJ 8 Apr 1915
"Juno" (Southwell). We do not find any disease in the bees sent. You should be feeding the bees with syrup now if they need it. Give the syrup luke-warm, wrap the feeder up well and the bees will be all right. There is very little that you can do to effectively prevent "Isle of Wight" disease. You may medicate the food with quinine and keep the hives supplied with disinfectants; also see that there is a constant supply of clean water. We cannot say exactly the number of bees, but probably between 2,000 and 2,500. A good stock now should cover about six frames. The first symptoms are the bees loafing about on the alighting board and the colony showing very little inclination to work. Here and there "crawlers" may be noticed; these increase in numbers as the disease progresses.

BBJ 15 Apr 1915
Bee Shows to come. June 29th to July 3rd, at Nottingham. Royal Agricultural Society's Show, Bee and Honey Section, under the direction of the BBKA. Prizes arranged in groups of counties for Associations affiliated to the BBKA. Schedules from W. Herrod Hempsall. 23, Bedford Street, Strand, WC. Entries close May 31st.

BBJ 15 Apr 1915
M. Brodhurst (Southwell) (1) Use Naphthaline and Apicure. (2) Yes.

BBJ 22 Apr 1915
Having more bees than I can manage, have decided to sell some hives and stocks; up to date and healthy. What can I book for you?   Uriah Wood, expert, Arnold

Bee-Keepers' Record May 1915
Hayes Densimeter
We have received one of these instruments from Mr. Hayes. They will no doubt prove a boon to judges for determining the comparative densities of bottles of honey in cases where there is keen competition. We append Mr Hayes' description.

"After hearing Mr Pugh's lecture on 'Judging Honey' at the meeting of the BBKA in October 1913, I determined to try and find a means of testing what is generally called the 'density' of honey, but which, in fact, is its *viscosity*. However, seeing the former term is more generally used, it will perhaps be best to keep to it, as it will be the better understood and for this reason I have called the instrument a *Densimeter* or measurer of density. After very considerable experiment I decided that this is the most easy and ready way of testing the density of honey.

These instruments are carefully adjusted in a scientific way to a standard and have been tested by several judges, who state that they are just what is wanted in close tests, or by a beginner or anyone who wishes to come to a definite conclusion by a scientific process regarding the density of any sample.

To Use: Hold the densimeter between thumb and finger and allow it to sink into the honey up to the figure 2 below the normal line, hold it there until you have noted the time; then allow it to sink of its own weight for not less than thirty seconds, or as much more as time permits, keeping it upright with just a touch from the finger, as occasion may require. At the expiration of the time allowed, grasp it with thumb and finger with an upward motion and read off the honey-line; which will be clearly seen on the tube. The deeper it goes, the thinner the honey. It should be quite clean, free from honey, and dry for each test and this can be easily done by washing in a tumbler of water and drying it with a soft cloth. Two densimeters expedite a large test".

BBJ 13 May 1915
Those bee-keepers intending to exhibit at the "Royal Show" at Nottingham must note that entries close on May 31st. Applications for schedules should be made, and entries sent in as early as possible, to the Secretaries, BBKA Office, 23, Bedford Street, Strand, WC. We trust that our readers and beekeepers generally will do their best to make the "Hives and Honey Department" at the premier show of the country a success by sending in a record number of entries, and thus hearten and encourage the BBKA in undertaking to keep up their department at the show this year. Do not forget there is the WBC Memorial Gold Medal, and that Colonel TH Jolly has again offered a WBC hive to be competed for.

BBJ 17 Jun 1915
Report from North Notts. I am sending you a short report from this district. I think this has been the best time for bloom that I have known chestnut, sycamore, and hawthorn in profusion.

My three stocks are doing well and ready for second supers. A neighbour has, however, beaten me, as I have extracted about 25 lbs. of honey for him to-day (June 10th) and it was only on Whit Monday that the supers were put on his bees and my own. It is grand honey - light, good flavour and very thick. What we need now is a good rain for about twenty-four hours to bring the clover on. I was sorry to hear the vicar of a village some few miles away has lost sixteen stocks from "Isle of Wight" disease. I hope it will not come any nearer here. We, all enjoyed the "Extracts from an Expert's Diary" last week. Well we remember the old motorcycle and the "sparking plug"!

Since writing the above I have extracted 50lbs. of honey from two supers, This is early for this district as we do not usually have any honey ready to extract before the last week in June.   T. Marshall, Ivy Cottage, Sutton-on- Trent.

The monthly meeting of the BBKA Council was held at 23, Bedford Street, Strand, London, WC. On June 17th, 1915. Mr. CLM. Eales presided, and there was also present AG Pugh.

The Chairman said how pleased he was to see such a good attendance in these times of exceptional stress and he sincerely hoped that every member of the Council would use his utmost endeavour to be present at the next Council meeting at the "Royal Show" at Nottingham.

Next meeting of the Council in the Secretary's Office, Hives and Honey Department, Royal Show, Nottingham, on July 1st.

BBJ 24 Jun 1915
We give our readers one more "reminder" that the "Royal Show" opens at Nottingham on Tuesday next. All who possibly can should make a point of attending the Show and paying a visit to the "Hives and Honey" department. Although we cannot at the time of going to press say with certainty what number of entries there are in the honey classes, we believe they will be equal in number to, if not exceeding, those at Shrewsbury last year.

The city of Nottingham is in a central position, with good railway facilities, and the attendance should be good, especially in view of the fact that there will be very few Shows this year. This will be apparent if a comparison of our "Bee Shows to Come" column in this issue is made with the corresponding issue of last year, where there is a list of twenty Shows. This year there are, so far, only two.

As bee-keepers have responded so well to the appeals made in our columns by making a goodly number of entries, it only remains to follow this up by a good attendance at the Show. We are looking forward to meeting even more old friends than usual in the native county of our Junior Editor and his brother, our Manager, who will both be present. We hope none of our readers will fail to pay a visit to the honey tent, and to use a somewhat hackneyed phrase, "Bring your friends with you."

BBJ 24 Jun 1915
BBKA Meeting. The Chairman said how pleased he was to see such a good attendance in these times of exceptional stress, and he sincerely hoped that every member of the Council would use his utmost endeavour to be present at the next Council meeting at the "Royal Show" at Nottingham.

BBJ 24 Jun 1915

Honey under a False Trade Label. A prosecution by the Board of Agriculture and Fisheries, before Mr. Mead at Marlborough Street Police Court on April 22nd, will no doubt be interesting and instructive to our readers.

FW Weitzel, trading under the name of The Globe Honey Company, was prosecuted for selling to Messrs. WS Chapman & Co., Ltd., honey under a false trade label, viz., "Pure Cambridgeshire Honey. Elsom & Co., Heydon Apiary, Royston."

Mr. J. Cornelius, one of the Board Inspectors, stated that he bought a jar of honey at one of the retail shops of Messrs. Chapman & Co. and submitted it to Mr. W. Herrod-Hempsall for his opinion as to whether it was honey or not and, if honey, its probable source. The opinion given was that it was a bad sample of foreign honey. He had also been present when a search was made of the defendant's premises when, amongst other things, labels giving descriptions of various kinds of honey were found; the foreign honey was also found.

Mr. G. Hayes, of Beeston, Nottingham, after answering questions as to identification of himself and position as regards his ability in the matter, was asked for his reason for concluding the sample of honey submitted to him was foreign and stated that:

1. The honey contained a large number of pollen grains from eucalyptus and was of that plant.
2. The honey contained other pollen grains which he had not found in British honeys.
3. The frequency of pollen grains was greater than in English honey.
4. There was a lack of certain pollen grains which are to be found to a less or greater extent in all English honeys.
5. He also submitted photo-micrographs of pollen grains from the honey in dispute, and also from a sample of what he knew to be Cambridgeshire honey.

BBJ 24 Jun 1915

HH Dennis (Notts.) (1 and 2) The queen has not mated. (3) No; they have not the room.

BBJ 8 Jul 1915

The monthly meeting of the BBKA Council was held in the office of the "Hives and Honey" Department in the Royal Show Ground, Nottingham, on July 1st. The Rev. FSF Jannings presided, and there were also present Miss MD Sillar, Messrs. G. Bryden, J. Smallwood, Association representative, WW Falkner (Leicestershire), and the secretary, W Herrod-Hempsall.

BBJ 8 Jul 1915

The "Royal Show" at Nottingham. It would be difficult to find a more suitable ground for this show than Wollaton Park, in which stands Wollaton Hall, the seat of Lord Middleton. The surface is level, and there is no lack of space, and the scenery is delightful. As a whole the show may be counted a success, the attendance being very good.

Bee-keepers are to be congratulated on the splendid display of honey on the stage. The trophy class was not so good as last year, only three displays being put up, two by Mr. R. Brown, of Somersham, one of which won first prize, Mr. J. Pearman, of Derby, taking the second prize. One other trophy was entered but could not be staged as the entrant was unable to procure bottles. The classes for wax, mead, vinegar, etc. were not so well filled as usual but the samples shown were of very good quality. The Rev. FSF Jennings, FES. staged a group of waxes of various kinds, mineral, vegetable, and insect, including several samples of beeswax, and the tray from a solar wax extractor containing wax just as it had dripped into it. This exhibit was a source of great interest to the numerous visitors and secured the first prize, the second in that class being taken by Mr. W. Dixon with a very tasty display of honey confections and medicines.

Mr. Watts exhibited his well-known Bee Escape and also his recently invented Hive Roof Escape, for which he was awarded first prize. Unfortunately, Mr. Watts will not be able to place the roof escape on the market at present as it is impossible to find a firm who can undertake their manufacture owing to the great amount of Government work in hand. This difficulty will vanish when the war is over, and bee-keepers will then be able to avail themselves of this very efficient device.

A splendid stock of hives and appliances were shown by Messrs. Jas. Lee and Son, Messrs. TN Harrison and Son, Mr. EJ Burtt, and Mr. P. Meadows. No money prizes were given in these classes but we hope these firms will have been rewarded for their loyalty in thus contributing to the success of the show by doing some good business. We are sure our readers will be pleased to hear that Mr. J. Pearman, of Derby, has this year won the WBC Gold Memorial medal. He has for many years been a regular and successful exhibitor at all the large shows, and a great number of the minor ones, but we venture to think no prize he has won will be as valued as this one.

The second prize was again won by Mr. R. Brown, of Somersham. Our readers - especially those who have met Mr. Brown "Dick Brown"' as he is known to his friends - at the various shows and enjoyed his genial company, will be sorry to hear that he has been for the past month, and is still, very seriously ill and was therefore unable to be at the show. This is, we believe, the first time for about thirty years that the "Royal" has been held and "Dick" has not been present. He was greatly missed, and by none more than ourselves.

The NBKA held their annual competition in connection with the BKA and made a most creditable display. The William Herrod Perpetual Challenge Cup was again won by Mr. Jas. North. They also held a Honey Judging Competition during the show, the results of which will be published later. Our Senior and Junior Editors were appointed judges but, to the disappointment of all, Mr. Cowan was unable to attend, his place being ably filled by Mr. E. Walker, Mr. G. Hayes was steward, and the secretarial work of the show was undertaken by Miss M. Dagmar Sillar and Mr. J. Smallwood.

The following awards were made:
Collection of Hives and Appliances, including a Suitable Outfit for a Beginner in Beekeeping - An Award of Merit was given to TW Harrison and Son, 5 and 7, Cheapside, Nottingham,
Honey
Class 538. Observatory Hive, with Bees and Queen - 1st, JT Wilson.
Classes 540 to 546 confined to Members of the NBKA.
Class 541. Six Sections of Comb Honey of any year - 1st, G. Marshall; 2nd, H. Merryweather; 3rd, RB Hutchinson; rn. and hc, JE Allsopp
Class 542. Six Jars of Light Extracted Honey of any year - 1st, JT Wilson; 2nd, James North; 3rd, William Lee; rn. and hc, H. Merryweather; hc, Charles E Smith.
Class 543. Six Jars of Medium and Dark Extracted Honey of any year (excluding Heather Honey) - 1st, James North; 2nd, CE Smith; 3rd, JT Wilson; rn. and hc, G. Houghton.
Class 544.Six Jars of Granulated Honey of any year - 1st, James North; 2nd, CE Smith; 3rd, W. Hopkinson; r.n. and hc, AG Pugh; hc, G Marshall
Class 545. Beeswax - 1st, JT Wilson; 2nd, W. Darrington; 3rd, G. Marshall.
Open Classes.
Class 554. Three Shallow Frames of Comb Honey for Extracting, gathered during 1915 - 2nd, JT Wilson.
Class 559. Exhibit of not less than 3lbs of Beeswax, the Produce of the Exhibitors Apiary - 3rd, JT Wilson
Exhibit of a Practical or Interesting Nature Connected with Bee-Culture, not mentioned in the foregoing Classes, including Candy for Bee Feeding, Articles of Food or Medicine in which Honey is an ingredient - Certificate of merit, JT Wilson

BBJ 15 Jul 1915
NBKA Honey Judging Competition at the Royal Show. Eight samples of honey were submitted for this competition and, as regards their source, the result of the microscopical examination was as follows:
Pollen Grains.
Sample No. 1. Clover, lime a few.

No. 2. A mixed fruit honey hawthorn, apple, bean, gooseberry, plum and a few clover.
No. 3. Clover, with just a few others.
No. 4. Apple, hawthorn, sycamore, chestnut, melilot, honey dew (tree honey).
No. 5. Heather blend, clover, heather, etc.
No. 6. Mainly clover, a few odd grains of heather, etc.
No. 7. *Calluna vulgaris* chiefly, with *Erica cinerea*, clover, etc.
No. 8. Clover, chestnut, *compositae*, etc.

Putting out of court any heather or heather-blend as per conditions, the placing of the remaining samples by the judges was as follows:

1st - No. 3   2nd - No. 6   3rd - No. 1

The prize-winners are as follows:

1st. Mr. W. Dixon, Leeds, 30 marks for source, 25 marks for placing; total 55.
2nd. Mr. JT Wilson, Shirebrook, 26 marks for source, 25 marks for placing; total 51.
3rd - Mr. J. Herrod-Hempsall, Luton, 25 marks for source, 25 marks for placing; total 50.
Highest possible - 40 marks for source, 30 marks for placing; total 70.

Geo. Hayes, Secretary.

BBJ 15 Jul 1915

JHA (Notts.) You can get tins to hold 7 lbs, 14 lbs or 28 lbs from any appliance manufacturers. Try and get some grocer or dairyman to take it, or advertise it in the "BBJ." A good idea is to put it up in "family" tins of 3 lbs or 7 lbs. If you are a member of the NBKA take advantage of their "Members' Exchange Mart."

BBJ 22 Jul 1915

NBKA Honey Judging Competition at the "Royal Show". I send herewith a few further particulars with regard to this matter, as we think they will prove both interesting and instructive. They are given for that purpose and not, as so many feared would be the case, to show up the competitor. It will be quite obvious that no one can say or know who were right and who were wrong - beyond the published names of the winners - except the individual competitor. It is hoped that each will compare the duplicate of his efforts - which he retained - with the actual sources as published, which should in some measure educate him in the matter.

No. 1 was a sample of clover honey, with just a flavouring of lime sufficient to make its presence known. Some of the competitors put this down to be from sycamore, fruit, lime, mustard, and apple wholly; whilst others considered it was from fruit and sycamore, fruit and charlock, fruit and clover, clover and sanfoin. Only one competitor gave its true source

No. 2. This was a sample of early honey gathered in the vale of Belvoir, near to Belvoir Castle, and was a fine sample of "mixed fruit honey." It was variously ascribed to the maple, hawthorn, bean, sycamore, and heather. Five competitors named its true source.

No. 3 was as pure a sample of clover honey as is to be found generally. Twelve competitors were correct as to its source, three stated it was mixed with sanfoin, fruit, or raspberry, two that it was lime.

No. 4. This was a sample of new honey gathered in the park at Nottingham, and was what is generally termed "Tree Honey" being obtained from apple, hawthorn, sycamore, chestnut, and other melliferous trees. There was also a slight amount of honeydew in it, which was noticed by some of the competitors. Although this had been extracted a fortnight before the show, and that at the time of the show the lime blossoms had not opened, no less than five competitors considered it was from that source. Three only were correct in their diagnosis.

No. 5. Heather-blend. Dark in colour. Five competitors named this correctly whilst some put it as wholly from the heather, others from sycamore and fruit, hawthorn, gooseberry, clover and bean, tree honey, blackberry, lime, and chestnut.

No. 6 was practically the same as No. 3 with the faintest trace of heather. Only one arrived at its true source, whilst eleven put it down as clover, one as lime, and others as clover and some other mixture.

No. 7. Heather from both *Calluna vulgaris, Erica cinerea*, and a few other sources, a stronger heather mixture than No. 5. With one exception all attributed this to the heather, either wholly or partially, the one dissentient named it as lime.

No. 8 was a rather puzzling sample, which was obtained from Devonshire, and was chiefly from the clover, but there were a good few dandelion and other compositeae which imparted a yellow colour to it. It was also partially granulated, and bore a faint trace of carbolic. Only one named its true source, the others considered it from lime, dandelion and mustard, sycamore, apple, sanfoin, buckwheat, hawthorn, etc.

BBJ 22 Jul 1915
PWD (Beeston). The bees are suffering from "Isle of Wight" disease.

BBJ 29 Jul 1915
"DoNNo" (Chilwell). Clearing Supers. To get the bees out in anything like that time you would have to use a very strong solution of carbolic for the cloth, which would taint and spoil the honey. We should prefer to take the super a distance away from the hive and brush the bees off using a feather or a twig from a bush to remove those that refused to be dislodged by shaking.

BBJ 29 Jul 1915
PJS (Notts.) Under the circumstances we think your best plan would be to drive the bees out of the box and tie as much as possible of the comb containing brood into frames, re-hive the bees on them in the new hive and feed up for winter.

BBJ 12 Aug 1915
MM (Mansfield). The honey is spoiled by honeydew which gives it the dark greenish

colour. It is worth 3d or 4d per lb. in bulk. In future please do not place a letter next to the bottle. There was no cork wad in the cap and the honey had leaked out and saturated the letter which we have had the greatest difficulty in deciphering, in addition to being extremely messy. Will other readers please note.

BBJ 9 Sep 1915
"Enquirer" (Mansfield). No. 1 is rather thin and the flavour spoilt by ragwort. No. 2 is also poor in density but the flavour is good.

BBJ 11 Nov 1915
XYZ (Notts.) No. 1 is mainly from clover and a little from mixed sources. It is quite suitable for show purposes. No. 2 is from clover. There are signs of fermentation in it.

BBJ 6 Jan 1916
The great reaper is appallingly busy in many fields, and his scythe has also taken its toll from our ranks Mr. AG Pugh has suffered the loss of his wife.

BBJ 27 Jan 1916
The monthly meeting of the BBKA council was held at 23, Bedford Street, Strand, London, WC, on January 20th, 1916. Mr. CLM Eales presided. Also present was AG Pugh. It was proposed by Mr. Pugh, seconded by Mr. Eales, and carried: That the committee consist of the following members: Messrs. WF Reid, JB Lamb, GW Judge, J. Smallwood, A. Richards, J. Herrod-Hempsall and the Secretary, W. Herrod-Hempsall. The committee to have full powers to act, three to form a quorum, the meeting to be held within the next fortnight.

BBJ 27 Jan 1916
A New Theory on "Isle of Wight" Disease.
I am working late now on "munitions," this being the fashionable name for murderous articles of all sorts which are sent to "somewhere in France."

After this experience I shall be able and capable of measuring bee-hives to one thousandth of an inch, as the limit of variation we are allowed is two thousandths. I tramped up all those stairs of yours at the BBJ Office once last year but you were out. I wanted to discuss the "Isle of Wight" disease - some folks only want to cuss it. Now you must be aware that of late years the Isle of Wight has been a very favourite rendezvous for the supposed honeymooning Hun; they swarmed there. The English bees have evidently been feeding on some of this honey (not the moon), and hence the disease which has proved so disastrous to them is a pure cultivation in the digestive organs of the bacillus *Luna Hunni*, and being English bees it cannot be expected to nourish them, so they die! Of course you will add this explanation to your examination papers next year alongside the F.B. questions. The remedy is to clear out all the Huns from everywhere.   FJ Cribb, Sand Rock House, Retford.

BBJ 16 Mar 1916
Roll of Honour
Major Thos. Henry Denman, Retford. 3rd Supernumerary Co., 2/8 Notts and Derbys.
Dr. Thos. S. Elliot, Southwell. Surgeon, 2/8 Notts and Derbys.
Captain GH Black, Gunthorpe. 7th Battn. Sherwood Foresters.
Captain FM Donne, Nottingham. 2/5th Sherwood Foresters.
Captain and Brigade Chaplain Rev. Wallis Sidney Hildesley, Colwick.South Notts Hussars.
Captain and Brigade Chaplain Rev. Ed. John Powell, Cinder Hill. 3rd Mounted Brigade, Notts and Derbys.
Captain Bernard Jessop, Kimberley. 8th Yorks A.P.W.O.
Lieut. Norwood N. Howard, Sherwood. 1/5th Sherwood Foresters.
Sergt. WH Mellors, Norton Cuckney. Notts Sherwood Rangers. KiA
Lance Cpl. Tom Walter Fletcher, Kimberley. "A" Co., 1st KRRC.
Trooper Cyril Fredk. Riley, Normanton-on-the-Wolds. South Notts Hussars.
Private Frank Fletcher Fidler, Hucknall. 10th Lincolns.
All the above are members of the NBKA.

BBJ 16 Mar 1916
The Annual Meeting of the NBKA was held on March 4th, at the Peoples' Hall, the Mayor (Mr. JG Small) presiding over a large attendance. Amongst those supporting the Mayor were Messrs. GE Skelhorne, Sneinton; Wm. S. Ellis, Sherwood; AG Pugh, Beeston; FG Vessey, Newark; RW Darrington, Eastwood; A. Riley, Beeston; MH Fox, Kirkby; G. Smithurst, Watnall; TN Harrison, Nottingham; W Adams, Mansfield; HM Riley, chairman of the Leicestershire BKA and others, including a good number of ladies.

The Annual Report read by the Secretary, showed that there were now 247 members, a slight reduction; but nevertheless a gratifying number when the present circumstances are taken into consideration. With the aid of the usual grant from the County Council (£25) the Association had been able to give expert instruction all over the county, and eleven experts had visited over 200 apiaries and inspected 807 stocks. Fourteen per cent of bees examined were found to be suffering from disease, which was the highest Mr. Hayes had known during his connection with the Association.

During the year shows were held at Newark and Beeston and also at the Royal Show. The "Isle of Wight" disease had made great inroads during the season, and they feared that by now many stocks would be found to have succumbed to the disease. Unfortunately as yet they had been unable to give any cure, no remedies having been found to be reliable.

Referring to the value of bees, Mr. Hayes stated that without bees certain fruit trees would bear no fruit at all. There was, he believed, a great future for bee-keeping.

It was reported that Mr. A Riley (Beeston) had gained the final diploma, and Mr. E Hollingsworth (Heanor) the intermediate diploma.

The financial statement revealed an increased balance in hand of £16 16s 6d. The report and accounts were adopted, and certain alterations in rules were sanctioned.

The Duchess of Portland was again elected President, and Mr. Hayes, after twenty years' service, was re-elected Secretary and Treasurer. A list of members at the Front was read, and it was decided that they be retained as members though their subscriptions might not be paid.

Replying to a vote of thanks, the Mayor said the ignorance of the value of the bee to the community seemed to have spread to the Board of Agriculture, for had they done what they should at the proper time the bees of the country would have been saved in a far greater measure from the ravages of disease, though it might have occasioned hardship to some bee-keepers.

Following the meeting tea was provided, competitions arranged, and an enjoyable conversation held.

One of the most notable features of the meeting was the exhibit of honey of various colours and grades from Great Britain, France, Canada, America, Australia, New Zealand, Hungary, India, California, Hawaii, and Jamaica, comprising no fewer than 80 different samples. Flowers were also shown from which the honey was chiefly obtained. Mr. Doleman gave a paper on "Are we to return to old-style Bee-Keeping?"

The winners in the single-bottle classes for honey were:

Granulated: 1st, Mr. JT Duckmanton, Langwith; 2nd, Mr. J. North, Sutton-in-Ashfield

Liquid: 1st, Mr. JT Duckmanton: 2nd Mr. AG Pugh, Beeston.

There were also the usual prize drawings for members at tea, and for all who had paid their subscription for 1916.

The meeting went so well that the evening appeared to be far too short.

BBJ 30 Mar 1916

BEEKEEPING BETWEEN TWO QUEENS

The monthly meeting of the BBKA Council was held on March 10th, at Shearn's Restaurant, Tottenham Court Road, London. Mr. WF Reid presided, and there were also present AG Pugh and Association Representative G. Hayes (Notts).

The following Association nominated a delegate, which was accepted Notts - Mr. G. Hayes.

You know at the last Annual Meeting we had a very interesting lecture for which we were indebted to Mr. G. Hayes. The last remark the Council make is their gratitude to the Editors of the BBJ. Many associations have given up their reports. Such reports are a means of keeping members of an association together, and here in our association we can always rely on the BBJ to publish our reports, etc. I am sure we are very much indebted to them for their courtesy in spreading news through the country. I think these are the main points. Our thanks are also due to Mr. Smallwood for taking so much trouble with regard to the finance of the Association.

Mr. Pugh: All previous reports have had, if I remember rightly, a list of experts. I think it should not be dropped; but it is probably saving paper.
Mr. Herrod-Hempsall: For economy sake the list was omitted this year, but it will not be dropped permanently.
Mr. Bryden: May I move a vote of thanks to the officers who gave their services during the past year?
Mr. Pugh seconded, and it was carried unanimously.

BBJ 6 Apr 1916
A. Wells (Newark). It is "Isle of Wight" disease. Destroy the contents of the hives - bees, combs and quilts - but the hives may be disinfected. Scorch them out with a painter's blow lamp, or paint them out with a solution of Izal, or Calvert's No. 5 Carbolic Acid, one part to two parts of water. Place out in the air for a few days till the smell has disappeared.

BBJ 20 Apr 1916
Leicestershire and Rutland BKA Annual Meeting. After the officers had been re-elected, with a few slight changes, the company adjourned for tea and later listened to an interesting discussion, opened by Mr. George Hayes, on "Isle of Wight" Disease.

BBJ 27 Apr 1916
Anxious (Notts). A queen that was reared but not mated last year will not mate now. The longest time we have known to lapse between a queen emerging from the cell and mating is six weeks and this only twice.

BBJ 4 May 1916

The monthly meeting of the BBKA Council was held at 23, Bedford Street, Strand, London, WC, on April 20th. Mr OLM Bales presided. Amongst those present was Mr AG Pugh. Elections to the Board of Examiners for Lecture List included AG Pugh.

BBJ 15 Jun 1916
Few grand 1916 fertiles, immediate delivery 5; No disease in apiary.
                                                Lowe, expert, Park-road, Chilwell.

BBJ 22 Jun 1916
The monthly meeting of the BBKA Council was held at 23, Bedford Street, Strand, London, WC. on June 10th. Mr. WF Reid presided, and also present was AG Pugh.

NEP 6 Jul 1916
Bee-Keeping in Nottingham. Financial assistance is being granted by the City Authorities to stamp out "Isle of Wight Disease." Any person whose bees have recently died, or show signs of the disease, should communicate immediately with Mr G Hayes, Secretary of the NBKA, 48, Mona Street, Beeston.

BBJ 6 Jul 1916
One more "Royal Show" has passed and taking into account the times through which we are passing, it has proved wonderfully successful. Unfortunately the last two days were spoilt by the rain, which at times came down in a deluge, turning the ground into a quagmire. Had it not been for this the attendance would have been the largest since the show was last held at Manchester.

Class 11. Twelve Sections of Comb Honey, excluding Heather Honey, of any year, approximate weight 12 lbs – 3rd, G Marshall
Class 13. Twelve Jars of Extracted Medium or Dark-coloured Honey of any year, excluding Heather Honey, gross weight to approximate 12 lbs - 1st, T. Marshall
Class 14. Twelve Jars of Granulated Honey, excluding Heather Honey, of any year, gross weight to approximate 12lbs - 3rd, T. Marshall; r.n., G. Marshall.
Class 23. Exhibit of not less than 2 lbs of Beeswax, in two cakes only - 3rd, T. Marshall; r.n., G. Marshall.
Class 24. Exhibit of not less than 3 lbs. of Beeswax - 3rd, G. Marshall

BBJ 13 Jul 1916
The monthly meeting of the BBKA Council was held in the Secretary's Office, Hives and Honey Department, Royal Show Ground, Manchester, on June 29th. Mr. AG Pugh presided and having welcomed those delegates who had not attended a Council Meeting previously the meeting closed.

BBJ 27 Jul 1916
Roll of Honour

We print a further name to those sent in, and shall be pleased to have other names as soon as possible.

     2nd AM. W. Hopkinson, Wellow   RFC

NEP 10 Aug 1916
Four beehives and various appliances for sale, cheap.   7 Lilac-grove, Beeston

BBJ 24 Aug 1916
Novice (Nottingham) Transferring troubles. We cannot say why the bees decamped as it very unusual for them to desert brood. Your method was much too rough and ready. A better plan would have been to drive the bees, cut the comb carefully out and tie that containing brood and as much of the other as possible into standard frames, which could then have been placed in the hive and the bees put on them As it was so late in the season the best thing would have been to winter the bees in the skep and allow them to work on to standard combs next year, as described in the " Guide Book." We advise you to join the NBKA - Hon. Sec, Mr. G. Hayes, 48, Mona Street. Beeston.

BBJ 24 Aug 1916
"Enquirer" (Beeston) A good sample of clover honey of moderate density. You have used rather too much smoke, which has slightly tainted the honey.

BBJ 31 Aug 1916
Two strong stocks of bees, in skeps, 15s each; guaranteed healthy
                     J. Moore, Bleasby

BBJ 31 Aug 1916
Dioxogen and "Isle of Wight" Disease. I tried this on my bees, having lost 45 stocks through this pest. I opened the hive early in the morning (14-frame hive), used 1 in 6 of Dioxogen, sprayed all frames, bees, sides, ends and bottom of the hive so that it ran out at the entrance - this was done with warm water so as not to chill the brood. I covered up quickly and left it until the next day; there were no crawlers about and by the third day they were working hard carrying pollen in. Up to the present I have seen no signs of the disease returning.

I may say Dioxogen is the same as peroxide of hydrogen, which will do, and is much cheaper.   W. Mountney, Expert, Southwell

BBJ 28 Sep 1916
The monthly meeting of the BBKA Council was held at 23, Bedford Street, Strand, London, WC. on September 21st. Mr. WF Reid presided. Also present was AG Pugh, amongst others.

BBJ 12 Oct 1916

Four first class rose comb Rhode Island Red cockerels, De Graff's well known strain, splendid winter layers, 10s each.   J. Moore, Bleasby

BBJ 19 Oct 1916

It is now just over a year ago since I wrote and told you how having tried many things to cure "Isle of Wight" disease and failed, I burnt my last stock at home. I was left with six stocks at my place of work half a mile away. I packed them in October, 1915, with plenty of stores.

On looking up early this spring I found four very weak, with plenty of food; the other two being swarms in 1915 were a bit stronger, so I took them home, as you said I might try again this spring. I had done all I could to kill any germs that might be about. I used a large bucket of strong Jeyes' fluid, so strong that it killed every bit of grass that it touched, and also dug lime into the garden. Well, by the end of April the four stocks had dwindled away and died. I cleared all hives away and covered the ground around with lime, so that I started this spring with the two stocks at home, neither very strong.

I gave them candy into which I had put Izal; they ate this up. Soon they were working for themselves with no sign of the dreaded "Isle of Wight" disease. These two were from Italian parents. Thinking that one strong stock would be better than two weak ones, I decided to build up one at the expense of the other. I got No. 1 hive very strong and as the weather was very dull, and the bees could not fly much, they swarmed.

Not having any standard frames ready I hived them on shallow frames until the next day when I took ten of the best frames out of the hives where the bees had died off. I scraped the bars, boiled the metal ends and mixed some Izal and water and, with a garden syringe, gave them a good washing. After one hour the bees were on them. In three or four days four combs were drawn out and filled with eggs. I noticed for a day or two the hive was covered with white excrement which I cleaned off.

The swarm gave me 30 lbs of honey: the parent stock 34 lbs. The season being very late all three got nearly enough food for winter, which means something with sugar at 6d a lb.

In March I gave a friend three miles away a sovereign and he put a strong stock of natives into one of my hives for me. I put them on a new stand at my work-place. They also swarmed but, thinking about sugar, I removed all queen cells and returned the bees. They were out again in nine days. I repeated the same thing again. After this they set to work and gathered 64 lbs of surplus honey which for this district is very good indeed. I am delighted to say that now, after all my trials I have been rewarded

BEEKEEPING BETWEEN TWO QUEENS

by not seeing a crawler about all the season.

Several things I have done which may seem simple. First, I have made it a rule every day to pick up every dead bee I have seen about, as I think they are no good lying around. Second, I have washed my hives with strong Izal and water. Also syringed all around the hives twice a week and sent some into the entrance of the hives.

I have been very interested in the correspondence in the Journal. I read every bit, often before I have my tea. Now I was so pleased the letter (in BBJ) on the use of Dioxogen for "Isle of Wight" disease that, knowing a friend who has a fine apiary of twelve stocks, who said something was wrong with his bees, I walked over the same night that I read the letter to let him also read it and advised him to try the remedy. I found several stocks very bad with "Isle of Wight" disease. My bees having had it I knew at once what it was.

Well, he has tried it and last week I went to see how he was getting on. He said he believed that his bees would get free from the disease again, as very few from one hive only were crawling.

Five years ago no one kept bees in our village and very few bought honey. I have talked it up, had a few bottles on show and told my friends to eat it not keep it. The result is I have sold my own 127 lbs, bought from my friends 60 lbs, sold that all retail at 1s 2d a jar, and my customers state they want more. Many people bring jars back as they are so dear. I have plenty of the tie-over jars at 8d a dozen.

Wishing you and your little paper success. I forgot to say that a well-known firm of chemists are making a show in their windows here with "New Season's Honey," 10d lb jar; 6d $\frac{1}{2}$ lb. jar. Feeling a bit put out at this mean show I went in and asked the manager if he would guarantee it English. He asked what I wanted to know for. I said: "If you will I will have a lot of it," but he said he could not. I opened a jar, and had a smell, although I am not an expert, but I told him that it had not even the smell of honey, to say nothing about the taste.   G. Ward.

BBJ 26 Oct 1916
The monthly meeting of the BBKA Council was held at 23, Bedford Street, Strand, London, WC. on October 19th. Mr. W. F. Reid presided. There was also present, amongst others, AG Pugh,

BBJ 9 Nov 1916
A conference of members of Leicestershire and Rutland BKA was held at the Vaughan College, Leicester, on October 21st. Mr. AE Biggs presided. Amongst others present was Mr. George Hayes, Secretary, NBKA.

The subject chosen for discussion was: "How the Association can best help those members who have lost their bees through "Isle of Wight" disease." This was ably opened by Mr. WE Moss, of Hinckley, who pointed out that it was their duty as an association to help the members. The disease had been prevalent in Leicestershire for the last three or four years, and in some cases took a very virulent form. As a county they were situated in the very heart of the country, and had by no means the worst bee flora. He hoped some scheme might be developed whereby the members who wished could have their apiaries replenished at a nominal cost.

BBJ 23 Nov 1916
The monthly meeting of the BBKA Council was held at 28, Bedford Street, Strand, London, WC. on November 16th. Mr. W. F. Reid presided. There was also present, amongst others, AG Pugh.

After considerable discussion, introduced by Colonel Jolly, Mr. Richards proposed, Mr. Smallwood seconded and it was carried unanimously, "that the BBKA draw up a re-stocking scheme for the benefit of those associations who wish to carry out such work and that a sub-committee be formed to deal with the matter." The following was elected to the subcommittee to deal with the matter - AG Pugh with others.

Earlier in the afternoon, Mr. Judge, on behalf of the Committee, reviewed the work of the Re-stocking Scheme, which had just been brought to such a successful conclusion. The Crayford Association was the first in this country to take steps to counter the effects of the so-called "Isle of Wight" disease among bees, by introducing a practical breeding scheme.

Ninety per cent, or more of the bees in the district had been destroyed through its ravages a few years ago, and by the successful conclusion of the scheme which Mr. Judge initiated in 1914, and put into effect with the assistance of Mr. Bryden and Mr. Barnes, seventy colonies had been produced and distributed to subscribing members in all parts of the county this season.

The scheme had created considerable interest among other associations, so much so that the NBKA amongst others had each asked to be furnished with full details. NBKA had decided to put schemes of a similar character into operation next year. Bees are necessary in the production of home supplies of food, not only of honey, but their usefulness in the fertilisation of fruit is of such importance as to make it imperative not to delay re-stocking.

BBJ 23 Nov 1916
For Sale, young liver and white Spaniel, well trained, and a good worker.
                        Apply, A.Beachen, 2, Bottle-lane, Mansfield

White Leghorn Pullets. April hatched, near laying, splendid birds, 5s each; owner joining up.
                                        Hall, Highfield Lodge, Newark

BBJ 21 Dec 1916
For sale or exchange, young Liver and White Spaniel dog, well trained, and good worker, good to children; owner going to the Front, and obliged to part with dog.
                                        Mr. A. Beachen, 2, Bottle-lane, Mansfield

BBJ 8 Feb 1917
Wanted, drawn out Combs, in wired shallow frames, drone base preferred.
                                        Walter Trinder, Edwinstowe

NEP 4 Mar 1917
The Dowager Countess of Carnarvon is inaugurating a common weal campaign in the east of the county, one section of which will deal with the food campaign and war time cookery. She has obtained two vans for the purpose of touring the county, one of which will be stationed at Newark on the 14th inst. for 10 days, while the other will visit Balderton, Collingham, and Sutton-on-Trent making a three days' stay at the two former villages and two days stay at the third. The programme each will include demonstrations in war time cookery, bottling of fruit, horticulture, bee-keeping, potato spraying, etc. Tuxford will then become a centre for one and Retford for the other. Should the venture prove the success that is ardently desired the campaign will be continued, and other places selected as centres.

Elizabeth Catherine Howard (1856-1929 Italy), 4th Countess of Carnarvon 1878-1890 then Dowager.

BBJ 8 Mar 1917
Wanted, two Stocks of Dutch or English Bees, free from disease; state price.
                                        J. Brooks, Winthorpe

BBJ 22 Mar 1917
The AGM of the NBKA was held in the Peoples' Hall on March 3rd, when, owing to the unavoidable absence of the Mayor (Councillor JE Pendleton) the chair was occupied by William S. Ellis, Esq. There was a very good attendance of members and friends.

The report stated that the year commenced with 247 members, and 18 had joined during the year. Owing to the fact that no show had been held there was a slightly increased balance, which would enable the committee to do more for members during the coming twelve months. In connection with the scheme being prepared, which would absorb the greater part of the money available, the County Council had very greatly assisted by grants, and the City Council had again, after a lapse of years, renewed grants for technical education.

Great havoc had been caused by the "Isle of Wight" disease; in fact, never of late years had the country been so denuded of bees, with the result that fruit crops had been very deficient and imperfect. All known cases of the disease had been dealt with by the experts, who had burnt up all infectious matter, and it was felt that whilst no cure had been discovered for the malady, the future was brighter in that palliatives were being experimented with and giving satisfactory results. There was a balance in hand of £35 17s 10d, as against £16 16s 6d last year.

"We must never say die," said Mr.Ellis, proposing the adoption of the report. All things considered, he thought they had every reason to congratulate themselves, for undoubtedly there was strong evidence of vitality in the Association. (Applause)

Her Grace the Duchess of Portland was cordially thanked for presiding over the Association for the past year and it was unanimously resolved that she be asked to continue. The office of General Secretary and Treasurer was again reposed in Mr. Hayes. The District Secretaries were accorded a hearty vote of thanks for their past services, and were all re-elected. Mr. Riley was thanked for his services as auditor, and re-elected. The existing Committee was re-elected *en bloc*, as were also the representatives to the meetings of the BBKA. At this point the meeting was adjourned for tea, to which a large number sat down.

About 6pm the meeting was resumed, and a drawing took place for those present at tea, and the prizes were awarded to the successful competitors in the classes for honey, which was to be given to the hospitals.

Mr. E. Hollingsworth, of Heanor, having gained his Final Diploma as Expert of BBKA, this was publicly presented to him by the Chairman amidst hearty applause.

Later in the evening Mr. AG Pugh introduced the Bee-stocking Scheme. This he explained in full detail both as regards the advantages of such a scheme and the difficulties that had to be met in carrying it out. Mr. A. Riley also pointed out why it was called a Bee-stocking Scheme and not a re-stocking scheme - which was that after the re-stocking of apiaries under the present shortage of bees, it was intended to continue the apiary as a model and experimental apiary where members could obtain bees of reliable quality at a minimum of cost and where lectures and general apicultural instruction would be given.

After several questions had been put and answered, it was unanimously decided that the scheme be adopted, and that a sum not exceeding £30 be voted to the Bee-stocking Sub-Committee for the initial outlay on the undertaking. The usual prize drawing concluded a well-attended and enthusiastic meeting.
[We knew our native county was noted for hosiery but little did we think that they

would go so far as to make stockings for their bees. In war-time, too.Eds.]

John E Pendleton was Mayor in 1918 (not 1917). He was awarded ther 'Order of Leopold' by the Belgian government for his work with the Belgian refugees in Radcliffe-on-Trent.

BBJ 29 Mar 1917
The monthly meeting of the Council was held at 23, Bedford Street, Strand, London, on March 15th. The following Association nominated a delegate, and was accepted - Mr. Hayes (Notts)

BBJ 29 Mar 1917
Enquirer (Notts). The bees were affected with "Isle of Wight" Disease. Thanks for your suggestion; we will do so if possible.

BBJ 26 Apr 1917
The AGM of Leicestershire and Rutland BKA was held on April 14th, in the Vaughan College, Leicester. Mr. HM Riley mentioned that the NBKA had secured a grant of £30, which was used for compensating beekeepers whose hives had become diseased. He wished the (Leics.) County Council would give them some help. In the evening prize drawings were held, and a lecture on judging of honey was delivered by Mr. AG Pugh, of Beeston.

BBJ 3 May 1917
Wanted, stock guaranteed healthy Bees; also Swarm.
                                    Crosland, Atherstone House, Wilford-lane, W.B

BBJ 21 Jun 1917
For Sale, Stock of Bees. For particulars - Brooks, Lincoln-road, Winthorpe

BBJ 21 Jun 1917
Wanted, swarm or nucleus British Black Bees, guaranteed healthy; state price when ready.   Brooks, Lincoln-road, Winthorpe

NEP 3 Jul 1917
The van campaign which was originated and inspired by the Dowager Countess of Carnarvon, and which has for its immediate aims the economy, increase and preservation of food, money, health, and strength, has made a most promising start. With a staff of experts lent by the Board of Agriculture and by the Notts. Education and Agricultural Committees, the campaign opened in Newark on June 14th, after preliminary conferences between the Countess Carnarvon, Inspectors of the Board of Education and the Newark Food Control Committee. An open space was secured and four tents were erected, two for demonstrations in war cookery, fruit and vegetable bottling, another for lectures on horticulture and the other for bee-keeping

and other purposes. Although the time for preparation was short, a distinct all-round success was achieved. On the same day, commencement was made at Balderton, where the school and a field in the heart of the village were placed at the disposal of the van and of the various lecturers, whose subjects included cookery, horticultural pests, potato spraying, and fruit and vegetable bottling.

Collingham, Sutton-on-Trent, Tuxford, Retford, and Worksop have, in turn, become centres for the surrounding villages and before the campaign is over, it is hoped that all the most important districts in the county will have been visited. The planned by inspectors of the Board of Education and then Miss Hood, who is acting as advance agent in a voluntary capacity, visits each proposed centre a few days before the van due in order to obtain the support of the people in the district most likely of service to making the visit of the van of the utmost use to the inhabitants.

BBJ 12 Jul 1917
Four young surplus Dutch Queens, guaranteed healthy, 4s. each, registered post free.
J. Moore, Bleasby

BBJ 2 Aug 1917
Head your colonies with my 1917 young vigorous three-banded Italian queens, bred for honey production and disease resisting.
Hudson. The Apiary, Abbey-street, Worksop

BBJ 9 Aug 1917
A Dorset Yarn. Bees were working the lavender flowers largely during the past week. If one could get all the honey they obtain from them in one section, the taste would be very sweet, I surmise; but it is mixed with honey from other flowers in the cells. Mr. George Hayes, of Nottingham, would be able to tell us what percentage of lavender is in it.

BBJ 16 Aug 1917
Strange Mortality among Bees. Much alarm is felt by beekeepers in North Devon, where whole swarms of bees have been found dead. It is presumed that they died in consequence of having worked among the flowers of potato plants which had been sprayed with a patent mixture to keep off disease.
From the *Nottingham Evening Post*.

BBJ 16 Aug 1917
Pure Italian nucleus and hive, 4 frames, 1917 queen; 27s 6d.
Beeson, Southwell

BBJ 13 Sep 1917
Bees have done fairly well in this locality. One of my stocks has beaten the record for this village. I have bottled off 144 lbs. taken from six racks of shallow combs. I had

four full ones on at once. I hope it will escape disease and survive the winter, as the hive is now packed full of bees and stores. Another stock I had at an outlying farm about $1\frac{1}{2}$ miles away gave me 45lbs. surplus but as I am unable to give it so much attention the bees swarmed or no doubt it would have given much more. We are very busy and like so many more, short-handed. Three of my sons have "joined up," so I have not so much time for the bees now. T. Marshall. Sutton-on-Trent

BBJ 11 Oct 1917
YYZ (Notts). Both queens bad mated. We could not find Nosema spores in the bees. We cannot explain the cause of other conditions.

BBJ 11 Oct 1917
Two fine 1917 Dutch queens, guaranteed healthy, 3s 6d each; registered free.
 J. Moore, Bleasby

MARRIAGE.
On October 11th at the Gunnersbury Congregational Church, Chiswick, AG Pugh, of Queen's Road, Beeston, to Catherine Edith Lundie. Mr. Pugh is well known as an enthusiastic bee-keeper, not only in Notts. but all over the country, and we feel sure our readers will join with us in congratulating him on his recent marriage, and we hope he and Mrs. Pugh may spend many happy years together.

BBJ 13 Dec 1917
Every beekeeper who has diseased stocks should therefore at his earliest opportunity test the remedial value of these antiseptics, whose prophylactic use must also be encouraged, inasmuch as "prevention is better than cure," and it is the only way to ultimate success. Both "Proflavine" (which is less toxic and cheaper than "Acriflavine") and "Dichloramine-T " (which is considered more efficacious than "chloramine-T") are manufactured by Messrs. Boots Pure Drug Co., Ltd., of Station Street, Nottingham. "Kerol" is prepared by Messrs. Quibell Bros., Ltd., of 153, Castlegate, Newark.

It is a pity that such noted manufacturing companies have no direct interest in bee-keeping, and have not yet been made to realise the many-sided gain in championing its interests.

In his book "Bee-keeping in Wartime" which he belatedly published in 1918, William Herrod-Hempsall, indicated what beekeepers could expect to pay to set up hives after the war. At that time extracted honey was selling in bulk at 1s 9d per pound. A copy of this book in electronic format can be found in the NBKA library.

	Lowest		Highest		
	s	d	£	s	d
Hive with 10 standard frames, super, etc	10	6	1	5	0
Feeder	1	3		1	9
Foundation 1 lb	2	6		2	6
Smoker	2	3		3	6
Super foundation	1	3		2	6
Extractor	19	0	2	10	0
Veil	1	0		1	6
Swarm	10	0	1	0	0
TOTAL £2	7	9	5	6	9

It should be realised that in 1918 agricultural labourers, the lowest paid manual workers, were earning 60-70s per week.

BBJ 14 Feb 1918
A few healthy Swarms for disposal, about third week in May, £1 each. Orders, with remittance, booked in advance, and supplied in rotation.   J. Moore, Bleasby

BBJ 7 Mar 1918
At Sutton House, Round Green, Luton, Beds. on March 2nd, Thomas Herrod-Hempsall, aged 76 years. He was the father of our Junior Editor and the Manager. We are sorry for any delay in answering correspondence, etc. during the last few days. The above, will explain the reason.

BBJ 7 Mar 1918
Wanted, good 10in. Lawn Mower with cylinder and grass box.
                        Wootton, Hope Street, Beeston

BBJ 14 Mar 1918
The NBKA AGM was held in the Peoples' Hall on February 23rd the Mayor (Councillor John G. Small) presiding. There was a very good attendance of members, who took great interest in the proceedings.

John Godfrey Small was Sheriff of Nottingham in 1914/15 and Mayor in 1917

The minutes of the previous Annual Meeting were read and confirmed.

Letters were read from Mr. Pugh, Mr. EO Vessey and others, regretting their inability to be present.

The annual Balance Sheet was next dealt with and as all appeared to be satisfied with the same, it was passed unanimously.

The Committee's report was as follows:
In presenting this, the 33rd Annual Report, your Committee observe with regret the scarcity of bees which prevailed in the county during the year under review - and according to reports, this was general throughout the country - not only for the loss of honey which was required to replace sugar, but for other food, especially fruit, in which the bee plays such a prominent part in bringing about fertilisation.

For three years the "Isle of Wight" disease has been prevalent in our apiaries, destroying a large number of stocks. We are, however, very hopeful as regards the future, for two reasons: firstly, we have in use some commendable disinfectants, which are at least helpful in keeping the disease in check; and, secondly, we hope to be able to supply a good number of nuclei from the bee-stocking apiary during the current season.

The summer of 1917 was good for the secretion of nectar, and bees gathered a fair quantity of high-class surplus. The present price of honey is higher than it has ever been in the memory of our oldest member.

A perusal of the Balance Sheet will suffice to show that we are in a good sound state financially.

The County Council and the Estates and the Public Park Committees of the City Council have again favoured us with their greatly esteemed grants, which have enabled our experts to make an autumn visit to those who possessed bees last year, as well as the spring visit.

That "Isle of Wight" disease is still with us there is no denying, although its virulence appears to be more localised. That it will be with us for some time we must expect, seeing that no definite cure has yet been found. The latest disinfectants have, in many cases, been found effective, and with these at our hands, it has made it much safer to recommence than it was previously.

The Bee-stocking Committee set out with the full intention of being able to supply bees during 1917, but regret they were unable to do so, owing to many unforeseen difficulties which they had to meet. Bees were exceptionally scarce, and we were only able to obtain three stocks. From these it was not possible to supply the 45 applicants with bees, so it was considered it would be best to stock the apiary as far as possible, to enable us to make a good start in 1918.

The apiary has been established on a plot of ground held by Mr. Pugh in an ideal spot for breeding bees except that it is a poor honey district. We have all we need for the season's work, except a few more stocks of bees and an apiarist and we hope that some of our members will be able to help us out of this difficulty.

In conclusion, we are full of hope that the coming season may have brighter things in store for us than has been the case for the last three or four years.

At the conclusion of the reading of the report it was resolved that the same be adopted and printed with the list of members.

Her Grace the Duchess of Portland was heartily thanked for allowing her name as President of the Association, and was unanimously re-elected for the current year. The secretary and treasurer were re-elected. The District Secretaries were thanked for their services and re-elected. After thanking Mr. Riley for his services as auditor, he was asked to again undertake the duty, and was re-appointed. The representatives to the BBKA meetings were re-elected.

As this finished the business for the afternoon, Mr. Smethurst rose to propose a vote of thanks to the Mayor for finding the time, amongst his multitudinous engagements, to come and preside over the meeting. This was very warmly seconded by Mr. Riley and supported by Mr. Thos. N. Harrison, and was carried with acclamation.

His Worship, in replying, said that he wished something more definite could be done to overcome the disease, so that the industry might revive. He himself had suffered loss for three or four years, but he was willing to try again. He saw that the Association was doing all that it could to battle with the disease, and to increase the stock of bees in the county by means of its bee-stocking apiary, and wished that it might be successful in doing this. He thanked the members for the welcome they had given him.

The evening meeting commenced at 6pm and there were many more present than in the afternoon. Mr. WP Meadows of Syston, was among those present, Mr. A. Riley, of Beeston, presiding.

A most interesting meeting was concluded with the annual prize drawing, and a vote of thanks to the Chairman.

BBJ 14 Mar 1918
The Junior Editor and Manager are deeply touched by the numerous letters expressing sympathy in the loss sustained by their mother and themselves. As these number some hundreds, and the pressure of work just now is enormous, it is impossible, much as they desire, to acknowledge each one separately. Therefore, to all who have written they hereby express their gratitude and appreciation for the kindly sympathy

and feeling expressed in the letters received.

BBJ 14 Mar 1918
Bee-keeper's know it pays to have high fecund stock! Grand hen White Wyandottes from Tom Barron's No. 8—260 to 287-egg—pen, Eggs 10s 6d per dozen (I paid 21s), eight acre grass run; also pen from birds trap-nested for 13 generations, mated to Barron cockerel, 7s 6d dozen. Inspection cordially invited. Order quickly, as only two pens. Stanley Wootton, Hurlingham, Hope Street, Beeston.

The Wyandotte is an American breed of chicken developed in the 1870s. It was named for the indigenous Wyandot people of North America. The Wyandotte is a dual-purpose breed, kept for its brown eggs and its yellow-skinned meat.

BBJ 28 Mar 1918
"Queens" (Nottingham) Effects of hybridising.
1. The first cross between Italian and Natives are usually gentle.
2. It is usually considered that disposition is inherited from the drone and working qualities from the queen.
3. Yes.

BBJ 4 Apr 1918
The following Association nominated a representative on the Council, and was accepted - Notts, G.Hayes.

BBJ 4 Apr 1918
"Isle of Wight" disease was the cause of death. The soiling outside the other hive may have been done by bees from the diseased colony. Wash it off with disinfectant and water and treat the bees with one of the advertised remedies.

BBJ 4 Apr 1918
Wanted, by a wounded soldier, few Stocks of Bees; Dutch or Italian preferred. Price to R. Flintoff, Carburton

BBJ 11 Apr 1918
Bees wanted few healthy Stocks. A. Pride, The Nurseries, Radcliffe-on-Trent.

BBJ 18 Apr 1918
JH Smith (Notts) Transferring bees from old hive. Your hive is evidently an old "Stewarton". We should prefer to transfer by the method given in "The Guide Book." We do not think either of your plans would be successful. The first would be almost certain to end in disaster, as all the flying bees would return to the old stand, leaving the queen with too few bees in the other hive. Your second plan would probably

result in the bees all going to the combs containing the brood, leaving the queen to perish.

BBJ 18 Apr 1918
Bees Wanted one or two healthy Stocks, 1917 Queens. Moderate price given. Apply Geo. Ward, Langley Mill

BBJ 2 May 1918

> To prevent and cure Bees of "Isle of Wight" Disease,
> JC Allsopp's B'kure (Registered).
> The Powder is simple to apply. Quick in action.
> Full directions on tins.
> Price 2/6 per tin. Postage 5d.
> JC Allsopp, 87, Gertrude Road, West Bridgford

BBJ 9 May 1918
Wanted, Hives, in good condition   Uriah Wood, Arnold

BBJ 6 Jun 1918
Strong Stocks of Bees wanted. Must be guaranteed healthy.   Trinder, Edwinstowe

BBJ 6 Jun 1918
A Dorset Yarn - JJ Kettle
"I have had a most interesting letter from Mr. Trinder, of Edwinstowe, re ventilation of hives and prevention of disease. There is a great deal of logic in it. I hope some of our experts will give us their wisdom on the subject. I have lifted up the brood chamber and outer case with a couple of sticks in some of mine, to see if this adds to the stamina of the young bees to stand the long damp winter and shall keep them up until robbery begins. Mr. Trinder suggests sheet perforated zinc by the sides of brood chambers, and I must try the same on some of my lot for next winter."

BBJ 20 Jun 1918
For Sale, one May Swarm from bar hive, two Nuclei, three and four frames, all from healthy stocks.   J Brooks, Winthorpe

NEP 22 Jul 1918
Food producing & saving. Notable exhibition at the University College. The Mayor Nottingham (Coun. JG Small), supported by the Sheriff (Coun. H. Offiler) and several prominent citizens, opened an Ambitious Food Producing and Food Saving Exhibition the University College this afternoon. The exhibition which, as the Mayor said, was organised first to stimulate the desire to produce and secondly to educate people as to how take care of that which they produced, includes displays of beekeeping, practical lectures and demonstrations by NBKA.

BEEKEEPING BETWEEN TWO QUEENS

Herbert Offiler was Sheriff of Nottingham 1917/18

BBJ 25 Jul 1918
Preliminary examinations were sanctioned at Nottingham.

BBJ 25 Jul 1918
JE Seaton (Notts.) Transferring from skep to frame hive. The skep should have been placed on the frames without a queen excluder under it, so that the queen could have used the new combs. When she had done this, the excluder should have been placed under the skep with the queen on the bottom combs. In 21 days the brood in the skep would have all emerged and it could have been removed. If the combs in the frames are drawn out and occupied by bees, drive the skep until the queen is secured, put her on the frames of comb, with the excluder over them, and replace the skep, leaving it for three weeks when it may be removed and the combs cut out, broken up and the honey strained from them.

BBJ 1 Aug 1918
Is it not about time that a word or two was said in favour of our old native British bee? The craze for foreigners has of late years apparently quite knocked the native out, but, in my opinion, it is, taking one season with another, still the best bee for our variable climate. Judging by nearly all the bee literature one sees, the natives are not in it.

The foreigners are puffed up as being lovely tempered, wonderful cappers of their comb, extraordinary honey gatherers, disease resisters, etc. until anyone not in the know, through not having had experience of them, must conclude that the natives are not worth having as a gift, while people are falling over each other in their haste to buy the foreign strains at five pounds a lot. The Dutch were almost guaranteed as being proof against "Isle of Wight" disease, But now one hears of their going down along with the others and, in addition, that they are just swarming pests.

In the days gone by when foreign bees were practically unknown, the complaint of the skepmen - and women too - was that the bees would not swarm enough. I have seen the bees "hanging out" week after week in a big cluster and the owner complaining that he or she was tired of waiting for them to swarm.

I have kept from twenty to forty colonies year after year without a swarm from the lot. Just had to pile up the section racks or boxes of frames for extracting, go to business and be away all day, knowing that there would be no trouble with swarming while away and that if there was any honey to be gathered the bees would store some for me.

Now this is all changed and it is a case, mostly, of almost constant overhauling, trying to prevent swarms coming off - and going off too. It used to be a treat to be able to go and lift three, four, and five racks of sections off a hive at a time, practically all filled and sealed. Now the experience seems to be a rack or two with perhaps a few sealed sections and the rest either not touched or partly filled; and a swarm which has often flown away.

Only three seasons ago I took about half a ton of honey from eight colonies - four originals, and four artificial swarms from them. The bees which gathered this were natives from off, or near to, the Yorkshire moors. This season I am having continuous trouble with swarming, or preparation for it, as the queens last season nearly all mated with foreigners. The foreigners, I'll admit, do breed more bees than the natives and are all that bee dealers may require, but the majority of us want honey as well as bees and to a man who is away at work all day the bee that will give honey with least trouble is the bee for him. There is no bee that will give a better looking piece of comb honey than the native.

They are as good-tempered as most and, in my experience, as good disease resisters, and if the queens are not quite as prolific as the foreigners, one that will fill solid a dozen combs with brood is quite enough for me. All bees are more or less uncertain in temper. I have known a stock of pure Italians to be perfect demons, and I have known many a stock of hybrids which could be manipulated almost without smoke or veil.

I have one such now, the queen of which is a wonder for breeding. Her progeny are good honey gatherers, too, but they have the one fault that just when they get into full swing they must prepare to swarm, although having plenty of storage as well as breeding room.

In forty odd years I have had only two seasons in which my natives did not gather sufficient for their winter stores. One season was in 1879; the other was 1886. One other year I had a good dozen Italian colonies, in addition to a number of natives. The Italians all had to be fed up for winter. They gave me not a pound of honey from the lot.

The natives stored sufficient for themselves and forty pounds each for me. The Italians were dethroned after that as a matter of course. It seems to be the fashion to re-queen with outsiders so as to avoid inbreeding. Now I cannot see where the fear of inbreeding comes in for we all know that the strain of bees in any neighbourhood is continually having fresh blood brought in by stray swarms and the question often arises in my mind, does not Nature provide against this in another way? All old hands know that it is almost impossible to keep any strain of bees pure, for the queens will mate with drones from a hive other than their own.

Most of us know what a keen sense of smell bees have, and, allowing for this sense alone, is it not very probable that the young queen out for a mate will avoid the drones from her own hive, and by the same rule the drones avoid their sister princess; or, if they do not avoid, prefer those from other colonies. This seems to be Nature's way all through animal and bird life, therefore why not insect life? There are two points more in favour of natives: one is that they are not so given to robbing as are some of the "foreign" strains, and the other is that they can be shaken off their combs instead of having to be brushed off. If real natives, one or two good jerks will clear a frame of comb, or a section, of every bee. "Robin Hood" (Presumably a Notts Beekeeper)

Capt. (A/Maj.) Thomas Stokoe Elliot, RAMC.served in the RAMC 1914-1922. The citation for his Military Cross in 1918 reads "For conspicuous gallantry and devotion to duty. This officer was unremitting in superintending the dressing of the wounded during four days' fighting. He was in charge of three advanced dressing stations, each of which was in turn destroyed by enemy shell fire, but he managed to evacuate all the wounded. Officers, personnel and patients were all encouraged by his cool example." In the Honours List for 1918 he was awarded a Bar to his MC for services in France and Flanders. He died in Salisbury, Southern Rhodesia in 1922.

BBJ 15 Aug 1918
For Sale, 10 new bars Bees, 1918 Queen £3.   C. Bryan, Kirkby

BBJ 19 Sep 1918
Sell or Exchange, a Double-barrel Gun, maker Kennington & Son, top lever, left choke, pistol grip, in perfect order. £7; or exchange for two Stocks Italian Bees. Must be healthy.   Thos. Parkin, 21, Muschamp Villas, Warsop

BBJ 17 Oct 1918
KMW (Retford) The cause of death was "Isle of Wight" disease. Better suffocate the few remaining bees and burn them and the contents of the hive. The hive should be disinfected.

NEP 17 Dec 1918
The efforts which are being taken to get soldiers back to the land were mentioned the meeting of Notts. War Agricultural Committee to-day. Coun. Minivers asked what was being done the matter and Mr. Hoald, the Labour officer, replied that every man with the colours was being officially circularised, whilst at the same time the Smallholdings Committee of the County Council had been working out the details.

Ald. Heath drew attention to the amount of overlapping now going on between the various authorities and expressed the opinion that it was high time the bodies concerned in this instance defined their positions. Mr. JH. said that in Lincolnshire the County Council committee had already taken action. Ald. Heath referred particularly to the overlapping with regard horticulture and bee-keeping. The Education Committee had been dealing with these matters for a long time, and were not prepared to give them up.

Mr. John Wilson, a bee-keeper of many years standing, passed away at the village of Besthorpe, near Newark, on December 19th, 1918, aged 78 years.

Mr. Wilson, who was a very capable gardener, was one of the old type of family servants, having been in the service of the same family for 61 years. He was one of our oldest bee-keeping friends, and was well known as a most capable bee-keeper when we first commenced bee-keeping. His services were much in demand as a judge at local flower shows, especially those that had classes for honey, and we were indebted to him for several hints on showing honey and wax when we first ventured to try what we could do on the show bench.

Always bright and cheerful and with an ever-smiling face, he was always - like all true bee lovers - ready to impart any information he was able to give. In those days - over 30 years ago - the great dread of all bee-keepers was an attack of foul brood, and we well remember him telling us of his efforts to combat it, one plan that he tried being to go through the hives every few days, and put in each diseased cell a drop of carbolic acid, but this failed to eradicate the disease.

Mr. Wilson had also a gift for mechanics, and when over 60 years of age learnt to drive a motor car, acting as chauffeur until well over 70. He was held in the highest respect and esteem in the village and neighbourhood. His death leaves another gap in the ever narrowing circle of our old friends. May he rest in peace.

NEP 4 Jan 1919
Honey, finest quality, 3s per pond, free post or rail.   J Moore, Bleasby

BBJ 30 Jan 1919
There are to my knowledge several keen bee-keepers who would greatly welcome an occasional good lecture and the mutual help to be derived from a live association and a local section thereof, while there are great possibilities of vastly increasing the numbers of those who keep bees.

The other day I read with considerable envy the report of an association (I believe it was the Notts) which had organised active independent sections in every small town and district in the county. What can be done in one county can surely be done in another, and I for one should be only too pleased to do what I could towards obtaining a similar state of affairs in Berks. I should not, however, care to be a party to poaching on the preserves of an existent association by starting another association in the county, or anything of that sort, and should be quite satisfied if this letter resulted in thoroughly waking up the Berks BKA. I should be pleased to hear from any other bee-keepers in the county interested in the matter.

Rippon, Springfield House, Abingdon

BBJ 6 Feb 1919
N. Wood (Arnold).Rearing queens from eggs. At the time of removing the queen from the colony put an empty comb in the centre of the stock from the queen of which it is desired to breed. Three days later go through the combs of the queenless colony and cut out all queen cells. Shake all the bees from the comb put in the other stock, which should now contain eggs and place it in the queenless colony, when the bees will utilise some of the eggs in it for rearing queens.

BBJ 20 Feb 1919
I read with much interest remarks in December respecting work being done by our friend, Mr. J. Tinsley, at the West of Scotland Agricultural College, Glasgow, and at their experimental apiaries at the Holme Farm, Kilmarnock. The report referred to was written last year and as I was recently in Scotland and made a visit to the apiaries at Kilmarnock, in company with Mr. Tinsley, I am pleased to be able to state how the hee-keeping schemes have progressed since that report was published.

Mr. Tinsley, as many of your readers are no doubt aware, is not a "canny Scot," but our friend of past years, with whom we have had many pleasant experiences amongst the bees and "shows" in Staffordshire. He has been for some four years the bee-keeping expert, lecturer and apiarist to the West of Scotland Agricultural College. He is very popular as a lecturer in the extensive area covered by the College activities, and is in close touch with all bee-keepers who report any disease in their apiaries, travelling long distances to investigate, and giving advice and assistance gratis to all who are in difficulties.

He is, with other professors, deep in the study of bacteriology and they certainly seem to be getting masters of our dreaded foe, "Isle of Wight" disease, for I was delighted to learn that nearly a 100 colonies have been distributed amongst beekeepers during the past season, and there are still about 80 stocks in the model apiaries with which to carry on the good work next year.

It was too late in the season to make an inspection, but I was pleased to note an utter absence of dead bees, or debris, which unfortunately one has become so accustomed to find on the alighting boards, etc. during recent ravages of "Isle of Wight" disease.
I enclose a photo of an apiary from which it will be seen several classes of hive are being experimented with, and I was delighted to find how well the authorities

supply Mr. Tinsley with everything needed to enable him to give the large number of pupils passing through his hands such tuition that will enable them to pass the severe technical examination in both practical and theoretical bee-keeping which the College provides for their students.

The study of bee anatomy, diseases, etc., by means of the microscope is a special feature, and the fact that the experimental apiaries have, notwithstanding the actual importation of disease for experimental purposes, produced very large quantities of most excellent honey, in addition to sending out such a large number of stocks, whilst retaining such a goodly number for future operations, is certainly very cheering to those of us who have had such disappointing results to report from other re-stocking apiaries.

Mr. Tinsley has already had a number of disabled soldiers pass through his hands with a view to teaching them modem beekeeping, and it is hoped to do further good work in this direction in the immediate future.

I feel that the work being done for our industry by the West of Scotland Agricultural College, Glasgow, is, indeed, commendable, and hope their researches will be rewarded by the discovery of that which we are all anxiously waiting for - a real cure for "Isle of Wight" disease. AG Pugh

BBJ 27 Feb 1919
Wanted, by the NBKA. Expert male or female, for season 1919. State qualifications and salary required to Geo. Hayes, Secretary, Mona Street, Beeston.

BBJ 27 Feb 1919
"Anxious" (Retford) The bees had died from" IoW" disease.

BBJ 28 Feb 1919
Notts Horticultural Sub-committee. A Conference of Beekeepers and others interested in beekeeping, will be held at the Peoples' Hall on March 1st at 6pm. An address will be given on the re-stocking scheme by W Herrod-Hempsall, Esq.

BBJ 13 Mar 1919

The NBKA Annual Meeting was held at the Peoples' Hall on March 1st, and was presided over by the Rev. A. Thornley, when over 100 members and friends were present, including several council members from the Leicestershire and Derbyshire Associations.

It was stated that 50 new members were enrolled during the season, making a total membership of 246. The honey season had not been all that could be desired, though existing bees did fairly well, and in regard to finance they were in a satisfactory position. The County Council and the City Council had favoured the Association with grants which had enabled experts to take instruction to 146 apiaries.

The examination of candidates for the preliminary diploma as expert was held at Beeston in August, when the following satisfied the examiners: Miss Darney (Retford), Messrs. G. Smithurst (Watnall), W. Sharp (Kirkby) W. Jackson (Nottingham), G. Ward (Langley Mill). S. Dodsley (Heanor) and AH Hanson (Ilkeston).

The Bee-stocking Committee reported the failure of the whole of the Association's stocks as a result of "Isle of Wight" disease and Mr. Skelhorne (Nottingham) said that in spite of this disaster, bee-keeping associations would have to go on struggling and doing their best to increase stocks. It would never do to stand aside and let the disease beat them. Pressure should be brought to bear upon the Government to investigate the disease, for unless the industry was kept going food production would suffer considerably.

All members present signed a petition asking the Government to introduce legislation for the protection of bee-keepers.

The Rev. A. Thornley, replying to a vote of thanks, expressed the opinion that the Government should vote sums of money, as in the case of the poultry industry, in order that investigations might be made, with the view of finding a remedy for the "Isle of Wight" disease.

The officers were all thanked for their services, and re-elected for the ensuing year.

After those present had partaken of tea, the meeting was resumed, and in the absence of Mr. W. Herrod-Hempsall, who should have addressed the meeting, but was prevented through illness, Mr. Hayes and Mr. Pugh together explained the Government Bee Re-stocking Scheme. They further stated that the Horticultural Sub-Committee had requested the Notts. Association Bee Stocking Committee to take over the bees and queens they had ordered and work them to the best advantage.

The Bee Stocking Committee had considered this, and now recommended the Annual

Meeting to vote a sum of £35 for the purchase of the bees and queens, to enable them to carry on the work of their Bee Stocking Apiary. The scheme put before the meeting was considered very satisfactory, and it was decided most unanimously to adopt the same; and authorise the Bee Stocking Committee to proceed as they suggested.

A vote of sympathy was expressed to Mr. Herrod-Hempsall, with a wish for his speedy recovery. The meeting was concluded by the usual draw for various useful articles, including hives, etc. in which 17 prizes were awarded.

G. Hayes, Honorary Secretary

Rev Alfred Thornley, MA, FES, FLS born near Preston in 1855. President of the Nottingham Naturalists' Society, 1897-99; Consulting Entomologist and Lecturer to the Midland Dairy Institute, Kingston, formerly Curate of Kimberley and St. Anne's, Nottingham; Vicar of South Leverton from 1885

BBJ 27 Mar 1919
The forty-fifth Annual Meeting of the BBKA Association was held at the Central Hall, Westminster, London, SW. on March 20th. Mr. AG Pugh was voted to the chair in the temporary absence of the vice-chairman at another meeting. There was a very large attendance of members, the room provided being full to overflowing.

The monthly meeting was held immediately after the Annual Meeting. Mr. WF Reid presided, and there was also present - AG Pugh amongst others. The following Association nominated a representative on the Council which was accepted - Notts. G. Hayes

BBJ 3 Apr 1919
Several surplus WBC Hives for sale, nearly new; would accept part payment strong skep.   Thompson, Muschamp Villas, Warsop

BBJ 3 Apr 1919
Wanted, by the Notts BKA, Expert, preferably female, April-August, 1919, to work the re-stocking scheme. State qualifications and salary required to G. Hayes, 48, Mona Street, Beeston

BBJ 17 Apr 1919
Wanted, Stock or early Swarm; guaranteed healthy.   Lowe, Lilac Villa, Chilwell

BBJ 1 May 1919
The BBKA monthly meeting was held immediately after the Annual Meeting on March 20th. Mr. WF Reid presided, and there present was, amongst others, AG Pugh who

proposed the election of Mr. Cowan as Chairman, eulogising the splendid work which he had done, and although not able to attend many meetings, is still doing for the Association and bee-keeping generally. Mr. Eales also, in seconding the proposition, paid tribute to the work of Mr. Cowan. The resolution was carried by acclamation. The Association nominated representative on the Council for Notts was G. Hayes.

BBJ 8 May 1919
For Sale, Marvel Extractor, little used, no gear, 18s.   Strickland, Aldercar

TO CLEAR.
WBC Ends for brood frames, 3s gross.
Ditto for shallow frames, 3s 9d gross; postage 5d.
Sections, $4^{1}/_{4}$ two and four-way split and grooved,
100, 7s 6d; postage 1s.
Excluders, 2s. 3d each; postage 6d.
Metal Dividers, for 3 sections, 2s doz.; postage 9d,
Wood Dividers, 1s doz.; postage 4d.
List on Application.   Established 1878.
WALTON & Co.,
MUSKHAM Works, Newark.

This advertisement appeared several times over the beekeeping season in 1919.

BBJ 15 May 1919
Queens, 1918 (Italian preferred), wanted at once.   Smith, 11, Woodbeck. Retford,

BBJ 22 May 1919
A meeting of Sheffield and District BKA was held on May 8th, at the Tontine Cafe, Sheffield. Mr. CH Chandler, President of the Allotments Federation, presided and Mr. George Hayes, Nottingham, gave a very interesting lantern lecture on "The Pollination of Flowers." The lecture was a fascinating revelation of the way in which flowers adapt themselves to secure the assistance of bees and other insects in bringing about fertilisation. Great interest was shown in the lecture, at the conclusion of which a hearty vote of thanks was adopted, on the motion of Councillor Bashforth, seconded by Mr. CM. Hansell.

BBJ 22 May 1919
Queens, 1918 (Italian preferred), wanted at once. Smith, 11, Woodbeck. Retford

BBJ 26 Jun 1919
Nucleus Italian Hybrid for sale. 6 bars, £3 5s   C. Bryan, 5, Urban Road, Kirkby

BBJ 3 Jul 1919
Can spare a few 6-frame Stocks, headed by Italian Queens, very prolific, £3 3s
                                                                Uriah Wood, Arnold

BBJ 17 Jul 1919

Miss Darney (Notts). Cambridgeshire is a good county for bee-keeping.

BBJ 31 Jul 1919

For sale, two Hybrid Italian Stocks, 1919 queens, nine bars, never had any disease, £3.   C Bryan, 5, Urban Road, Kirkby

BBJ 7 Aug 1919

Several healthy 8-frame Italian Heather Stocks, all headed by this season's raised queens, and packed with young bees, £4 each.

                                 Hudson, "Sunny Vale Apiary," Rockley, Retford

BBJ 14 Aug1919

Bee-keepers requiring a supply of sugar for feeding should apply at once to the Secretary of the Committee dealing with this matter in their respective counties for a registration form, which must be filled in and returned to the source from which it is obtained. A certificate will then be issued entitling the holder to 10 lbs. of sugar per stock any time up to December 31st, 1919. This must be presented to the Local Food Committee, who will issue the necessary coupons for the amount allocated from his grocer. The address from which the registration forms can be obtained is The Secretary, Notts. Agricultural Executive Committee, Milton Chambers, Milton Street, Nottingham.

BBJ 11 Sep 1919

Stock Geese. Pure White Embden, pen two geese and unrelated gander, £3; crate 16s, returnable; cash or deposit.   J. Moore, Bleasby

Embden goose is one of the oldest domestic breeds, with records of it dating back about 200 years. The origin of this breed is not clear but most probably, the Embden goose was originated from the town of Embden in Germany. The Embden goose are the most common goose breed used for commercial meat production, because their large size, white feathers and fast growth rate.

BBJ 18 Sep 1919

After 30 years of bee-keeping I almost decided to give it up, after losing 64 stocks with the "IoW" disease, leaving me only four stocks of old English bees. I tried almost everything recommended, but "Flavine" was the best so-called cure. This year I allowed my two best stocks to gather supplies, from the other two I have been making new stock; my count is now 12. From the two stocks I took 180 lbs of honey - 72 shallow bars.

I bought a hive of bees on March 31st. This year I have made five new stocks from this, as well as selling two queens. This bit of luck has revived my lost spirit.

BEEKEEPING BETWEEN TWO QUEENS

There is another thing I should like to mention. A friend of mine who had lost all his bees through the disease left two hives with the frames of comb and dead bees in. A stray swarm went in last summer, the bees wintered well and are alive to-day and healthy. He has taken 120 lbs. from this hive.

Take another case. I had this year some empty hives with all the old combs in. A stray swarm went in and I have got 16 combs of honey from it, also a big swarm as well. What about the "Isle of Wight" disease in these two cases? Can anyone tell me why the bees did not take the disease? W. Mounting, Southwell

George W Mounting was a watchmaker born in Newark in 1863. He lived with his younger sister, Alice W, born in Newark in 1869.

BBJ 25 Sep 1919
The monthly meeting of the BBKA Council was held on September 18th. Mr. WF. Reid presided and among those present was AG Pugh.

BBJ 30 Oct 1919
The Diary Show. The stand of honey and wax, though considerably shrunken in size as compared with pre-war shows, was visited by a goodly proportion of the large crowds which thronged the Gilbey Hall throughout the show. The duty of judging was undertaken by Mr. CLM Eales. The list of awards included:
Twelve jars of medium-coloured extracted honey (other than heather) - vhc, Mr. W. Trinder, Edwinstowe,
Bees' wax, not less than 2 lbs. in two cakes - hc, Mr. W. Trinder.

BBJ 6 Nov 1919
Will you please permit me a small space in your valuable paper, to reply to an article written by Mr. Manley, regarding too much attention being paid to enrol bee-keepers instead of looking to the already too many, or the few who would like the bee industry to themselves.

I wonder what would become of the bees (if there are any left to look after) when the "would be's" die, because they cannot take them with them. I for one am affected by his article being enrolled only last August. I would like to state it is not only learners that get and spread the "Isle of Wight" disease, for the stocks of friends from whom I have bought bees are affected and they are old bee-keepers.

I am pleased to say the disease is killed but it has left his stocks so weak that be will have to unite to go through winter. This is not the first case of "Isle of Wight" disease that has been cured.

Although my bees are free, others all round are affected. I have three stocks healthy and intend to get more next spring, say a "bee farm," but not to monopolise the bee industry. If any novice or other reader has got a stock that is strong and affected with "Isle of Wight' disease, or two weak stocks to unite and thus make one strong stock, send postage, and I will supply them with the same remedy that has cured several.

CJ Beecroft, 95, Sutton Road, Huthwaite

BBJ 20 Nov 1919

On September 9th I was at Norwell, near Newark, with an old bee-man who is well known at all Notts Honey Shows. He had one stock working vigorously. He said, "What do you suppose they are working on?" "By the smell of the honey, I should say it is red clover." He pointed to a field that was dead-looking and said, "They are not getting it there. "Well," I said, "I bet there is some somewhere."

In a bit I left him and went to Ossington. At a point well known to our junior editor, where the road's to Ossington and Carlton meet, the corner field was a sight to behold. At first glance it looked just like the heather looks from a distance when it is well out - a purple mass of bloom.

"Oh!" I said, "Here is where the red clover honey is coming from, and a good mile away." But I got a surprise. I had noticed his bees were dark, with odd ones having a dark golden band. Here they were at work on red clover - lots of them. There were also a few heads of alsike clover among it. And working on this alsike were some very light-coloured Italian bees.

I watched a while, but I never saw a light coloured bee go on the clover. I had always understood that Italians did work the red kind. I saw a hive or two at Ossington - don't know if they were light Italian; but I heard the Vicar of Weston had a hive of Italians. If they were his and he reads this account, he will be interested to know where I saw them at work, for they must have been over two miles from home; and he will also be interested to know that I saw a hive less than a mile from Weston badly affected with "Isle of Wight" disease, being robbed out by wasps on the same day, and that they are all dead now.

That was the first I saw of the disease this year, and I had been thinking how well I was escaping it; but I was not many days before I saw it quite plain in one of mine, the best lot even and I have since seen it in another place (five miles from Clay Cross).

I tried putting some sulphur in the smoker and giving them a few puffs in the early morning, and - well, I had seen a few score crawlers before, but that day they came out in hundreds, till there were very few left in the hive at night, so I put more sulphur in the smoker and finished the job then. Oh, it's a sure cure; I've seen no crawlers since. But I don't recommend it, as it seems a dangerous remedy. I have three stocks left, and they have shown no sign of it up to now.   Tom Sleight

BBJ 20 Nov 1919
The Standard Frame. The majority of your correspondents on this subject - if not all - seem to have overlooked the fact that what suits one district will not suit all. Each one naturally writes of his - or her - experience in the one particular district.

Now I, having been somewhat of a nomad, have had a pretty well all-round experience, and, if I was keeping bees for honey only in a good fruit, sainfoin, and/or clover district in the south, I would not have the Standard frame at any price but would most certainly go in for a deeper one.

But what about the heather districts? Beginning at the South of England, and going to the North of Scotland, with the Derbyshire, Yorkshire, Northumberland and other moors intervening and also a great part of Wales. For most of these districts a deeper frame would be utterly useless for bees working on heather will look after their own larder first (I might state they will do this anywhere during a glut), consequently in a moderate, or poor heather season, the brood combs are filled as fast as the young bees emerge and if deep frames were in use very little, if any, honey would reach the supers and even in a good season only half a crop of super honey would be obtained. Of course the surplus in brood combs could be pressed out but that means destroying the combs and then, again, there is no comparison between pressed honey and honey in the comb, saying nothing of the difference in price obtainable.

Before the advent of the "Isle of Wight" disease, I daresay there would be as many bees kept in the heather districts as in all the southern flower districts and it is fair to suppose that such will be the case again; therefore I cannot see how we are to get away from the present Standard frame for these districts.

In a great part of the northern districts, heather honey is the only honey thought of as honey, all other flower honeys are merely looked on as feeders for getting colonies into condition for the heather. And how are we to get away from the commercial side of the question - trading in bees on frames of comb, stocks and nuclei. Not many bee-keepers of any account nowadays carry all their eggs in the honey basket; the majority, I think, go in for trading in bees as well and if not selling, have to buy occasionally. What is more awkward than having frames come into your apiary which will not fit the hives used? I have had some, so I know.

I had for years three sizes in use, the Standard, a deeper one and another still deeper, say, about a dozen of each. This was in a good clover district, with just enough fruit bloom to keep breeding going, and I always found that the colonies on the deepest frames came out strongest in the spring: were ready for supering - or swarming - a week or two earlier than bees on the Standards, were not so apt to swarm and did not require nearly so much feeding up for winter. Often those on the Standards would all require feeding up, while the majority on the deep ones required none.

As a number of our Southern friends seem to have in use the deeper frames and will, no doubt, have made converts through this correspondence why not those amongst them who have standards in use and are about to try the deeper frame, keep to the standard length?

Where hives are WBCs a shallow eke underneath can then be utilised by taking off the bottom bar, will give the required depth and where not WBC's a shallow lift to fit the top of the brood compartment for frames to hang in will do it. Combed standard frames may need on each end of it a $1/4$ in. thick strip of length required to deeper frame, then to these two strips tack inside other strips about $1 1/2$ in. longer; these longer strips tack to inside of bottom end of the 8-in. sidebar.

This will make frames strong enough for any ordinary purpose and no waste, except that a good comb must be cut up for making strips to fill the new space: one good comb will make three or four strips. With all due deference to Mr. Manley and others, I think the 14-in. long frame is quite as good for a broodnest as the 16-in. one and, in tiering up, it would be much more than an ordinary wind that would upset the hive if placed on a stand 18 to 20-in. square.

Robin Hood.

BBJ 27 Nov 1919
About a month ago I was cycling from Tuxford to Ossington and at Moorhouse I came across the most primitive set of wooden hives I have ever seen.

Two years ago the owner had 14 stocks go under. He had never heard of "Isle of Wight" disease so he described how they died. It was the same old crawling. I said, "Don't you ever read the Bee Journal?" He said he had never seen that bee, did not know what kind it was. He did not think I meant a paper till I told him. He had kept bees 12 or 14 years and only once in all that time had he come across a man who knew aught about bees, who told him how to make those curious wood boxes he had.

He had one stock in the spring, and it swarmed on May 13th. It filled a super and some sections, but he had no dividers in the rack, so they were about all grown together. I got one, a fair one - the honey fine, a straw colour, and very thick. In all my travels I never saw a more cosy, ideal place for a dozen hives.

Tom Sleight, Clay Cross

BBJ 7 Dec 1919
In connection with the use of the brood hatching chamber, I made one of these, on

the lines of Dr. Abushady's article appearing in the BBJ, which introduced it, some time last winter, and experimented with it during the late spring and early summer this year. It had several modifications in detail owing to the scarcity of materials but the main principles were adhered to and followed out.

At the outset I was disappointed in not being able to give the contrivance a good trial, through the loss by "IoW." disease, of six out of ten stocks during the late winter which included the best and strongest. The cause being traced to the importation, by a neighbour, of an apiary from an infected area into a healthy district inasmuch as we had been free from "IoW" disease for two years: a case which would not have occurred under the protection of the legislation now suggested.

The best remaining stock was not by any means strong, so therefore had to wait a considerable time whilst building up to ten combs.

The difficulty which presented itself to me was the finding of five frames of comb which were well covered with capped brood to put into the hatching chamber, as on the best was a large proportion of eggs and larvae, in various stages of their development and though leaving a few adhering bees on each comb to attend to the wants of the young larvae. It was with some timidity I boxed them up above the nest. The suggestions of Dr. Abushady were followed as closely as possible and the vacant spaces left in the brood chamber below were filled with drawn-out combs, so that the queen could get to work right away.

In examining the hive after about twelve days had elapsed I found most of the brood, and certainly all that which was capped over when the frames were put in, had emerged and most of the bees had descended into the brood chamber, but there was still some newly capped brood to hatch out, which I distributed amongst other hives.

Not being satisfied with this first trial, owing to the stock being not so strong as would be desired to give results worthy of the use of the appliance, I proceeded to set it up again for another trial, as now the stock was in a condition that would warrant the placing on of a super and therefore, I judged, in a fit state to prove the utility of the hatching chamber for the purpose I had in view in making the experiment, viz., the possibility of intensive breeding.

The brood chamber of this stock had now produced three well-covered combs of capped brood and two others not so well covered. These latter I put into other hives in exchange for better ones, thus making five ideal combs for the purpose which were placed in the hatching chamber after the vacant spaces in the broodbox below had been filled up with new frames of foundation and allowing not more than fifty bees to remain on the capped combs.

Unfortunately, I have lost the notes I made at the time as to the dates and the condition of weather but I remember well being very dubious of the results as the weather became cold and windy, with very few bees flying, and although the box was well covered and made as snug as possible, I had the strong conviction that without the usual clustering bees on the brood there might not be sufficient warmth to hatch the bees out properly and I was therefore anxious when the opportunity arrived to examine them. The results were better than anticipated.

The box contained a large quantity of young bees, for practically the whole of the brood had emerged, the bees evidently remaining instead of descending through the escape, probably on account of the unsuitable weather detaining the older bees below in their brood nest and there being little or no commotion through the departing and returning workers. I was particularly struck by the extreme docility of these young bees as compared with the older bees of the same hive.

These youngsters remained there several days without attempting to get out, apparently well content with the home in which they were born, but eventually they were transferred bodily to another hive to form a nucleus. In the meantime the bees in the brood box below had fully drawn out three of the combs and filled them with brood, eggs, and larvae, the other combs being partly drawn out.

It is, perhaps, right to mention that a number of these artificially hatched bees were found to have distorted wings and were unable to fly but I should not infer that was caused by their somewhat unnatural brooding, or even their voluntary detention in the brood hatching chamber, as the same bees might have been similarly affected in the brood chamber proper, but would not be so easily detected.

Now as a result of these somewhat preliminary trials I came to these conclusions:
1. that, although the weather was unseasonable, the bees during that time were occupied in useful, active work, instead of remaining inactive (except for covering the brood, and a little attendance on the queen), as they would undoubtedly have done, most of their comb space being occupied by brood which would remain there for about ten days before hatching;
2. that the queen was able to continue laying, practically uninterrupted, when otherwise there would have been fewer empty cells available for depositing eggs, thus making an ultimate increase in the number of bees, because I am convinced that under the same conditions, had I employed the means to increase as used in previous years, by placing a super of shallow brood frames directly over the brood nest for the queen to ascend, she would not have done so, owing to the unseasonable weather and the fact that my bees

have not always readily taken to this procedure;
3. an increase of five to my stock of drawn cut brood combs.

These conclusions alone are quite sufficient to satisfy me that the appliance is a useful adjunct to a hive, and will repay the trouble of making and using, apart from other conveniences it may be put to, and I fully intend to continue the use of it.

LW Walton, 34, Holme Road, West Bridgford

BBJ 20 Dec 1919
The monthly meeting of the BBKA Council was on November 20th. A letter of regret at inability to attend was read from AG Pugh. The Secretary reported that Mr. Pugh had been very ill but was slowly recovering. The Council passed a vote of sympathy with the wish that he would soon be fully recovered.

NEP 24 Dec 1919
Photographic studio, 24ft x 12ft sides, ends, would make good greenhouse, 300 feet glass, also three standard beehives, accessories. Owner having to clear out.

Ellis, 48, Bright-street Radford.

Advert (throughout the first half of the year)
"No Spring-Feeding, but Again First. The W.S. bees on eleven 16 x 10 framed were full and boiling over at the middle of May, and were the first to enter supers out of 23 stocks; and last, but not least, they required no spring feeding." Uriah Wood, Arnold

BBJ 1 Jan 1920
The monthly meeting of the BBKA Council was held on December 18th. Mr. WF Reid presided, and amongst others there was also present AG Pugh. The Chairman expressed the pleasure of all those present at having Mr. Pugh back amongst them again after his serious illness and that his repartee proved he had apparently recovered his usual health and spirits.

BBJ 22 Jan 1920
I am glad to see from your columns that native bees have still a few admirers, if not champions. Is not one of your correspondents mistaken in thinking that the brown variety were introduced not so many years ago? Some may have been then introduced but I think they must have been in force long before, for in our village on the edge of Sherwood Forest - nigh on fifty years ago, they were all "browns," and there were a many of them; and some few years later, when I was doing expert work, and afterwards, they were all brown natives between Nottingham and Kiveton Park in Yorks. I believe I saw them in every parish between and a good few each side, especially on the eastern side as far as the borders of Lincolnshire. The darker ones which we used to see at shows were mainly south-country bees, I believe.

The darkest - almost blacks - came from Truro in Cornwall. These "browns" were as good in all points as could be wished for. The queens were very prolific, as they would keep ten standard frames of comb full of brood from top to bottom, and corner to corner for several months during the height of the season, and were practically non-swarmers if given room enough.

In 1887, from thirty-three colonies I took over 33 cwt. honey, about half of it in sections. During several years previous to and after 1887, I sent out several hundred queens. 1886 was a bad honey season but a good breeding one. I had more than one swarm that season, the biggest of which scaled just over 13 lbs. This lot left six racks of sections (126 sections) and ten combs of brood, even outside combs being brooded right up to top and down to bottom bars. This speaks for the prolificness of the queen. Most of the above-mentioned sections were not even drawn out, but all before swarming were simply packed with bees.

If "Isle of Wight" disease has not cleared them out there should still be some of these "browns" in the Dukeries, for many of the old oaks there were occupied by stray swarms of them.

Now, as to the native bee being more susceptibly to "Isle of Wight" disease than others, is this really so? The Dutch were imported as being "resisters," but are found on experience not to be so. The Italians - I scarce dare write it - are, in my opinion, doubtful, and I know I am not alone in that. With me for several seasons the first to show the disease were Italian hybrids. It appears to me that where a neighbourhood is fairly rotten with colonies which have died of disease where a number of such disease centres are in roofs and hollow trees, to say nothing of hives in similar condition left open, no strain is more immune than another.

Three years ago I had three lots of driven bees from Pickering. I put them on combs containing honey from colonies which I had destroyed through showing unmistakably the "Isle of Wight" disease. In November these driven lots got at a pile of combs of similar quality and cleared them out. At the time there were within a few yards of them two twelve-frame hives which had for some little time shown the disease.

I did not have time to attend to them for about a fortnight, and when I did go to them found both lots dwindled to a few hundred bees in each. Being just about "fed up," I said, "Here goes," and sweeping every bee off carefully I put two combs from each hive, which each contained a patch of hatching brood about the size of my hand, into the centre of the hives of driven bees, two combs going into one, one into each of the others. The bees were hatching out before my eyes at the time. These three colonies came out healthy and strong in the spring following, and did remarkably well that year. Through having to move I sold all out.

One of them was left in the village and gave a fair quantity of honey and three or four swarms and the year after its swarms practically set the whole lot of bee-keepers in the village on their feet again. Unfortunately the "Isle of Wight" disease came round again last autumn and cleared nearly all out once more.

I am not grinding my axe in writing in favour of natives, for I have no bees now. One trait they had which nowadays many would consider a bad one, and that was as soon as the honey season was over, say, end of July, breeding would be curtailed, even if queen was a late young one. It was surprising the small quantity of stores which would carry a stock through. If a skep weighed anything like fourteen pounds in September it was considered safe for wintering. I have known many a one weigh pounds less than that even and still live through and swarm early without feeding.

The Pickering bees were pure natives, of a rather darker strain. Good honey gatherers all. From one colony of same strain and an artificial swarm from them. I, in 1916, extracted 280 lbs honey (not so bad for the despised native) and left sufficient for winter and spring stores.
Robin Hood

BBJ 19 Feb 1920
Bees. Healthy Stocks, Swarms, Nuclei, early delivery. Chickens, hatching weekly; best strains. Particulars, stamp. J. Moore, Bleasby

BBJ 26 Feb 1920
Conference of Bee-Keepers re Legislation. Present at Meeting of Bee-Keepers, etc., to discuss Bee Diseases Bill, at Surveyors' Institution on February 6th, 1920: Geo. Hayes, NBKA and C. Taborn, Notts. A. Ed. C.

BBJ 26 Feb 1920
What offers for Observatory Hive? Bees preferred. Particulars, F. Hopkin, Eastwood

BBJ 26 Feb 1920
HM Lowe (Chilwell) Testing sugar.
1. Lump, or loaf, sugar is not all pure cane.
2. It requires special apparatus, and analytical skill to determine if sugar is beet or cane.
3. A rough and ready test is to partly fill a bottle with a syrup of sugar and water, stand in a warm place for about 48 hours, keeping the bottle tightly corked. If the cork is removed at the end of that time pure cane sugar will have a sweet wholesome smell; that of beet sugar will be foul.

BBJ 18 Mar 1920
The 35th Annual Meeting of NBKA was held in the Wesley Hall on February 28th, Mr.

Wm. S. Ellis, of Hawksworth, presiding, there being a large attendance of members and friends, amongst whom were representatives from the adjoining counties of Leicestershire and Derbyshire, also Mr. C. Taborn, the horticultural organizer for Notts, Mr. W. Herrod-Hempsall, Mr. J. Herrod-Hempsall, and others.

A satisfactory record of progress came under review, it being reported as follows: During this period we have enrolled 48 new members and at the present time the number of members who have paid their subscription is higher than for the two previous years, viz., 240. We are also pleased to note that some few have increased their subscription to enable us to meet the present high cost of everything. We welcome back those of our members who have returned from the Forces, and trust we shall see them active in the more peaceful pursuit of bee-keeping and lively members of our Association.

The Balance Sheet, when carefully examined, will show that we have been able to carry our finances through in a way which we hope will prove satisfactory. The cash balance is small owing to the grant made at the last Annual Meeting to the Re-stocking Committee, which we think all will agree was money well expended, as it has been of great value to the cause of bee-keeping in this county.

The price of sugar came under review and it was stated that negotiations were in progress between the Ministry of Food and the Royal Commission on Sugar Supplies, with a view to its reduction in price for the purpose of feeding bees.

Mr. Taborn stated that they wanted to see four or five times as many bee-keepers in Notts, for getting four or five times as much fruit, etc., and that the County Council had in mind the starting of another re-stocking apiary for the production of bees to this end. (Hear, hear.)

The Duchess of Portland was re-elected president. The committee were re-elected, with the addition of Messrs. W. Trinder, Edwinstowe, and AE Goodlad, Mansfield. Mr. G. Hayes was again appointed secretary and treasurer, and he mentioned that at the next Annual Meeting he will have completed a quarter of a century in office and that he would then have to ask to be allowed to retire. Mr. A. Riley was re-elected auditor.

Mr. Riley reported that Mr. Pugh was unable to be present owing to illness and it was requested that a letter be sent him stating the meeting's regret at the cause of his absence, and wishing him a speedy recovery.

An adjournment was made for tea, to which a large company sat down.

A conference of bee-keepers and others was held in the evening, when a very lucid, instructive address on 'The best methods of increase' was given by Mr. W. Herrod-Hempsall. It was listened to with rapt attention, and at its conclusion a most enthusiastic vote of thanks was accorded him.

NEP 19 Mar 1920
Fertile Italian queen bees are being distributed by the authorities as widely possible to beekeepers throughout the country, and good supplies are available. The bees cost 8s 6d each.

BBJ 25 Mar 1920
The AGM of Leicestershire and Rutland BKA was held at the Museum Buildings, Leicester, on March 13th. Mr. GW Dunn presided over a very good attendance, which included AG Pugh (Notts).

BBJ 20 May 1920
Wanted, a few 1919 Queens of good strain. Cages sent if desired.
J. Moore, Bleasby

BBJ 20 May 1920
My Champion Strain of Hybrids. 1920 3-frame Nuclei, immediate delivery, 45s; box and carriage free; Fertiles, 8s; Virgins, 4s. Guaranteed healthy.
HM Lowe, Park Road Apiary, Chilwell

NEP 1 Jun 1920
Fruit Bottling Outfit for Sale also three Beehives and Patent Honey Extractor.
T3 Evening Post

BBJ 3 Jun 1920
Swarms. A few strong ones for delivery during June, £2 each.   J. Moore, Bleasby

NEP 9 Jun 1920
Popular Demonstrations in Bee-Keeping will be given under the auspices of the NBKA at the Arboretum, Nottingham, at 6pm on s, July 17th and 24th. Come and see the Wonders of the Hive and hear how to keep bees.

BBJ 10 Jun 1920
The monthly meeting of the BBKA Council was held at 23, Bedford Street, Strand, London, WC2 on May 20th. Mr. WF Reid presided. The following Association delegate, as nominated, was accepted – Notts. Mr. G. Hayes.

BBJ 10 Jun 1920
Miss EH Darney (Notts) The queens were virgins.

BBJ 10 Jun 1920
Pure Carniolan Alpine Queens, imported direct, 12s 6d each. Orders in rotation.

J. Moore, Bleasby

NEP 15 Jun 1920
Four WBC Beehives and Appliances.
Crossland, Atherstone House, Wilford-lane, West Bridgford

BBJ 24 Jun 1920
Will exchange good Hives for 'Swarms'.   Presley, Bakestone Moor. Whitwell.

NEP 1 Jul 1920
Four good beehives for sale   163 Gregory-boulevard, Nottingham

BBJ 1 Jul 1920
Considering the far-reaching range of the Apis Club and the newness of its constitution, the attendance at the first Annual Conference at the Central Hall, Westminster, on the last afternoon in May, was not altogether disappointing; many members having travelled long distances to be present at the initial meeting, with an eagerness that betokens enthusiasm and a keen interest in the welfare and future proceedings of the Club. Yet it must be acknowledged that the rate of one out of every twenty is a very meagre proportion, and members should look to it, and feel that their presence, even if their voices are silent, is required at the next meeting, whenever and wherever that may be.

The gathering was certainly to be congratulated in having in its midst such an able and capable chairman as Mr. JB Lamb, to preside over their meeting and the fullest confidence of all present went with the representative committee elected towards the close.

As may be imagined, from the commencement the whole of the proceedings were centred around and dominated by the financial affairs of the undertaking, which, when made known, put aside every other consideration for the time being; and the members were visibly nonplussed by the overwhelming task Dr. Abushady had borne with such perseverance and with such inadequate support for a whole year in bringing forth that finest publication in beedom - the "Bee World" – apart from the enormous work in the formation of the Club.

One very cheering note was the fact that new members were still being enrolled, and at the average rate of two or three every day; and on reviewing the discussions the writer feels that this fact is an extremely hopeful foundation to build upon, and he would like to make the suggestion that his fellow members do their utmost to encourage and increase this enrolment if possible, by each canvassing his or her bee-keeping friends with a view to inducing them to join. Lend them any copy of the "Bee World" (they are too precious to give away) for their perusal, if they have the craft at heart, a glimpse through its pages will be sufficient inducement; in this way, we can swell the membership and thereby greatly encourage the work of the representative committee authorised by the conference in their deliberations with the founders.

We have hitched our wagon to a star; our aims are high and noble, and we cannot retract or reduce them. But their attainment can only be fulfilled by each doing his share, and that for the moment seems to be to help to swell the numbers. We cannot do less.
LW Walton, West Bridgford

Dr Ahmed Zaky Abushaby (1892-1955) was an Egyptian physician living in England. He founded the Apis Club in 1919 which was an international organisation of individual beekeepers – it lasted until 1958.

NEP 1 Jul 1920
Four good beehives for sale.   163 Gregory-boulevard, Nottingham

BBJ 1 Jul 1920
For Sale, several surplus Stocks and Nuclei, Dutch-Italians.   Brooks, Winthorpe

BBJ 8 Jul 1920
The monthly meeting of BBKA Council was held in the Hives and Honey Department, Royal Show Ground, Darlington, on July 1st. Mr. AG Pugh presided.

BBJ 15 Jul 1920
Strong Stock Italian Hybrids on 10 frames, 1919 Queen, £3 10s; healthy; purchaser to send box.   Sharpe, Bentinck, Kirkby-in-Ashfield

NEP 16 Jul 1920
Wanted a WBC Beehive, in good condition   WW Evening Post.

BBJ 22 Jul 1920
Can Spare a few 6-frame Stocks, headed by Italian Queens, very prolific, £3 10s; carriage paid. A customer says: "Bees are doing very well. Send three more stocks."
Uriah Wood, Arnold

BBJ 29 Jul 1920
Bee-keepers requiring a supply of sugar for feeding should apply at once to the Secretary of the Committee dealing with this matter in their respective counties for a registration form, which must be filled in and returned to the source from which it is obtained. A certificate will then be issued entitling the holder to 14 lbs. of sugar per stock any time from August 1st to December 31st, 1920. This must be presented to the Local Food Committee, who will issue the necessary coupons for the amount allocated from his grocer. All needing sugar must register, whether registered before

or not. The address from which the registration forms can be obtained - The Director of Education, Shire Hall, Nottingham.

BBJ 29 Jul 1920
Queens, English or Italian, 1920, fertile 10s, virgins, sent by post. Neither strain has disease.   Jackson (Certificated Expert), Post Office, Bleasby

BBJ 12 Aug 1920
Two Stocks good healthy Bees, 8 and 10 frames, 7s 6d frame, carriage paid; box 12s, returnable.   Beeson, Southwell

BBJ 12 Aug 1920
Six Stocks of healthy Bees for Sale on 6 or 8 frames, £2 10s and £3.
                                       J. & J. Willatts, 20, Howitt Street, Hyson Green

BBJ 19 Aug 1920
For Sale, a few 5, 6 and 7-frame Stocks, Italian Hybrids, 6s per frame; one on 10 frames, 55s   Mrs. JE Walker, Winthorpe

BBJ 16 Sep 1920
The flow is over in this part of the country. I say "flow," but as a matter of fact there has not been one here this season. Strong stocks did well in May but since then there has been nothing doing owing to the bad weather, which has been one long spell of cold and wet. Stocks that had nothing in the supers have had to be fed for some time now. It has been the worst season I have known in all my experience and many beekeepers will have a loss on the year's working.

I can't help envying friend Kettle of the weather conditions he has been enjoying in Dorset. If he had had the same sample that we have here, the bees would not have stored anything in shallow combs, let alone sections, for the simple reason that they have not been able to get anything to put in them. However, "Hope springs eternal in the human breast," and we are hoping for better things next year.

A sharp look-out should be kept for robbing now that bees are flying freely and nectar is scarce. A gentleman asked me to go and look at his bees last week as they were robbing. One stock had been cleared right out of stores and the population reduced to less than half, through fighting.

This vice is not easy to stop once it gets real hold and the best policy is to "nip it in the bud ' by reducing entrances according to the strength of the stock inside. Above all, don't leave even the scent of any syrup or honey about in the daytime and do all the feeding at dusk. Hives should not be opened in the middle of the day if it can be avoided.

BEEKEEPING BETWEEN TWO QUEENS

I have noticed one or two little "scraps" in my apiary, but nothing of a serious nature has occurred. I paid dearly for my experience some years ago and am now always on the alert when the "flow" is declining.   H. Morton Lowe, Chilwell

BBJ 29 Sep 1920
Grocers' Exhibition. As a whole this exhibition is the best for some years but, so far as quantity is concerned, the same cannot be said of the honey department. It was, however, a most creditable display, taking into consideration the very unfavourable season.
Class 67. 12 1-lb jars light-coloured honey - 5th, Mr. W. Trinder, Edwinstowe

BBJ 23 Dec 1920
On October 23rd I went a cycle ride down Trent-side way and what a lot of corn there is out there, whole fields of barley to lead (gather in) and some to cut but the nice sunny days that followed would enable them to get it in. There seems to be very few bees that way; my informant said he did not think there was a bee left in five villages round Normanton-on-Trent. My old friend Marshall, at Norwell, had got eight good lots and, what is more, he had taken some good clover honey this year. The 'Isle of Wight' disease takes some reckoning up.

At Woodhouse (Mansfield?) three years ago a stray swarm took residence in a hive. The combs were stained and full of dead bees. They cleaned it out themselves, stored it and wintered, cast two swarms the next summer and got a lot of honey. This year they have taken 12 stones (168 lbs) of honey from the lot and sold three swarms. Now they have five lots and the mistress told me the bees had cleaned all the hives out themselves. It is really amusing, when one reads and sees what some people have done in the way of disinfecting hives; it makes one wonder is it all necessary?

October has been out and out the best month of the year, only wet on four days, thirteen sunny days, and two part sunny: no wind, only one good drying day, bad for corn drying: and such foggy nights: wind south on only four days, the prevailing wind seems to be east to south-east, temperature 65°F. November 5th.

Clover blooms freely here some seasons with nothing in it the same as he says. I lay down on a little mound one sunny afternoon in July when one would have thought clover would yield honey freely. It was white over with heads and three or four bees were busy in them. How long one had been there I cannot say, but I kept my eye on it twenty minutes and it went round hundreds of heads in that time. Then I lost it but it did not fly toward the hives then. I cannot think when there is a good flow on from clover that it takes bees all that time to load up.

November 7th. A lovely sunny day for time of year. I took a cycle ride round by Hardwick Hall. Could see plenty of ivy in bloom but a bit too cool for any bees to be at work on it. I also saw six fields of corn that wanted leading yet.

What a long drawn-out harvest this year; from starting in some counties in July till November and quite green now it is cut. In many places it is killing the clover with the stocks standing on it so long. Tom Sleight, Clay Cross

NEP 17 Jan 1921
January and February frequently prove the most fatal months to bees, owing to the risk of starvation. This is especially likely to be the case this year as many bee-keepers practised economy feeding their colonies last autumn owing to the high price of sugar. In September large numbers of colonies were not supplied with the proper amount of stores. Food in the form of well-made candy should given without delay.

BBJ 27 Jan 1921
Wanted, Stocks guaranteed pure English Black Bees. Price, particulars - Brooks, Winthorpe,

BBJ 13 Feb 1921
The mild weather has to-day induced my bees to leave the combs and come out into the open. Noticing one hive unusually busy about 3pm, I took up a position near to it merely out of curiosity. However, I soon found out the cause on observing the queen go down the alighting board and back, up the hive front a few inches, when she took to the wing, flew round three times and returned to the hive. She only got up a few feet above the hive, so that I was able to observe the flight from beginning to end.
Clement Smith, Collingham.

BBJ 3 Mar 1921
For Sale, six Bee Hives, standard size, with frames, queen excluders, feeding boxes, section racks, smoker veil, and sundry lots, painted, and all in good condition, £7, or near offer; owner giving up. MW Swallow, Claypole

BBJ 10 Mar 1921
Pure imported carniolans. Stocks, Nuclei, Queens, May onwards. Particulars, Stamp. Large number of May and June strong natural Swarms. Offers wanted.
J. Moore, Bleasby

BBJ 17 Mar 1921
The 36th AGM of NBKA was held in Nottingham on February 26th, when Mr. A. Riley, Beeston (in the absence of the Sheriff), presided over a representative gathering of over 100 members and friends, including some from Derbyshire and Sheffield Associations.

The committee's report, read by the secretary, stated that the past year had been a very successful one for the Association, a large amount of work had been done in the way of lectures, demonstrations, shows and visiting and especially in the re-stocking apiaries, all applicants for nuclei having been supplied. This work had been greatly assisted by the grant from the County Council, and with the cordial co-operation of the Horticultural Organiser, Mr. C. Taborn.

Her Grace the Duchess of Portland was thanked for her services as President and re-elected for the ensuing year with acclamation. The auditor, Mr. Riley, was cordially thanked for his services, and unanimously re-elected, as were also the district secretaries. The executive was re-elected, with the addition of Messrs. EL Walton, WE Collishaw and GR Bostock.

As the committee had been unable to find a suitable person to take up the secretarial duties, Mr. Hayes consented to continue for one more year and he was cordially thanked and re-elected. Messrs. Pugh and Hayes were re-elected as representatives to the BBKA.

It was resolved that in future the subscription should be not less than 5s. per annum, except in special cases to be submitted to the local and general secretaries.

After an adjournment had been made for those present to partake of tea, the meeting was resumed as a conference at 6pm, when Mr. Joseph Price, the expert for Staffordshire, gave a very interesting and instructive address on his work in that county, which was well received, many questions being put to him which were answered; afterwards a very hearty vote of thanks was accorded him.

There were also the usual prize drawings at this meeting, which was one of the best we have had for years, and augurs well for the revival of interest in the industry.

John H Freckingham was Sheriff of Nottingham in 1920/21 and Mayor in 1926/27.

BBJ 24 Mar 1921
You are asking bee-keepers for anything of interest, but I do not see anything from this district; I am about one mile south of the Trent. It is not one of the best districts as the honey flow is finished with the limes. We have no heather here, but we have a nice lot of clover from which we get most of the honey.

I lost my six stocks four years ago with the "Isle of Wight" disease, but started again with one six-comb lot last April. I did fairly well, considering it was such a bad season. I took forty-seven pounds of honey at end of June, then the weather broke up, and it was rain, rain, rain!

The bees started building queen cells, so I decided to divide them and made up five stocks which all have wintered and I see to-day, February 21st, are carrying pollen in.

I have retained the old queen to rear my queens from this year. I am inclined to think we have been rearing our queens from eggs from queens that are too young. In poultry-keeping we say do not set pullet eggs, as we know the chicks are smaller and not so strong. If this is so with poultry why not with bees? Perhaps more of our bee friends will say what they think of this.

A. Pride, The Nurseries, Radcliffe-on-Trernt

[Our correspondent is quite right. Strong, vigorous queens are more likely to be obtained if they are raised from the eggs of a mature queen.]

BBJ 24 Mar 1921

"Dirtywax" (Notts). Cleaning cage of 'Gersler' wax extractor. Scrub it with hot water and soda, or Fels Naphtha soap, using a brush with fairly stiff bristles, or try rubbing it with a cloth dipped in turpentine.

BBJ 14 Apr 1921

With reference to Mr. A. Pride's comments in the Journal for March 24th and the Editor's foot-note with regard to queen-rearing, I do not see that an old queen can be better than, or equal to, a young queen for the rearing of the best queens. I do not see that chick rearing and queen rearing form a good comparison as the development of a bird is quite a different thing to that of an insect.

I take it lhat maturity is that period in the life of a living thing when all growth and development has been completed and it stands at the apex of its existence. Every day after that period marks a gradual falling away from that period of completed strength.

A chick - if a female - reaches maturity when she becomes a hen, an insect when it has passed from the egg through the larval and pupal stages, emerged from the cocoon a perfect insect, mated, and - if a female - laid her first egg. After then, no further development or growth takes place.

Surely, then, a queen reaches maturity when she has laid her first egg, and every day after that means, through her efforts when ovipositing, a gradual drain on her powers and a consequent gradual falling away from her period of maturity. Granting, then, that the best queens are reared from mature queens, should not they be reared from one recently commenced laying, and not from one more than half-wornout with two or three years' almost continuous work?   JB Claxton, 80, Spalding Road, Nottingham.

BBJ 14 Apr 1921

BEEKEEPING BETWEEN TWO QUEENS

Wanted, spare Queens, any breed; must be healthy. Offers to J. Moore, Bleasby,

BBJ 28 Apr 1921
The monthly meeting of the BBKA Council was held at 23, Bedford Street, Strand, London, WC2. on. April 21st. Mr. WF Reid presided. The following Association, amongst others, nominated representatives on the Council and was accepted - Notts.

BBJ 28 Apr 1921
Owing to removal. Twelve fine, healthy stocks of Italians, highly disease-resisting strain, very prolific, excellent honey gatherers, and all headed with 1920 Queens.
Hudson, "Sunny Vale," Rockley, Retford

BBJ 28 Apr 1921
Italian-English hybrid, strong, healthy stock, £4, carriage paid; box returnable; also strong Stock of Italians with imported Penna Queen.
Sharpe, 8, Bentinck, Kirkby-in-Ashfield.

BBJ 5 May 1921
Four Lee's Holborn hives, telescopic lifts, sweet and clean, well painted, cheap as dirt, 17s 6d. each. Compare list price.   Lowe, Chilwell

BBJ 12 May 1921
Having no time during the coming season to attend to my Apiary, I have decided to sell Bees, Hives, Appliances, including an up-to-date Geared Extractor, Ripener, etc., etc., by Auction on Saturday, May 21st, at 3 o'clock, at my Garden, five minutes from tram terminus. There will be some good bargains. No disease.   Uriah Wood, Arnold

BBJ 26 May 1921
There are many other things, profitable and otherwise, to be derived by keeping bees besides honey, and without much hesitancy one could tick off each of his fingers with a distinct benefit to the bee-keeper. There is an interest and occupation for the mind; the deepening of one's knowledge in the entrancing study of the open book of Nature; a keener and quicker sense of judgment; all these in time reflect in the character of one who pursues the craft.

Then, coming to the more general benefits, we may include the resultant health and pleasure obtained from an out-of-doors hobby, and the intermingling with its closely allied sister - horticulture; the humour in the many incidents which occur, and it is on these latter little-spoken-of benefits that I write of, for during the long winter evenings by the fireside, when one has finished every word in the current "Journal" and other bee literature, or perhaps when one is perusing the diary of last year's work in the apiary, many thoughts of past experience there recur to one's mind - the failures, successes, and the little incidents which come cropping up though the humour in them was not so apparent at the time; yet, on looking back, the bee-keeper finds they bring forth a quiet smile, if he himself has any humour left in his soul.

It has occurred to the writer many times that among the habits and instincts of the honey bee there is included a very practical, though little spoken of, sense of humour, which is often indulged in to his discomfiture but whether this humour is spontaneous on the unit's part, or the considered will of their particular community, is for us a matter of conjecture only.

One Sunday afternoon, on an after-dinner stroll, I met a friend who had several times expressed a desire of seeing my wonderful bees and, being a fine day and a good opportunity, I took him to the orchard-garden apiary where my bees dwell. He was naturally a little nervous, this being probably his first introduction to bee life, so I located him under a tree where I explained he could see the happy throng without fear of molestation on their part.

After a little manipulation and many explanations of their doings and the inner construction of the hive, which is a great mystery to the uninitiated, we sat and watched them policing at the entrance.

He was evidently impressed with my exhibition of coolness whilst handling them and foolishly I boasted of my fearlessness, but hardly had my words been uttered when, straight from nowhere and as swift as a bullet, one of those capricious little wretches collided purposely with the extreme tip of my nose with an impact which quickly brought tears to my eyes and made me eat my words, and as we walked quickly away the merry buzzing of those myriad workers, as they whisked in and out of the hive, seemed to develop to my stinging senses into a roar of derisive laughter; then I saw no humour in it, I could see nothing but my quickly swelling nasal extremity. But my friends did, and the bees - I wonder? Apparently there is a truth in the old custom of "telling the bees," and since then I have been careful what I say concerning them in their presence.

In the Nottingham Castle picture gallery is a noted picture which always attracts attention, and is entitled 'Busy Bodies and Busy Bees' by Lucy Ann Leavers (1845-1915) - a fine painting of several terrier pups evincing a great interest in a skep of bees which is shown placed on a bench. One dog is standing with his fore feet on the bench looking intently into the entrance, where the bees are seen going in and out; another is closely studying a bee on the floor. Of course, the picture is a splendid animal study and the expression on the dogs' faces is - well, all dog lovers know how they can look when anything greatly interests them; but to the writer the great point of the picture is the mischievous little imp slinking away with his bobtail tucked closely between his legs, and a bee perched on his back just above the top of it. His curiosity has abated for the time being.

BEEKEEPING BETWEEN TWO QUEENS

Dogs, as a rule, have a great respect for the near vicinity of beehives; my dog has but he has a different kind of respect, however, for the extractor. His great ambition is to be about when the extracting is in progress to lick up the many drops of honey which persist in reaching the floor during the operation and for months afterwards will "sit up" should he happen to be in the room where this utensil is kept.

No better or more useful watch dog could be desired than a hive of bees situated in an orchard or allotment where one tries to cultivate small fruit. One can feel quite safe if the bees are there though a million little boys, with that immense fruit craving they all possess, live near. He would get more because there would be less taken away (even his garden "friends" would be less frequent visitors) and his crop would be finer, on account of the extra fertilisation, as all bee-keepers know.   Trent

BBJ 2 Jun 1921
Italian hybrid queens. Can spare several 1921 Queens, fertile 10s, virgins 6s. (sent registered post); no disease. W. Jackson, Certified Expert, BBKA, Post Office, Bleasby

BBJ 9 Jun 1921
What offers for 12 large Canadian Feeders, 30 round Tin Feeders, 20 Zinc Excluders, four Super Clearers on board, five Watts' Super Clearers, one Smoker, two Veils (black), all in good condition?   Uriah Wood, Arnold

BBJ 23 Jun 1921
Being nice weather during the Christmas holidays, I thought a ramble out on the old bike round Worksop and Retford, just to see how the bee trade was flourishing in that part, would be grand.

Taking a route through from my home in Derbyshire until, on crossing the border into Nottinghamshire, I take the main Worksop and Mansfield road, a tarred surface covered with the drizzle of the previous night. Here I had the sun and wind at my back and cycling was a pleasure. After I had gone about a mile I looked back - cycling the other way would be anything but a pleasure. The sun glittered on the water on the road, making it impossible to see. There not being much more of interest I pass on through Worksop. When about a mile out I can spy a few hives on a hillside; further on there is an old apiary in a sheltered spot, where bees have been kept for years and where most years one can get as good a sample of honey as I ever tasted. But, alas, it has been missing this year. I call and go inside: "Glad to see you," said the bee-man. "I have just been reading 'Bee Notes from Derbyshire' I always lookout for them." "Well, and how's the bee trade this year?" I inquire "Oh, bad," he said. "Did nought but swarm, did not get a bit of honey and had to feed. In all my bee years I never knew a worse time and, what's more, I found three lots dead to-day. I have been cleared out three times and, if they go this time, I'll bother with them no more."

"Well," I said, "it's hard times but in such a lovely spot for bees as this, never say die."

From what he says the "Isle of Wight" disease is very bad in that part. I understood him to say a man close to him in North Carlton had given £9 for two stocks last summer and both were dead. As the shades of evening are falling I leave him and retrace my steps to Worksop, where I stay the night.

Monday, a lovely warm morning, so I set off to Retford. Near to Manton colliery I see them digging some beautiful new clover to plant houses. I thought, "Oh my! They are robbing somebody's bees of some good forage next summer." But there had been some bad foresight on someone's part to sow a field with clover to be dug up to plant houses, and they always seem to plant houses on the best land to be found for growing crops, while a mile or two further on there are hundreds of acres of forest land that won't grow anything, where I should think would be the ideal place to plant houses. I pass places on this forest land where I could dump a dozen hives of bees down, and it would be next to impossible for them to find any other flower but clover within a mile all round; thirty and forty-acre fields of it. I'll guarantee pure clover honey could be got there. But I ramble on past another lodge gate on the Clumber property.

A lonely house in a wood, the only house on the roadside for four miles as far as I could see. Bees kept there would have to live on the trees. There may be a few blackberries amongst them, but very few of them were good bee trees, but such giants. A ride on that two or three miles under the leafy shade on a hot summer's day would be grand. They must have been growing there for ages by the size of them. At last I come to the open country again.

With Checker House Station on my left, the old posting house of the Normanton Inn of stage-coach days a little to my right, I dive down an ideal country lane where, by the look of the bushes, blackberries had been picked by the bushel and the little bunnies had burrows in all directions. Being only a by-way to the fields, and more of a footpath than a road, cycling was anything but a luxury over that two miles, but one has to leave the more beaten track if one would get close up to Nature in all her moods.

Tom Sleight

Checker House Station was on the Great Central Sheffield - Worksop - Retford - Lincoln main line. The station was closed in 1931 and to goods in 1963.

BBJ 7 Jul 1921
Six hybrid nuclei, 4- and 6-frame, with 1921 Queens, £2 and £3 respectively; your travelling box or 10s. deposit (returnable); two Queens, 10s. each
W. Jackson, Post Office, Bleasby
BBJ 7 Jul 1921

BEEKEEPING BETWEEN TWO QUEENS

Now, this may not be all "bee gospel" as they say, but I guess there are old British bee keepers in outlying parts of our Empire to whom the notes of this ramble will be as the balm of Gilead to their souls. How they will devour this, especially if they happen to know any of the parts. If it does one poor soul good I shall be well repaid for the time it has taken me to write it. But I leave the old lane; in summer time it would be grand to walk down it and listen to the hum of the bees on those bramble bushes, that is, if they are kept near there. Two houses at the lane end, with each half an acre of garden to them, where farm labourers live and quite half a mile from any other house, one would think would be just the spot to keep bees but no bees are there. One reason why bees are not kept in such places as those is because the houses belong to the farm where the men work and consequently, if living in one, anything goes wrong and you have words with your employer, the first thing he would throw at you would be the bees and they can sting in more senses than one.

I go into Little Morton, where I get talking to a man who kept bees fifteen years ago I said, "Why did you give them up?" "Oh," he said, "For six Sundays running, just as I sat down to dinner the bees would swarm, so I outed 'em." I was sorry to hear it as bees could get clover honey there second to none, I am sure, for the land is farmed on the four-year system - clover, then wheat, turnips, then barley, then clover again - and as there are only five farms in Little Morton parish of some 2,700 acres, you may guess there is always plenty of white clover within a bee flight of there, and that was the centre farm. So I leave him and travel on past Babworth into Retford, not having passed a beehive by the wayside all the way from Worksop that I knew of, a journey of some eight miles.

Now having travelled the last 12 or 15 miles on what is known around there as the forest sand land, striking the Leverton road out of Retford, the land quickly turns to clay or red marl. I pass a house where I have a faint recollection there used to be some toll gates in bygone days. A little further on I leave Welham Hall on my left, as I rise the slope of Grove and Clarboro' Hills. The soil here is quite red and in a little cutting through the hill there juts out a slab of white stuff, two to four inches thick that, when ground into a powder, resembles Epsom salts. Anyway, it strikes a sharp contrast to the red clay above and below it.

The day which had been dull and warm, cleared off to a bright, sunny afternoon, which made one feel glad to be alive, for I am sure the thermometer must have been near 52^0F it seemed so warm. All nature seemed to be asleep but the leaves on the blackberries were still a bright green. How quiet and peaceful; habitations are few and far between about here. I can see Little Gringley across a field, where are three or four houses. A few bees used to be kept there in days gone by when I went through it; so I journey on, when suddenly the red clay turns yellow, and it was "some" clay, too. Someone ploughing it was turning furrows up that were all in a piece from end to end: not much about here for bees, I thought, on this stiff soil. But, oh! there is, for here is a field of beans, the first I have seen in my 40 odd miles' ride. A few more yards, and I have the whole Trent valley before me, and Lincoln Minster in the "teens" of miles away on the hill top the other side of it. Then I come into Leverton, where my journey ends for a while, as I call to leave two very old friends one of my best heather sections.

Never having seen heather honey before, they marvelled at the neatness and whiteness of the capping, and didn't they enjoy it for their tea? It was good to think one could give anyone such a treat and as they have half an acre of apple trees and bees not very plentiful about there, they thought of setting up a hive or two, so I had a look round. A better place would be hard to find than among those apple trees and there seems a great lot of fruit in the village beside.

"Why," I said, "you will get the price of a hive of bees with all the extra fruit you would have." There are also good clover fields about and that field of beans I passed was not over a mile away, while about 30 years ago a gentleman in the village planted a great many lime trees that are just getting to blooming age, so a few bees here would easily get their own living and, perhaps, a litltle honey for the table as well.

"And when should you reckon would be the best time to start?" "In April, by all means, then you get the full benefit of them on the fruit trees." Tom Sleight

BBJ 14 Jul 1921
The Derbyshire BKAs Annual Show, held in conjunction with the Royal at Derby, proved to be of a praiseworthy character. Entries were satisfactory and the quality of the exhibits being of a high standard of excellence. The judge, Mr. J. Tinsley, of Kilmarnock, experienced no easy task in the allocation of awards.
Six Sections of Comb Honey (7 entries) – 1st, G. Marshall, Norwell
Six 1 lb. Jars Light Honey (23 entries) – 1st, G. Marshall
One Cake Wax, not less than 1 lb or more than 2 lbs (7 entries) – 3rd, G. Marshall,

BBJ 28 Jul 1921
The monthly meeting of the Council was held at 23, Bedford Street, Strand, London, WC2, on July 21st. Mr. WF Reid presided. The following application for preliminary examinations was granted - Nottinghamshire,

BBJ 28 Jul 1921
"Management of Bees," by Thomas Wildman, 1768, "Female Monarchy or Government of Bees," by Rev. J. Thorley, 1744, both with copper plates; cash offers.
 Walton, Muskham

BBJ 4 Aug 1921

4-frame Extractor, 115s; large Ripener, 30s; never used.

Dawson, Market Hall, Newark

BBJ 11 Aug 1921

It is with profound regret, which we are sure will be shared by all our readers, that we have to record the death of Mr. AG Pugh, of Beeston. In him we have lost one more personal friend of many years' standing. He had been in failing health for some time and passed away after an operation. He was a member of the Council of the BBKA for many years and also a very regular attendant at the Council meetings, where his advice was always to the point and his help of the greatest value. Bluff and breezy, with a keen sense of humour, his personality will be greatly missed at the Council meetings of the BBKA and the committee and other meetings of the NBKA. He was secretary and treasurer of the NBKA, a position in which he did much good work, from 1888 to 1895, and held that position when we first commenced beekeeping and joined the NBKA.

He was followed by the present secretary, Mr. G. Hayes, also of Beeston, who sends the following appreciation:

A form and face well known amongst British Beekepers has passed from our midst in the person of Mr. AG Pugh of Beeston who had been in failing health for some considerable time, and those who were closely associated with him - and he himself—were aware that his time here was not for a much longer period, but the end came somewhat suddenly.

He had just lately been to Italy and after that to Belgium, in the hope that fresh scenes and fresh air would recuperate him. However, on his return he again had an intense attack of his complicated complaints and, calling in a specialist, he was advised to undergo an operation. He consented to this but he had been so worn out by the disease that the operation proved futile and after it he soon passed away.

His death took place on the evening of July 26th, and he was laid to rest in the Beeston Cemetery on July 30th at the age of 66 years. He leaves a widow and four children, two of the latter married and two unmarried, to mourn his loss.

Pugh memorials Beeston cemetary

Mr. Pugh will be greatly missed in many places, for he was very active and energetic, but especially will he be missed at the meetings of the Council of the BBKA, the conversaziones, and the various shows where he was a regular attender and where he was in constant demand as judge, etc.

He was a prop and stay to his own county association where he was always looked up to for advice in all matters of difficulty, and to which he was ever ready to give ungrudgingly all the help he possibly could and here his loss will be felt most acutely.

A wreath of flowers was sent and two representatives from the NBKA followed his remains to their last resting place

BBJ 1 Sep 1921
The troubles of the bee-keeper are not confined to modern times, when the carelessness of some people with regard to disease leads to a desire for legislation. Doubtless, however, there is greater honesty in the craft nowadays than there was in the time of Queen Elizabeth.

An Act passed in the year 1581 shows that the honey and wax trade was a flourishing one, for it begins as follows
"Where by the Goodness of God this Land doth yield great Plenty of Honey and Wax, as not only hath and doth suffice the necessary Uses of the Queen's Majesty and her Subjects to be spent within this Realm, but also a great Quantity to be spared, to be transported unto other Realms and Countries beyond the Seas, by way of Merchandize, to the great Benefit of her Majesty and the Realm. "

Unfortunately, the growth of the trade had apparently led to the growth of fraud, for "a great part of the Wax made and melted within this Realm has been found to be of late very corrupt, by reason of the deceitful Mixture thereof, and the Makers and Sellers of Honey also have not only used to put the said Honey in Cask of deceitful Assize, but have used also deceitful Mixtures of the same," so the Act penalised every maker or melter of Wax who should "after the Feast of Pentecost next ensuing,"use or practise" any Manner of Deceit by mixture and mingling the same with Rosin, Tallow, Turpentine, or any other deceitful Thing, to the intent to sell and utter the same." The penalty was to be 2s. for every pound, "whereof the one Half to the Queen's Majesty, the other Half to the Party deceived, if he will sue for it." Wax melters were further required to stamp each cake of wax with their initials in order the better and the sooner to discover deceit.

As for honey, "all Barrels, Kilderkins, and Firkins filled with Honey by the Maker and Filler "were to have the man's initials burnt upon the head of the cask with a hot iron, with a penalty of 6s. 8d. for every cask not so marked, and honey-sellers who gave short weight were fined "five shillings of English money" for every half-gallon lacking, while honey found to be corrupted with any deceitful mixture was forfeited, one-half to be to the Queen's Majesty and the other half to him that should sue for the same. How nice it would be if British bee-keepers once more had such a great quantity of honey and wax that niuch could be spared for transport to other countries beyond the seas! (Miss) EH Darney. Retford.

BBJ 8 Sep 1921
The Annual Summer Conference of NBKA was held at Beeston, on August 20th, when about 50 members and friends assembled to talk bees and exchange notes. On arrival there were a number of objects of interest to view which well occupied the attention during spare moments.

At 3pm the meeting settled down to hear a discourse on the development and the wiring of frames but, before this was begun, Mr. Riley made a few sympathetic remarks on the loss to the Association caused by the death of one of its oldest members, viz., the late Vice-President, Mr. AG Pugh, after which his name was honoured by the meeting rising to their feet and standing in silence for a short time.

Many old frames were shown to illustrate their variations, also a number of various distance-givers, or spacers and ends.

Different methods of fitting up and wiring frames were shown and theories given as to the cause of bees building comb on the face of foundation instead of drawing it out. These matters created quite a lively discussion on various points, which lasted to the exclusion of other matter until it was time to partake of tea, which had been prepared by a small committee of ladies and to which all did justice.

After tea a short tour was made of Beeston as a little diversion. It somewhat filled the inhabitants with with surprise to see such a large company parading the streets.

The meeting was resumed at 6pm.

Competitions had been arranged for sections, extracted and granulated honey, and for wax, and these were judged openly to the audience for their education in honey judging. This appeared to be well received, and greatly appreciated by them. The awards were as follows:

Class I (extracted) – 1st, G. Marshall, Norwell; 2nd, W. Trinder, Edwinstowe.
Class II (extracted) – 1st, G. Marshall; 2nd, W. Trinder
Class III (granulated) – 1st, J. Ward, Inkersall; 2nd, W. Trinder
Class IV (beeswax) – 1st, G. Marshall; 2nd, W. Trinder; hc, Mr. Saddington, Ossington.

The conclusion of the judging with its comments brought to a close a very pleasant time spent together, which, it is believed, everyone enjoyed. There were other items on the menu, but time did not allow of them being dealt with.

Shows of honey, bees and wax were held at Newark on August 26th and at Aslockton on August 27th, both in connection with allotment holders' shows and proved very successful, there being a fair number of entries at each place. The awards were as follows:

Newark

Class for sections – 1st, G. Marshall, Norwell; 2nd, W. Trinder, Edwinstowe

Class for extracted – 1st, W. Trinder; 2nd, G. Marshall; hc, AW Broadberry, Collingham

Class for granulated – 1st, J Ward, Inkersall; 2nd, W. Trinder

Class for bees – 1st, G. Marshall

Class for wax – 1st, W. Trinder; 2nd, G. Marshall; hc, JT Baines, Caunton

Aslockton

Sections – 1st, G. Marshall; 2nd, W. Trinder

Extracted – 1st, G. Marshall; 2nd, W Trinder; hc, Rev. MWB Houston, Langar

Granulated – 1st, W. Trinder

Bees – 1st, GR Bostock

Wax – 1st, G. Marshall, Norwell; 2nd, GR Bostock

BBJ 15 Sep 1921

I leave the Lincoln road at Darlton and take an unfrequented country lane, where farms are in oddments, the land red clay, the hedges grown high with dog-roses that are in full bloom. Nature seems to run wild hereabouts; the fields nearly all grass with scarcely a clover nob to be seen. To look at it, for about three miles, it would (about) keep a bee to the acre. Normanton-on-Trent is the next village of note. Here you leave the clay for the Trent sand, and it is a regular market garden village.

Apples and plum trees at every house. There seems a fair crop of apples but the plums are "off." Outside the farm crops look a treat, particularly beans and peas, and rye six feet high. Half a mile further on is Grassthorpe, where I intend staying the night, as it is gone 9pm. It is another little village where plenty of apples grow, chiefly Bramleys Seedling.

June 18th. A bright morning and, feeling somewhat stiff with my previous day's ride, I set off again. Bees are a rarity hereabouts. In five villages just here not a bee left. I peep over into a garden, raspberry canes, red clover; black currants seem to thrive here.

BEEKEEPING BETWEEN TWO QUEENS

I just skirt Normanton and Marnham. Here are some very heavy crops of red clover. At Ragnall I see the best bed of 'Tripoli' onions I have seen anywhere. Making my way into Dunham (where they let me cross the bridge over the Trent for a penny) I cross the Trent to Newton. Here beans look a treat, and barley, too.

Leaving a road to the left where it said, '11 miles to Lincoln,' I take the Newark road. If charlock is a farmer's pest around Ashover, the red poppy is more so here, for miles along here the cornfields are a red mass. The land seems very light sand while here and there, by some freak of nature, red clay pops up and goes two or three hundred yards, and then disappears as suddenly.

Passing North and South Clifton I pass a man here that is doing fairly well with his bees. From five hives he had got thirteen, and sold one, and taken 50 lbs. of honey from a May swarm, but it looked a good place for bees there, clover and blackberries everywhere. Two miles further on lies Spalford, and I could see bees there would die in summer, for it was all sand and rabbits. It is sand that blows a lot, for I saw two fields of carrots where it had blown them up by the roots at one end of the field and buried them a foot deep at the other.

[Mr. Sleight should be there, and just outside Besthorpe on a windy day when the weather is dry. Clouds of sand can be seen from miles away, and we have seen drifts of it several feet deep on the roads. Eds.]

They grow a lot of carrots about there and where the wind had not got them they looked very well. The road along here floods when the Trent is over-full, and the fields, too. A man told me barley stood up to the neck in water for a week in August, 1912, and then he threshed seven quarters to the acre of it. For two or three miles I have not seen white clover by the wayside, but as I get into Besthorpe the roadside is a white mass for miles again.

Here, and to Collingham, clover seeds are dead in the fields and only want cutting and leading for they seem to be standing up dead. The sand land this hot day seemed to fairly burn my feet, for it was over 80^0F in the shade while I was about here. I spent about two hours in a clover field. I found it was cooler than riding. Here are dog-roses galore, nothing else for yards but the bloom of them on the hedges.

Corn crops around Collingham seem to be wanting rain, as bad or worse than it did on the forest sand. At Besthorpe I saw one small lot of bees and a few empty hives. I heard there was one man with about twenty hives at Collingham. An old roadman told me quite a lot kept bees years ago but they had about all gone now. So I leave Collingham to make my way past Langford, where I saw two or three pea-pickers pulling early peas for the market. At Langford I leave the Newark road and make for

Holme. For about a mile it is all meadows for mowing; the hay crops seemed fair.

At Holme Ferry, where I cross the Trent again, I put my bike on a boat, and a cool breeze from the north as we cross over brings a remark from the old boatman: "Ah, a north wind's no good to nobody in summer time; it never brings much wet." "No, but it suits bee-keepers; they do well with it."

Well, the old Trent did seem low as it did not seem far across. I crossed it thirty-seven years ago when it was top bank full; it seemed quite a long ride then. I should say it is about as low now as it was in 1887. I walked across it then, as did two or three more of us, clinging to the old 'milking boat' in Sutton Holme and a pretty object I looked for tar when I came out on the other side, as the boat bottom had been freshly tarred and the strong current carried me under it in a place or two. The other two, who knew the trick, walked on the low side. I cannot remember that it ever rained at all that summer till after all the corn was got. Tom Sleight

BBJ 15 Sep 1921
Cowan 4-frame Extractor, 90s; Non-swarming Hive, five drawers, four fitted foundation, £3; Taylor's non-swarming Hive, £2; two Burtt's Hives, 30s. each; guaranteed disease free. Dawson, Market Hall, Newark

BBJ 22 Sep 1921
Queens. Two 1921 Carniolans, mated Dutch, 6s 6d. each J. Moore, Bleasby

BBJ 29 Sep 1921
The experiment of holding the BBKA Annual Show at the Grocers' Exhibition proved a successful one. The exhibits were more numerous and of a better quality than for a number of years past. The Royal Show is held too early in the year for the current season's produce to be exhibited and, unless bee-keepers retain some of their previous year's surplus for the purpose of exhibiting, the exhibits are, and have been from this cause, both few in number and poor in quality. This year the show was held at the end of the season and the exhibits, therefore, comprised the cream of the season's products. We noticed also that many of the prize winners at local shows sent their exhibits; therefore there is no wonder that the judges, Messrs G. Brydon and CLM Eales, had a difficult task in awarding the honours.
Honey
Class 6. Six sections of comb honey, excluding heather honey of current year (5 entries) – 3rd, G. Marshall
Class 7. Six jars of extracted light-coloured honey of current year (18 entries) – 2nd, G. Marshall; 3rd J. Ward
Class 8. Six jars of extracted medium or dark-coloured honey of any year, excluding heather honey (7 entries) – 2nd, W. Trinder

Class 9. Six jars of granulated honey, excluding heather honey of any year (4 entries) – 3rd, W. Trinder

BBJ 29 Sep 1921
The monthly meeting of the BBKA Council was held at 23, Bedford Street, Strand, London, WC2 on September 15th, 1921. Mr. WF Reid presided.

The Chairman referred to the loss of one of the oldest members of the Council by the death of Mr. AG Pugh, of Beeston, who had worked ungrudgingly for many years on behalf of bee-keeping. He was a man of sterling worth, perhaps not always understood on account of his rugged and abrupt manner. In all his dealings he was straightforward, incapable of a mean action and his services and presence would be very much missed by his colleagues. A silent vote of condolence with the widow and family was passed, and the Hon. Secretary was instructed to write accordingly.

BBJ 27 Oct 1921
Bees. Cash offers wanted. Strong stocks in skeps; guaranteed healthy; March delivery.   J. Moore, Bleasby

BBJ 10 Nov 1921
Kilmarnock Honey Show. The exhibition of honey was the finest seen this year in Scotland and was visited by great numbers of bee-keepers. The entries totalled 121. Mr. Joseph Tinsley, Lecturer to the West of Scotland College of Agriculture, judged the honey and attended on both days of the Show to give help and advice to the various exhibitors. Some exceptionally fine honey was on view. The following is the list of awards:
Light run or extracted honey (22 entries) - hc, Walter Trinder, Edwinstowe
Granulated honey (12 entries) – 3rd, W Trinder; vhc, J. Ward, Ollerton
Cake of beeswax (10 entries) - hc, W Trinder

BBJ 10 Nov 1921
Peterborough, Oundle and District BKA's Show was held in the Victoria Hall, Oundle, in conjunction with the local Horticultural Society, Mr. Mackender, of Newark, acting as judge. The local and Association classes drew a good number of entries, the honey being of a very good quality.

BBJ 8 Dec 1921
These dull, foggy days not being much in my line for rambling. I will just take the readers of the BBJ on a ramble I had, beginning on April 29th, and lasting ten days. I describe here only those parts which appertain to Nottinghamshire.

On nearing Worksop I thought a turn to the right, down an old narrow lane, would give me something that was more interesting than the straight macadam highways were doing, but I had not gone far before I found I was retracing my tracks towards the north end of Welbeck tunnel. Now here was a house nicely isolated, surrounded with every description of bee forage, trees and fields of clover, a situation that I should have called ideal; but no bees were there, although there was a large garden and two ladies at work in it. I paused to admire it, and was just going to broach the question,"Where were the bees?" when they went off to the further end so I had perforce to pass on.

The monster trees around there must be hundreds of years old. The next two or three miles into Worksop were certainly among sylvan scenes. It was along this drive that I saw the first sprig of hawthorn cut; just fancy. I thought, May in bloom in April. I had more often seen it out in June. On going into Worksop the inner man wanted attending to, so seeing the shop of a pork butcher named "Bee". I patronised him for a pork pie, so I can truthfully say all bees don't gather honey.

As I got out of Worksop, on the Doncaster road a mile or two in a wood was a hawthorn tree in full bloom, and sycamore in flower, too, while a little further on is Carlton, where I pass the night. Tom Sleight

BBJ 15 Dec 1921
April 30th A most lovely morning, so I set off towards Retford but I had not gone far before I took a wrong turn, and came to an old grass or sandy lane that was in parts a yellow mass of gorse where I failed to find any kind of bee, only "Tommies" (humble bees, Eds.).

I wandered on into a wood that was pretty well set with sycamore trees and just through it was a white clover field that was about the best set clover field I saw this year, about 20 acres, and no rye grass in it. Didn't I just calculate what a few good hives of bees would be able to do planted on that wood side in two or three more weeks' time. I had got on to the forest sand in that wood and for a few mile I thought Retford is never on this road, so I had to get my road map out to find out where I was. It is peculiarly sheep country around here and farmers had led turnips into the clover fields to get them out of the way for barley sowing. But where they were still growing they were yellow over with flower, real early forage to any bees about there, while in among the clover was yellow trefoil in bloom.

At last I got the right turn for Retford and for the next five miles I was never away from a clover field, either one side the road or the other, and sometimes on both sides It was here I passed the pretty village of Barnby Moor, in the midst of all these clover fields, but I failed to see any bees about there. If there were none what honey must have gone to waste. I am soon in Retford, where I spend an hour or two looking around the market. It appears to be a cheap little town but not knowing any bee men about there I am soon out in the country again

Taking the Leverton road, under a clear blue sky, I got pretty warm by the time I got to the top of Clarborougb and Grove Hills but the air is so clear I spend half-an-hour viewing the country round about. I have heard say one can see 30 church steeples from there. I never thought to count but really the distance I could see was marvellous. I could pick out Stone Edge and Alice Head, on the other side of Chesterfield, some 30 odd miles away.

It was on Stone Edge where I once stood and picked out Grove Hall, that lay about a mile from where I stood now (such clear days don't often occur, for I have passed Alice Head about 30 times this summer since then but it has never been clear enough to see Grove Hall since, at least not on the day I was up there).

A liltle further on I was able to see the other way. Lincoln Minster stood out bold and clear, a lot different to how it looked when I saw it from the same place four months before. I was on my way to Leverton to give my lady friends a start with their bees, but when I got there they had decided not to bother with any this year so I had another cup of tea with them and passed on through Tresswell to Grove. What fields of beans I passed along that road that were just about bursting into bloom. My next rendezvous was to call on Mr. Hudson at 'Sunny Vale Apiary' at Rockley.

So getting out on to the North road I pass Eaton and Gamston and am soon at Rockley. That gentleman not being at home, or anyone else, I thought they might have gone to Retford market and, as the day was young yet it being only 5 o'clock, I decided to wait an hour to see if he turned up. Seeing what peculiar hive roofs they looked (he had wintered twelve hives of Italians), I was just examining one when the whole family drove up in a motorcycle and side-car. The roofs were flat, covered with thin sheet iron, which hung over all round 2in and to keep it on, without nailing, he had cut and turned an inch back at each corner, which clamped tight hold of the wood. No wet could ever get in. I thought it a very good idea for any rough wood would make roofs like them.

I was very glad I waited to see him, as we had two or three hours good bee chat and he showed me some very interesting photos of swarms of bees. The first was in a glass case. He said it was petrified and it looked it, too. The next was hung around a man's whiskers, another was hung from a man's elbow while the last was a lady holding a mop shaft, with two swarms hanging from the middle of it, down to her knee. He said there were 17 lbs. of bees on that shaft when he weighed them. She had a good nerve to hold them while he took a photo.

He said all his bees were for sale, except one lot, on large frames as he was going on a summer tour. He seemed to fancy the large frames and was going to use all that kind another season.

Hope he sees this in print, and gives us his experience of them at the end of next summer. I did not consider it a first-rate place for bees in the spring. There seemed nothing near, only red clay plough fields, which might have beans and charlock in; while for clover he stood well, a large field just over the road, and it may be alsike or red clover, it didn't look like white. Still, he had got a good sample of honey last year.

I had to bid adieu, as I had six or seven miles to go and the sun had set, so passing Markham Moor and Tuxford I was soon at Grassthorpe, where I spent the next week. In those two days I had ridden 60 or 70 miles. and only seen Mr. Hudson's hives by the wayside; there must be a great shortage of bees in that part of the country.

Tom Sleight

BBJ 19 Jan 1922

While spending a week by the Trent side early in May last year I had several runs round about but this one on May 5th was a long one, in a very flat country. There being no hills to get off to walk up I was very often walking on the level to ease myself. Starting out of Grassthorpe at 10am. I passed Sutton and Carlton-on-Trent then followed the North Road through Cromwell and Muskham to Newark. Crossing the Trent outside South Muskham, I see at last they are going to build a bridge there and do away with what has been an apology for one and a disgrace to the greatest highway in the land for so many years.

Muskham Bridge at South Muskham (B6325 Great North Road) was opened in 1922 by the Duke of Portland

Further on, a large beet sugar factory was nearing completion. I wonder what amount of attraction the sweets from the beets will have for bees in that part, and whether they will be able to get a special blend of beet honey?

In 1919, the sugar-beet industry was inaugurated, 5,000 acres of land at Kelham being purchased for the purpose at a cost of about £500,000 towards which the Treasury contributed £125,000. The sugar factory was opened in 1921.

I had a look around the old castle and grounds, where I saw some splendid beds of wallflowers, which were more or less all of a hum with bees, showing someone has a few bees in or near Newark. Yes, the old grounds looked a treat. I remember as a youth of sixteen taking a cow to the Cattle Market when it was held on what are now the Castle Gardens and how dead tired I was when I got there after chasing that beast about for five hours on the road from Crow Park, nine miles away. When crossing the GNR line there, it shot up by the gate and got wedged between a fence and the waiting-room while two trains went by - it surely was wild, that cow. Whoever has transformed that old market into the beautiful grounds they are now has great credit due to them.

I passed on up Stodman Street into the market-place where Proctor's showmen were busy fixing up for the May Fair hirings. I lingered by the Town Hall steps and pondered over the times I have stood there in those hirings years ago, wondering what and where my fate was going to be for another year - never two years in the same village. Perhaps that was how I learned to read the countryside like a book.

Leaving those old memories, I pass down Kirkgate and Northgate out into the country again. Never having been through Winthorpe, I turn off the Lincoln road for a while just to see what a lovely little village it is. By the look of the houses there I should think a great many tradesmen from Newark live there. I spied two hives in a nursery as I went along but although I went round the village, I failed to see any more. By what I saw of the fields of clover around, it ought to be a good place for bees. Just outside the village I saw a field of barley in ear or just coming in and I thought how early it was - only May 5th.

It is quite flat open country from there to Brough and Stapleford where I saw five empty hives. The lady said bees were too dear now to re-stock them; but dear, or not dear, they would have got some honey for just by there I saw the best clover field on that trip. I guess it is very good land about there for alongside the road to Norton Disney was a field of oats that I never saw equalled anywhere last year, about 25 acres, as level as a table top and a very dark green. Turning a corner at the end of this field I came to a bridge over the Witham.

Though the morning was rather dull, the afternoon, now 2.30pm, opens out into one of those days of sunshine and small clouds that are a treat to be out on, so I sat down on the bridge. As it is fenced in with iron railings I had a good view both ways, also up and down the river. As the sun went behind a small cloud about half a dozen fish, big chubb, would come out from under the bridge only to dart back as soon as the sun came from behind the cloud again. It made me wonder if they could not bear the rays of the hot sun because it was hot then.

Two cuckoos in a tree close by were squabbling and fighting, and as I made a meal off some bread and cheese I thought whatever the coal strike had done for some, it was certainly giving me a chance to ramble about the country as I had never been able to do before. It was here I missed seeing a big bee-man close by at Carlton-le-Moorland.

Turning to the left, I went to Bassingham, where I learnt a lot of bees used to be kept, and I should say a very good place, too; but they did not know of any there now. Leaving Bassingham, I cross the Witham again, and also pass the farm where I spent my first years as a farmer's boy; then on through Thurlby, where a small field of swedes in flower would have given early forage to bees but I fail to locate any in the hamlet of Thurlby. A little farther on a tractor was turning the clay clods over to be baked by the sun.   Tom Sleight

John Phipps, Editor of "The Beekeepers' Quarterly" commented when asked if this piece should be included because Tom visited Lincolnshire: "Regarding Bassingham, no problem as bees from both Notts and Lincs will be unaware of the border and flying between both counties. Important for Notts beekeepers to know what crops, etc. are within flying distance of their apiaries no matter if into Lincs".

BBJ 9 Feb 1922
I had read in the BBJ of Roper's at Thorpe-on-the-Hill and I was getting close to it. At last, I thought, I am going to have some free-wheeling if this place is on such a hill as the name makes one appear to expect. Well, it certainly is not in a hollow but if being 20ft. above the surrounding country constitutes it a hill, then I'll call it one, but I climb some hills around Ashover that Thorpe people would term 'Mount Everests' if they saw them. But anyway, there was a glorious view of Lincoln Minster, six miles away. The sun did sparkle on the windows, and I was near going to see it but was glad after that I did not, for before 10pm I had had enough for one day, it being 5pm then.

I found Mr. Roper a most interesting chap but he had lost heavily in the winter. From 109 stocks put down to winter, only 30 survived. Out of twelve of Claridge's guinea queens only one had lived and to that hive, in his angry moments, he had fed the combs from some of the worst crawling stocks, with the result that there were eleven combs of brood, and bees covered seventeen combs that day. They were in what was like a Wells hive, the brood box holding twenty combs. He said, "What can you make of a complaint like that, when what one lot of bees have died on won't kill another lot? I tell you what, Mr. Sleight, in my opinion "IoW" disease will come and go, and nobody will know what cures it." He showed me the remains of his Dutch stocks in their quaint old skeps; some of them had died with "IoW" disease and those left were busy on a field of white turnips in bloom a little way off. He said how heavy the skeps had got since he picked one up a few days before. I had a look in those turnips - they were a yellow mass of bloom 6ft. high.

Mr. Roper is a young bee-keeper who is going to have some bees some day and some honey, too, if only he can get some stocks on to Lincoln Heath, as they call it. He said he was always pleased to have a chat with any beeman but he was very busy at his trade, having orders to get off that night. I did not like to hinder him any longer so I sped on my way to Eagle.

Eagle seems to be a place like Thorpe, on another little knoll of a hill, while all the country around is quite flat. From there to Spalford, where the sand is like the sea sand, on lanes where biking is a glorious treat but would not be on dark, foggy nights as ditches on each side are too deep. A good many fields seemed to grow nothing but gorse and rabbits and what a yellow mass the gorse was! If there is honey in gorse surely bees could have got gorse honey here; there were such quantities of it. That land does not seem to grow clover, as I never saw any on it - a poor place for bees when gorse has flowered, I should say.

Travelling on, I come to South Clifton, through there to North Clifton, where I ended up in a farmyard. I asked, "Is Dunham Bridge anywhere here?" "No. you will have to go back up the street and take the Gainsborough road to Newton before you get to Dunham."

Just as I turn on the Gainsborough-road I spied a board up, 'Honey for sale'. Hello, I thought, some bees here; I'll just see over the gate how many hives he has got and the bee-keeper stood inside. I said. "Well, how's the bee trade?" He replied. "Are you a beeman?" "Yes." "Well, come and have a look at my little lot."

I have never come across a better arranged apiary in all my travels: everything was arranged to get some honey and I know he has got some since then. There were five hives supered then, and 3 lb. of honey in one he showed me on May 5th." Why," I said, "We shan't have honey in supers till June."

When I had honey in supers from clover on the moors I went again to look in that hive and he had taken 130 lb. from it alone and mine were just starting. He said, bad as the season was last year (1920) he took 210 lb. from two hives. It is surely a good early part for honey. He proved a very interesting companion for an hour's chat and his garden, of that seaside sand, was clean and well looked after. He said that the notice-board sold nearly all his honey, it being on the Newark to Gainsborough turnpike; people going by in motors called for 5 and 10lb at a time.

I have an idea his bees crossed the Trent to Marnham for most of their honey, for there seemed little forage growing near his house, only red poppies. I had to get on, as it was turned 8pm and I had a long way to go, round Newton, over Dunham Bridge to Ragnal, where along that side of the Trent the hedges were one mass of May bloom. Passing Fledborough Station and Marnham as darkness set in, I reached Grassthorpe after one of the loveliest days out on a bike, having done fifty miles on a very nearly dead flat in just over twelve hours. I was indeed glad that I had not gone to Lincoln, for I had had quite enough.   Tom Sleight

BBJ 23 Mar 1922

The task of the Bee Re-Stocking Committee in assisting apiarists to make good the losses brought about through "Isle of Wight" disease has been successfully accomplished, and it is considered that there is no longer any necessity for the continuance of the scheme.

Mention of the matter was made at the Annual Meeting of the NBKA, which was held in the Albert Hall Institute, Mr. C Taborn, horticultural organiser of the County Council, presiding over a large attendance. In presenting the committee's 37th Annual Report, Mr. G. Hayes, the hon. secretary, said the membership now totalled 222.

Last year stood out prominently as a most abnormal season, the yield of honey in some localities being exceptionally good, whilst in others it was only moderate, though the quality in all cases was exceptionally pleasing. Nine experts had been employed to give advice and assistance to owners, no fewer than 150 visits having been made. The following candidates had passed the junior craftsman's examination: Messrs. CJ Bond (Elston), GR Bostock (Aslockton) and Miss W. Freeman (Edwinstowe).

The financial statement showed a credit balance of £26 7s 7d and it was decided to grant £12 to the Bee Re-Stocking Committee to clear off its deficit, it being pointed out by Mr. A. Riley that the scheme of the latter was never intended as a commercial enterprise but was formed with the object of meeting the requirements of bee-keepers. Regarding a proposal to establish a central instructional apiary depot for bee-keepers, the Chairman said it was hoped to incorporate a scheme which could be submitted in concrete form at the next Annual Meeting.

The Duchess of Portland was re-elected president, a tribute being paid to the work of Mr. Hayes, who consented to serve as hon. secretary and treasurer for another year. Messrs. Hayes and A. Riley were chosen as representatives to the BBKA and the district secretaries were thanked for their services and were re-elected *en bloc*.

The following were elected to the Executive Committee: Mr. E. Hollingsworth, Heanor (chairman) ; Mr. AE Goodlad, Mansfield (vice-chairman) and Messrs. MH Fox, Kirkby; G. Smithurst, Watnall; FG Vessey, Balderton; TN Harrison, Nottm.; GE Skelhorne, Nottm.; G. White, Sandiacre; W. Adams, Mansfield; W. Trinder, Edwinstowe; LW Walton, Nottm.; WE Cowlishaw, Mansfield; GR Bostock, Aslockton and W. Sharpe, Kirkby.

It was decided to hold several conferences in other parts of the county during the summer and autumn so as to reach the more outlying districts.

BEEKEEPING BETWEEN TWO QUEENS

The meeting then adjourned, and members and friends partook of tea.

A conference in the evening was opened by an address by Mr. Hayes on "The Treasures in a Pound of Honey, and How They May Be Found." Dealing with the synthesis of honey, the speaker emphasised the important part played by the flower, the sun and the bee in the production of honey and referred to the food and minerals which formed the medicinal qualities. Later he dermonstrated a few chemical experiments showing the analysis of honey.

BBJ 30 Mar 1922
The BBKA Council meeting was held at Pritchard's Restaurant, 79, Oxford Street, W., on March 16th, 1922. The chair was taken by Mr. TW Cowan, and there was also present G. Hayes (Notts). The following Association nominated representative on the Council was accepted - Nottinghamshire

BBJ 17 Aug 1922
On August 15th the members of NBKA, with their friends, made an excursion to the Agricultural and Dairy College at Sutton Bonington where they were received by the Principal, Dr. Wm. Goodwin. The party, numbering over 70, were shown round the college and grounds and partook of tea which had been prepared in the spacious Dining Hall to which all did justice and which was greatly enjoyed.

After a short breathing space the company assembled in the large Lecture Hall to hear a discourse from Dr. AZ Abushady on "After the Honey Flow." This was listened to with keen interest as the speaker brought out point by point the things it is necessary to observe and to do at the period named. This brought on a lively discussion and many questions were put and answered and everyone appeared surprised when they were told the meeting would have to be closed to enable them to catch their train.

Nearly all expressed their satisfaction at the enjoyable, interesting and instructive outing in which they had participated and, like Oliver Twist, asked for more such times so one has been arranged to take place on October 7th, at Nottingham.

Geo. Hayes, Hon. Secretary.

BBJ 9 Oct 1922
A well-attended conference of members of NBKA and friends was held in the Albert Hall Institute on October 7th, under the chairmanship of WS Ellis, Esq., of Hawksworth. The proceedings were opened by an excellent, well-delivered lecture on "Diseases of Larvae and Adult Bees" by Mr. J. Herrod-Hempsall, FES, who gave a minute description of the most important diseases the bees were subject to and how these diseases were to be recognised and dealt with to prevent them spreading and so far as possible to bring about a cure.

This was followed by three other short papers on:
1. "The Abnormal Season of 1922 and some of the lessons it taught," by Mr. W. Sharpe, Kirkby.
2. "The Pollination of Flowers by the Bee," by Mr. W. Trinder, Edwinstowe.
3. "The Production of Heather Honey," by Mr. D. Wilson, Belper.

Each paper raised considerable discussion, and a large amount of interest was shown in each.

Competitions for honey and wax (open to all members):
Class 1. Six 1-lb sections of comb honey produced in any year – 1st, 7s 6d, E. Saddington, Ossington; 2nd, 5s, G. Marshall, Norwell
Class 2. Six 1-lb jars of extracted honey produced in any year – 1st, 7s 6d, G. Marshall, Norwell; 2nd, 5s, E. Saddington
Class 3. Six 1-lb jars of granulated honey produced in any year – 1st, 7s 6d, G. Marshall; 2nd, 5s, E. Saddington
Class 4. Best sample of beeswax to approximate 8 ozs. – 1st, 4s, GR Bostock, Aslockton; 2nd, 3s, G. Marshall

The exhibits were judged by Messrs. Trinder, Dolman and Goodlad. They were followed by Mr. Riley, who finally decided the awards.

NEP 21 Mar 1923
The Rev. William Towers, appointed to the living of Bratoft from South Thoresby, was formerly curate at St. Stephen's, Nottingham. He is a well-known beekeeper, having been a successful exhibitor at numerous shows in Lincolnshire.

NEP 11 Apr 1923
At the monthly meeting of the Local Committee of the Notts. Agricultural Show, which is to be held at Worksop on May 23rd and 24th, it was reported that in addition to the agricultural section, which will great attraction in the local classes, there will be bee-keepers' competitions.

NEP 23 Apr 1923
Beekeepers. Two WBC Hives, Appliances, one shocked, nearly new.
                                                                        Ivy Cottage, Bunny.

NEP 18 May 1923
Cheap, six Beehives, complete with frames and supers.
                                        Davidson, The Mews, Mapperley Plains.
NEP 9 Jun 1923
Paralysed Sailors' and Soldiers' Home, Ellerslie House, Gregory Boulevard. The

BEEKEEPING BETWEEN TWO QUEENS

President, and Committee acknowledge, with grateful thanks, the following gifts during May: Their Graces the Duke and Duchess of Portland, usual weekly hamper; Colonel WH Blackburn, flowers, magazines and invitation to Gedling House; Lt. Colonel Clifton, invitations Clifton Hall; Mr. Bradley, part-gift Beehive;

Ellerslie House was set up in 1917, at the instigation of the Nottingham Sports Club, and by private subscription. The house, on the corner of First Avenue and Gregory Boulevard in Nottingham, was purchased by the 6th Duke of Portland and donated to a committee established to provide long-term care for back and other paralysing injuries among ex-servicemen.

NEP 31 Jul 1923
Twelve beehives, well made, honey ripener, wax extractor, frames, feeders, cheap
55, Julian-road, West Bridgford.

NEP Sep 1923
The bee-keepers of Nottinghamshire have had a disastrous season and they have been unable to reduce the price of honey.

Some Recollections
Extract from an article by Joseph Chambers for inclusion in the booklet celebrating 100 years of NBKA.

My interest in bees was first aroused in 1924 by an elderly neighbour who kept two stocks to obtain a little honey for himself. He talked to my sister, Annie, and myself of the doings of these marvellous little creatures and eventually promised to give us a swarm when he next had one. That settled, the day came when we saw him carrying thwe swarm in a straw skep across the fields to our home.

NEP 14 Feb 1924
An interesting presentation took place to-day at the LMS Railway Company's engineer's office, Queens-road, Nottingham, when Mr. George Hayes, on the occasion his retirement from the service of the railway company. The employee was presented by Mr. Dixon (superintendent) with a fitted oak smoker's cabinet on behalf of the members of the staff. Mr. Hayes had been in the railway service for nearly 50 years and of this time had been Superintendent's Chief Clerk at Nottingham for over 40 years. He was secretary and delegate for the engineers' department hospital fund, which contributed considerably to the local hospitals. For over a quarter of a century he has also been secretary of the NBKA.

NEP 22 Feb 1924
A branch of the NBKA has been formed in Retford.

NEP 3 May 1924

For Sale. Three-tier beehive, all fittings. 34 Waterloo-road.

NEP 16 May 1924
Beehives, Swarm Boxes, honey and wax extractors, drawn out super and brood Frames, Beekeeper's Outfit. 16ft. Flagpole, 26-stave Ladder, cheap.
55, Julian-road, West Bridgford.

NEP 16 May 1924
Bees. A few strong, healthy stocks for sale, can be seen any day after five.
W. Dennis, Rempstone.

NEP 20 June 1924
Sale new Hives and Bees complete, may be seen working.
Scrimshaw, Prospect Villa, Gunthorpe.
Beehives, drawn-out frames, bee appliances, 16ft. Pole, cheap.
55 Julian-road, West Bridgford.

NEP 24 July 1924
Bee-hives, non-swarming, lifts, supers, etc. good condition, half current price.
Potter, Tollerton Rectory

NEP 25 Aug 1924
Three Beehives, good healthy stocks, for Sale, Incubator wanted.
Wright, Northfields, Barnstone.

NEP 9 Sep 1924
Wanted, good Book on Beekeeping, cheap. Z9 Evening Post.

NEP 12 Sep 1924
Nottingham Journal allotments exhibition and flower show opened in the Drill Hall, Nottingham, to-day. It is the fourth year that the function has been held. Very interesting are the exhibits relating to the apiary on the stand of the NBKA.

NEP 17 Dec 1924
An interesting and instructive lecture on the "Romance of Bee-Keeping" was given to the members of the Social Guild in connection with Nottingham Belgrave-square Presbyterian Church last evening, by Mrs. GE Skelhorne. The chief points touched upon were the hive, its construction and inhabitants and the production of honey.

NEP 9 Mar 1925
According a statement of the secretary the AGM on, the work and importance of the NBKA are not sufficiently known. "Ours is not an association of faddists, or people with a hobby," said Mr. George Hayes."We want people to realise that it is an industry that is trying to advance." The meeting was held at the Albert Hall Institute, and the Sheriff (Mr. RA Young) presided over the gathering including enthusiastic beekeepers from all parts of the county.

BEEKEEPING BETWEEN TWO QUEENS

Robert A Young was Sheriff of Nottingham in 1924/25

NEP 10 Mar 1925
In connection with NBKA honey competition the awards were as follows:
Class 1. – 1st, J. Ward; 2nd, G. Ward
Class 2. – 1st, Hopkinson; 2nd, G. Ward

NEP 14 Apr 1925
What Listeners will hear during May:
Below is given the programme of talks to be broadcast during May, and from which will be seen that the subjects are in keeping with those arranged for the present month, which have so far proved the most enjoyably varied list yet put on May11th, 7.40pm. Mr. George Hayes, Bees and Bee-keeping.

NEP 30 May 1925
The demonstration apiary at Arnot Hill Park, Arnold, was opened to-day. Mr. G. Hayes, secretary, NBKA will lecture.

NEP 6 Jun 1925
Radford Bee Swarm. Mr. TWN Harrison writes: The very interesting report of this incident in the Post shows the narrow line which so often divides success and failure. Since my success last season taking a swarm of bees off a GPO pillar box in Sneinton-road, I had not had the stage so nicely fixed for demonstrating how the modern beekeeper manipulates bees, but *"the best laid plans of mice and men gang aft agley."*

The stage was set, the whole neighbourhood assembled, the swarm of bees numbering at least 15,000 was ready to be taken, but the position against the glass pane of an upper bedroom window was awkward in the extreme but mounted on the top of a ladder I did my best and but for the fact that amongst the 10,000 bees I shook into my swarm box I failed to secure the queen, explains the failure, as when I returned later with suitable appliances the bees had flown. The stinging of the bees is difficult to explain, as when swarming bees are usually docile but I received more stings yesterday than I had previously received during the last ten years.

NEP 9 Jun 1925
Beehives, two, with complete outfit.   Redgate, The Hut, Sandford-road, Mapperiey

NEP 30 Jun 1925
A swarm of bees alighted on the front of one of Messrs Boots' delivery vans at the top of Derby-road yesterday.

NEP 15 Jul 1925

The only country workers who rejoice in the hot weather are the beekeepers, who are finding that their honey crops are better than ever this year.

NEP 31 Jun 1925

A demonstration of beekeeping took place at the Apiary, Arnot Hill Park. Daybrook yesterday by Mr. G. Hayes. After dealing with the queen's travelling cage. Mr. Hayes proceeded to demonstrate the way of extracting honey from the hives. The straining, ripening, and bottling of honey was dealt with and the demonstrator went on to explain the method of preparation of section honey taken from the hives. How to deal with wax was also explained.

NEP 26 Aug 1925

Notts. Education Committee. Bee-Keeping. On August 29th, 5.30pm at the Demonstration Apiary, Arnot Park, Arnold. All interested are invited to attend.

NEP 4 Sep 1925

Allotment Produce. The allotment and garden holders competition organised by the Nottingham Journal, opened today at the Drill Hall for the fifth successive year. Though there was a slight falling off of entries due to the prolonged period of drought, the display as a whole was excellent. The NBKA and Notts. Education Committee had interesting stands.

NEP 17 Sep 1925

Bees. To encourage beekeeping. Strong stocks hybrids on ten frames sectional hives 20s each complete. 75 Lees-road, Mapperley

More recollections of Jospeh Chambers from the booklet produced to celebrate 100 years of NBKA.

My interest in NBKA was through listening to a talk about bees on the original BBC Radio Nottingham early in 1926. It was given by Mr Geo. Hayes at the close of which he said that further information could be obtained from him. As a result of this talk, both my sister and I became members of NBKA and attended our first AGM on 6th March 1926. The subscription at that time was 5/-.

The BBC opened the first regular public broadcasting station in the world on 14th November 1922 in London. In October 1926, a 25 kW long wave station was opened at Daventry. However, listeners would have had to buy a new radio. In August 1927, the first high power medium wave transmitter was opened, also at Daventry, replacing Birmingham and Nottingham.

NEP 6 Mar 1926

BEEKEEPING BETWEEN TWO QUEENS

The Annual Meeting of the NBKA was held at the Albert Hall Institute to-day, when officers for the ensuing year were appointed.

NEP 8 Mar 1926
It was reported at the Annual Meeting of the NBKA which was held at the Albert Hall Institute on Saturday, that 57 new members had been enrolled during 1925 making a total membership of the association of 264. The Mayor (Ald. Foulds) presided over a large attendance and expressed pleasure that the association had experienced a much better year and that the bees had been much freer from disease. He also alluded to the interest in the subject shown by the County Council and said he would do his best to get the Nottingham Council interested the subject of beekeeping.

Charles Foulds was Mayor of Nottingham in 1925/26.

NEP 14 Mar 1926
For Sale, Beehive, Cold Frame. Zinc Wheelbarrow, Hose Reel with short length of hose.   "Post."

NEP 15 Apr 1926
Beehives for sale. What offers?   Robinson, School house, Bestwood.

Further recollections of Joseph Chambers for inclusion in the Centenary booklet produced in 1984.

As a result of *Acarine* disease all stocks were terribly weak and obviously in no state to gather any surplus that season. An advertisement appeared in the BBJ offering three-pound packages of bees from the south of France for thirty-five shillings. (A week's wages for a working man!) We ordered one of these which arrived by rail on 8th May 1926. These were hived in the evening on eleven frames of new foundation with one pint of sugar syrup on top.

In two weeks time they had drawn these combs out which were full of brood and food so a first super was given – nine frames of new foundastion. One week later a further new super, followed by a third; all combs being drawn out and filled.

We actually extracted ninety pounds of light honey from them and then we took them to the heather where they put twenty pounds in the super asnd fed themselves up for the winter in the brood combs.

The worst feature of all this was that they very soon proved to be absolutely <u>wicked</u> stingers. Many other beekeepers had the same experience with them and they soon earned the name 'French devils'. They were such good workers we persevered with them, accepting stings and rearing young queens from the more docile stocks which we built up.

NEP 10 May 1926

Sales by Auction. N. East Stoke Hall. Edward Bailey and Son are instructed by Capt. A Bromley, RN to sell by auction, garden tools and outside effects including beehives.

NEP 18 May 1926
Cheap. Six beehives with frames and supers.
                              Davidson, The Mews, Mapperley Plains.

NEP 1 Jun 1926
Fruit Bottling Outfit for sale, also three Beehives and Patent Honey Extractor.
                              T 3 Evening Post.

NEP 21 June 1926
Notts. Education Committee. Lectures in Bee-Keeping will be given in the Shire Hall. Nottingham, Wednesdays, 6.30pm commencing June 23rd. Lecturer: Mr. G. Hayes, The course will include practical demonstrations. Full particulars may be obtained from the organiser, Mr. C. Caborn, Shire Hall, Nottingham.

NEP 30 Jun 1926
Five good beehives for sale.   163 Gregory-boulevard, Nottingham.

NEP 8 Jan 1927
For Sale. Two beehives, one fowlhouse.   Address I 67 Evening Post.

NEP 21 Mar 1927
The AGM of the NBKA was held at the Albert Hall Institute, Nottingham, on Saturday, Ald. C. Foulds presiding in the absence of the Mayor. The report of the committee showed an income of an adverse balance of £10 10s and a membership of 239 - a decrease of 25 on the year.

NEP 1 Jun 1927
Two beehives for sale. Cheap and accessories.   32 Lancaster-road, Nottingham.

NEP 9 Jul 1927
Beehives. Several well-stocked also this years swarms. Cheap.   F Clifton, Bulwell.

NEP 8 Aug 1927
Broadcasting. Miss Yates: Personal Experiences and the Pleasures of Bee-keeping.

In "Bee-Keeping Old and New" William Herrod recounts making hives and fittings for them from old scraps of wood no doubt 'rescued' from the carpenter's shop where Joseph was an apprentice. Four of the frames, shown left, the brothers "retrieved in our native village in 1928, forty years later, from our old friend T Marshall."

NEP 15 June 1928
Advantages of Co-operation. Bee-keepers are a fraternity organised in district and county associations, in affiliation with the BBKA. They have meetings for reading papers and discussion of their engrossing pursuit, lectures, excursions to apiaries, periodicals, lending libraries and examinations for Preliminary, Intermediate and

Expert Certificates. Many are the advantages enjoyed by bee-keepers through the social instinct, strong as it is, is in their bees; but whereas with the bees the community is everything and in its fierce, united energy the community is inexorable and remorseless to the individual, the qualities which make for success in the management of bees are the quieter, more passive virtues, gentleness, patience, tact, which pertain rather to intelligence and moral character than instinct. Perhaps this explains the success of ladies who make bee-keeping one of their occupations.

NEP 16 Jun 1928
Beekeepers. For Sale, two up-to-date Bee Hives 24 drawn comb, extractors, &c. what offers. Apply Mrs. Gifford, The Red House, Ollerton.

NEP 13 Jul 1928
Glorious weather favoured the fourth day of the "Royal Show" at Wollaton to-day, when local interest was intensified by the incorporation in the programme of the Notts. Agricultural Societys' county classes. Mr. W. Herrod-Hempsall, hon. secretary of the BBKA gave lectures and practical demonstrations with live bees.

NEP 28 Oct 1928
Solid Mahogany Dining Table, 4ft 6in square, three leaves, total length 10 feet.
                    Hammersley, Beech House, Queen's-road, Beeston.

NEP 8 Nov 1928
Nottm. Chrysanthemum Show. Nottingham's premier show, promoted by the City and County Chrysanthemum Society, opened to-day in the Albert Hall. Mr. Walter Trinder, of Edwinstowe, secretary to the NBKA, had a stand containing liquid honey, sections of honey in the comb and bees' wax.

NEP 16 Apr 1929
Six beehives, geared extractor and ripener for sale.
                    Richardson, 52, Edward-road, West Bridgford.

NEP 17 Apr 1929
Two WBC beehives for sale. Nearly new. With fittings.   Park House, Gunthorpe.

William Herrod-Hempsall writes in his book, "Bee-Keeping Old and New"
"We had the pleasure this year (1929) of paying a visit to that veteran bee-keeper, Colonel Sir Lancelot Rolleston, KCB, at his seat, Watnall Hall, Notts. He is keen on preserving for posterity the handiwork of our forefathers and as a result, on his lawn stands an original Nutts Collateral Hive, over 100 years old. No one would imagine from his alert and erect figure the gallant Colonel is 83 years of age."

NEP 3 Mar 1930

Increase in Proportion of Healthy Stocks. Founded some 45 years ago, the NBKA continues to flourish. And now has a membership of nearly 200. This satisfactory position was revealed on , when the Annual Meeting of members was held (for the first time in the history of the association) at University College, Highfields.

Col. Sir Lancelot Rolleston (a vice President) presided and warmly congratulated the association the satisfactory progress it was making. He alluded to the value of beekeeping, and said it was a matter in which the County Council were ready to give their support.

A report of the committee in a reference to the general condition of beekeeping in the county, stated the proportion of healthy stocks showed an increase. *Acarine* disease in particular was decidedly waning.

The Duchess of Portland was unanimously re-elected President and Mr A Riley was elected Chairman.

Mr Riley spoke in appreciative terms of the valuable assistance which the association had received from the Ministry of Agriculture, the County Council and the City Council.

Mr W Trinder (Edwinstowe) was re-elected general secretary and treasurer and a tribute was made to his work. The district secretaries were re-elected with the addition of Messrs Segar (Retford), Coppin (Beeston), and Straw (Worksop).

At the conclusion of the business tea was served and the members were afterwards conducted over the College. Later, the members re-assembled in the Great Hall when the the William Herrod Perpetual Challenge Cup was presented to Mr. B. Watson (Wellow Green), who once again took premier honours in 1929 in connection with competitions for honey at various shows.

An interesting talk on the general management of an apiary was given by Mr W Herrod-Hempsall (technical adviser in beekeeping to the Ministry of Agriculture).

NEP 5 Aug 1930
Rain marred the 38th annual exhibition the Southwell and District Horticultural and Industrial Society, held at Southwell yesterday. The adverse conditions of the season were reflected in decrease in the number of entries, but, in spite of this, there was a satisfactory display. NBKA's classes were well patronised.

NEP 1 Nov 1930
Practical work in the senior schools formed the subject of the address delivered to members of the Nottingham and District Education Study Society by Mr. EFD Bloom, HM. Inspector of Schools, at the University College, Shakespeare-street, to-day. Mr. Bloom reviewed the possibilities of handicrafts, book-binding, agriculture, poultry keeping, bee-keeping and other practical subjects.

NEP 1 Apr 1931
After 52 years of married life, a Radcliffe-on-Trent couple, Mr. and Mrs. Richard Rose, who lived on the Nottingham-road at Lamcote, have passed away within few hours

of each other. They had been ill about a week. Mrs. Rose died on Monday night at the age of 78, whilst Mr. Rose passed away the following morning, aged 82. Mr. Rose, until his retirement about 18 months ago, carried on the business of a chimney sweep in Radcliffe and the surrounding villages. Previous to this he was ferryman at the old Radcliffe Wharf, in the days when coal was brought up the river and sold to the villagers by Mr. Rose. He was a trustee of the Radcliffe Primitive Methodist Chapel, and a very keen amateur gardener and beekeeper.

NEP 24 Apr 1931
Beehives and Honey Separator for sale. Apply Elridge, The Lodge, Beeston Fields.

NEP 27 Apr 1931
The death has occurred of Mr. Thomas Nettleship Harrison, well-known Nottingham ironmonger and a member of one of the oldest local families. The business which he controlled was established 25 years ago by his grandfather in Cheapside and was carried on at that place until the property was demolished in connection with the building of the new Exchange. Mr. Harrison retired from active participation about four years ago. For many years he was a prominent member of the NBKA. He lived in Birkin-avenue but prior to going there resided in the Carrington district for over 30 years.

He was born in 1862 in Nottingham and married there in 1897.

NEP 27 May 1931
The death has taken place at "The Orchard," Uttoxeter of Mr. Henry Purcell Day, for many years a solicitor in practice in Nottingham. He was interested in bee-keeping and canary breeding.

NEP 13 Jul 1931
Wolseley car for sale, £10, taxed. Incubator, beehives.   A Rose, Radcliffe on Trent.

NEP 22 Sep 1931
Federation of Womens' Institutes. Country Produce, Drill Hall, Derby Road. Sep. 23rd & 24th 1931, at 2.30pm. Working Beehive demonstration.

NEP 5 Oct 1931
"The principles of beekeeping" formed the subject of a lecture given by Mr. W. Hamilton, bee-keeping instructor at Leeds University, to the Autumn Conference of the NBKA at the Albert Hall Institute on Saturday.

BBJ 18 Feb 1932

Obituary Notice by Joseph Herrod-Hempsall
Mr. R. MACKENDER.

There is naturally always a feeling of sadness and a sense of loss when we have to record the passing of our friends, especially those who have, been of the brotherhood of bee-keeping.

This feeling is multiplied a hundred-fold in announcing the death on February 1st, of the guide, philosopher and friend of our early bee-keeping days, Mr. Robert Mackender, of Newark.

Whatever success we and our brother, William, have attained in the pursuit of bee-keeping is due to our feet having been firmly planted in the right path from the beginning by Mr. Mackender and to the kindly help and advice he was always willing to give in the following years.

It is over forty years since we first met Mr. Mackender when he came as gardener to the Manor House at Sutton-on-Trent. A warm friendship was struck between our two families from our first acquaintance, which has continued unbroken up to the present day.

Mr. Mackender, who was 82 years of age, was born at Mildenhall, in Suffolk, and was for many years head gardener at Staunton Hall, near Newark from whence he went to Sutton-on-Trent and on leaving there he went to Newark where for 40 years he has carried on a seeds and bee appliance business, his shop at 43, Northgate being well known to all bee-keepers and gardeners within a wide radius.

As one of the oldest members of the NBKA, he was probably known to the majority of beekeepers, not only in the county but beyond its borders. He has been on the NBKA Committee for many years and rendered valuable assistance at shows, lectures, etc. No bee-keeper who went to him for help and advice was sent empty away. A man of kindly and generous disposition, a Christian gentleman, it is safe to say he had not an enemy, but all those who knew him at all intimately held him in the highest esteem. "A man he was to all the country dear"

He was wonderfully hale and vigorous for his age and took a keen interest in all beekeeping matters, the last important function in which he took part being at a gathering of Notts and Derbyshire bee-keepers at the house of Mrs. Huse at Stanton-by-Dale, Derbyshire, where he gave an address.

Mr. Mackender was an elder of the Baptist Church, and at a memorial service the minister spoke in the highest terms of his worth and services. His life was a model and example of what a true Christian's should be, and as one of his sons said: "He died as he had lived - nobly."

Although Mr. Mackender kept his methods and practical bee-keeping up-to-date, he evidently had a kindly feeling for at least one old superstition, for we read in a local paper he desired that at his death the old Custom of "telling the bees" should be observed, and soon after he had passed away this ceremony was performed by his grandson, Mr.R. Mackender, "telling the bees" in the six hives in the garden at his home that their master had passed away.
"Bees, sing softly; bees, sing low; He is dead who loved you so."

We are sure all our readers will join us in expressing our deepest sympathy with Mrs. Mackender and the family in their bereavement.

NEP 19 Mar 1932
Five beehives, two containing stocks, with all spare parts, most have 16 shallow frames drawn out, leaving district.   Hooton, Station-road, Lowdham.

BEEKEEPING BETWEEN TWO QUEENS

NEP 28 Jun 1932
Several beehives, sections, Frames and Extractor cheap to clear.
       Gilmore, Netherdale, May-avenue, Wollaton Village.

More recollections of Joseph Chambers from the booklet produced in 1984 to commemorate 100 years of NBKA.

During the winter of 1932/3, Mr M Bartle of Mansfield Woodhouse (Science master at the Brunt's School, Mansfield) who held a Junior Craftsman Certificate of BBKA, decided the time was ripe for other members of NBKA to become more highly qualified. He therefore generously threw his house open on Saturday evenings for any members to join in further studies with a view to obtaining certificates with the result that several of us obtained "Craftsman" status during 1933.

A further course of more advanced classes was arranged to be held in the County Technical College in Mansfield and given by the Beekeeping Adviser to Staffordshire County Council, Mr Joe Price. Mr Bartle and I were successful in obtaining our "Expert" ceritcates during 1934. My sister, Annie, also obtained hers at a later date.

NEP 15 Sep 1934
Beehives, extractor, Ripener Section, racks, frames; giving up.
       West, 'Lorelei', Kenrick-road, Porchester-road

NEP 15 Jun 1935
Three good beehives for sale, cheap.   1 George Street, Long Eaton

NEP 18 Jul 1935
The 'Three Rs" have been joined by many more letters of the alphabet in Nottingham schools. Among these are 'B' and 'T' at the Wm. Crane school They signify bee-keeping and tomato growing, as the scholars showed by sending a jar of honey and a basket of tomatoes to last night's meeting of the City Education Committee

EDWARD VII

He became king on the death of George V on 20th January 1936. He abdicated the throne without being crowned in December of the same year.

NEP 2 May 1936
Nottingham and Notts. Paralysed Sailors' and Soldiers' Home, Ellerslie House. The President, Patients and Committee acknowledge with grateful thanks the following gifts received during April: NBKA per Mr. GR Lucas, 9 Jars Honey

NEP 5 May 1936
Bees for sale.   72 Ragdale Road, Bulwell

NEP 5 May 1936
Beehives for sale. Eight hives all in good condition, the lot £6.
                              Mynors, Black Lion Hotel, Radcliffe on Trent.

The Black Lion, originally situated on the opposite side to its present location and dates back to the 18th century. It moved over the road in 1928 on the site of Buxton's farm and village pinfold.

NEP 15 May 1936
Bees, 10 Frame Stocks with New Hives for Sale. Apply Creed, 122 Eltham-road, West Bridgford

NEP 10 Aug 1936
Beekeeper. You could probably obtain the address from the secretary of NBKA, Mr GR Lucas, Inglenook, Edmonton-road, New Clipstone

NEP 30 Oct 1936
Mansfield Chrysanthemum Show. Prize Honey. The NBKA staged a show of honey and beeswax, prizes being awarded to the following:
    Light honey – 1st, CH Hone. Kinoulton: 2nd, Miss A and JJ Chambers, Teversal.
    Medium or dark honey – 1st, Miss A. and JJ Chambers; 2nd, CH Hone.
    Granulated honey – 1st, CH Hone.
    Shallow comb – 1st, Miss A. and JJ Chambers.
    Cake of beeswax – 1st, CH Hone; 2nd, GR Lucas, Clipstone.

George VI

Albert Frederick Arthur George was affectionately called "Bertie". He was also Duke of Saxony. He ascended the throne upon the abdication of his brother, Edward VII on 11th December 1936. Edward's coronation had been planned for 12th May 1937. So the coronation of George VI took place at Westminster Abbey on that date.

NEP 5 Feb 1937
To Seek Information on Rural Water Supplies. It was at the request of NBKA that Mr. SF Markham put the question which drew an answer from the Minister of Health, in the House of Commons yesterday, promising legislation to safeguard pure honey. Mr. Markham will ask Sir Kingsley Wood next Wednesday whether he can state the total number of villages supplied with water and the total amount expended in so supplying them under the provisions of the Rural Water Supplies Act of 1934; whether the work under the Act is still in progress and, if so, when the programme is likely to be completed.

Sir Sydney Frank Markham (1897-1975) was elected MP for Nottingham South in 1935. Sir Howard Kingsley Wood (1881-1943) introduced PAYE.

NEP 6 June 1937
Bees, beehives, etc. Sale cheap. Leaving district. 45 St Helens Road, West Bridgford.

NEP 2 Jul 1937
Two good beehives, various appliances. 35/- offers. Maycock, Cliff Drive, Radcliffe.

NEP 17 Jul 1937
Morley Senior School, Wells-road. Parents who attended the Open Day yesterday had the opportunity of seeing splendid display of woodwork, bookbinding, alabaster carving, needlework, etc. Recently, beekeeping was started at the school.

Morley School was opened in 1922 as a senior school for boys and girls. It changed over to a Secondary Modern School in 1944. In 1966 the senior children transferred to the new Elliott Durham School on Ransom Rd and the old Junior school (Board School) closed down and moved up to the Morley site on Wells Rd and was then called The St Ann's Junior School.

NEP 15 Sep 1937
The report of Mr. Caborn, horticultural organiser for Notts., to be presented at to-day's meeting of the City Education Committee, shows that there were 127 students at the evening classes for horticulture and 16 for beekeeping (jointly with the county authority).

NEP 24 Jan 1939
For sale cheap beehive and stock of bees.	Gamekeeper, Shortwell, Trowell

NEP 27 Mar 1939
NBKA has invited the British Beekeepers' Convention to hold the event in Nottingham this year and it will take place at University College, Highfields, on October 28th. The acceptance of the invitation by the Convention was disclosed at the Annual Meeting of the NBKA held at the Albert Hall Institute on Saturday. This is only the second time that the Convention has been held away from London, the previous occasion being at Birmingham.

It was also announced that the first joint meeting of the Leicestershire, Lincs., Derbyshire and Notts. Associations would be held at the Agricultural College, Sutton Bonington, on July 8th. The Report stated that the year had been satisfactory for the association. The balance at the end of 1937 of £37 had been increased to £45. Three exhibitions and various demonstrations and lectures were held.

The following officers were elected: President, the Marchioness of Titchfield; chairman, Mr. M. Bartle; vice-chairman, Mr. A. Worth general secretary and treasurer, Mr. GR Lucas; auditor, Mr. JH Trease; delegate to BBKA meetings, Mr. WG Nutting. Mr. M. Bartle presided over the meeting and after tea a lecture was given by Mr. L. Illingworth, secretary of the BDI Ltd., and of the Apis Club, on "International Beekeeping." Mr. Illingworth spoke of his experience of the pursuit of beekeeping in Switzerland and invited members to accompany him to the International Conference at Zurich in August.

In 1943 the Midland Dairy and Agricultural College's principal HG Robinson of the Agricultural College became its first Professor of Agriculture. In 1946 it was agreed that the buildings and faculty at Sutton Bonington would be transferred from the County Council to the University College. On 1 April 1947, the Midland Agricultural College became the Faculty of Agriculture of University College, Nottingham. It became a part of the new University of Nottingham in 1948.

The Marchioness of Titchfield from 1887-1922 became the Duchess of Portland in 1943 and retained the title until 1977. The ancestral seat was Welbeck Abbey in Nottinghamshire. When she died she was buried, as was customary in the family, in St. Winifred's churchyard in Holbeck.

Mr M Bartle was a science teacher at Brunt's school, Mansfield. He obtained his BBKA "Craftsman" certificate in 1933. He is commemorated by a cup which is competed for in honey shows held throughout the year.

It is with great regret we announce the death, with devastating suddenness, of Richard Trease at the end of May. Our heartfelt sympathy goes out to Mrs Trease and her family. In his quiet way, he gave great encouragement to the production of BEEMASTER – the newsletter of NBKA. He made mead as part of his beekeeping, was a keen conservationist and spent some of his time after an early retirement giving talks in schools. He will be much missed as a force for good. BM July 1989

NEP 12 May 1939
Beehives, etc. for sale.	35 Hazelgrove, Mapperley

BEEKEEPING BETWEEN TWO QUEENS

NEP 9 Jun 1939
Beehives wanted, must be cheap, state price.   P98 Evening Post.

NEP 9 Jun 1939
Wanted to buy – strong swarms of bees.
                                   Shrimpton, 18 Margaret-avenue, Long Eaton

NEP 8 Mar 1940
Kent BKA members will look after the hives of beekeepers in the County who are called up.

NEP 29 Mar 1940
Bees in Demand. Sugar rationing is largely responsibly the unusual demand for bees in country areas. Realising that honey can help the sugar deficiency, and that beekeeping is a profitable way to supplement the larder, many people are now becoming amateur beekeepers. Beehive manufacturers are having difficulty in obtaining enough suitable wood to meet the demand.

NEP 16 May 1940
Stock and Bees for Sale, ready for super.   72 Ragdale-road, Bulwell

NEP 23 May 1940
Notts Food Production. As regards beekeeping in wartime, it has been decided to divide the county into five areas with an adviser in each. During the season each adviser will give four demonstrations of practical beekeeping.

NEP 11 Jul 1940
Bees, four hives with honey, must sell.   70 Warren-avenue, Sherwood

Bee-Keeping. We have very large stocks of Beehives and Accessories. Inquiries ground floor for show rooms.
                       Pearson Bros (Nottm.) Ltd. Long Row Tel 45761 (9 lines)

NEP 18 Jul 1940
A postcard written in Manchester on May 3rd 1877 saying that some beehives would shortly be despatched, was yesterday delivered to Tiverton, Devon.

NEP 28 Aug 1940
The death has taken place at Scar Banla, Levens, Lake District, of Mr George Hayes, a native of Beeston. He was for over forty years Superintendent of the Beeston Valley Mission and for a similar period an official of NBKA. He was 82 years of age and went to Levens in 1929. He was a keen gardener and often judged at the local flower show. Mrs JR Dunn, his daughter, with whom he lived, was President of the Womens' Institute.

In 1887 Humber & Co, the cycle manufacturers, allowed their premises to be used for a Sunday School in Beeston and George Hayes served as Superintendent. As the number of children grew a bigger premses was needed and this was found on Queens Road. The Valley Mission was opened in

October 1900.

After George Hayes death in 1940 a lectern was bought in his memory, which is still used in the parish church. In the 1960s the congregation began to dwindle and the Valley Mission closed in 1971 and the building was sold.

NEP 23 Mar 1941
Beekeeping. The AGM of the NBKA at the Albert Hall Institute on 29th March, There will be a discussion on "Common Errors in Beekeeping." Chairman: JN Derbyshire JP. Open to the Public at 5.30pm. Business Meeting at 2.45pm. All interested in beekeeping are cordially invited.

NEP 28 Mar 1941
Advertiser prepared to buy strong healthy Stock Bees, reasonable price.
Phone Long Eaton 419.

Beehives, joiner made, WBC, once used, brood, shallows, frames, excluders, seen any day. Beehive, Spring-lane, Mapperley Plains.

NEP 11 Jul 1941
Beehive, new, with brood frames. 30/-. Sharman, Epperstone-road, Lowdham.

NEP 22 Jul 1941
Beehives from 10/- each; Honey Extractor, £1; Frames, Feeders, all guaranteed healthy. Bailey, Ben Buie, Shelford-road, Radcliffe.

NEP 1 Aug 1941
Two Beehives, Brood Boxes, Supers, as new, £2/10/- each. A 68 Post.

1942 It was reported that Col. Sir Lancelot Rolleston left an estate of £57,000 gross. Born in 1847 he died in 1941 aged 91 (Basford registry Office records).

NEP 10 Mar 1943
Wanted, Beehive and Appliances. Ward, Robin Down-lane, Mansfield.

NEP 11 Mar 1943
Beehive Wanted, WBC, good condition.
Harrison, 24 Girton-road, Hucknall-road, Sherwood.

NEP 24 Mar 1943
The Committee of the Horticultural Section of the Notts. Services and Prisoners' of War Comforts Fund met at the Guildhall last night to arrange the schedule for the Wollaton Park Show, to be held on July 31st - August 2nd. There will be 134 classes for flowers, fruit and vegetables; classes for allotment societies and for women and schoolchildren. Nurserymen and the NBKA will be invited to stage displays. Mr. J. Icke, organising secretary, will have charge of the arrangements and the classes.

NEP 26 Mar 1943
Mr. A. Worth, well-known over a wide area in beekeeping circles and for several years secretary of NBKA and the East Midlands Federation, has been appointed

instructor in beekeeping to the Dorset County Council. He is succeeded in the Notts, organisation by Mr. Bartle, 23, Park-road, Mansfield Woodhouse, who has also taken over the course of lectures in advanced beekeeping which are being held each Saturday afternoon in the Shire Hall, Nottingham

NEP 4 June 1943
The Nottingham Parks Committee apiary at Woodthorpe Grange, which is one of the few municipal apiaries in the country, to be opened on Wednesday by Mr. W. Herrod-Hempsall, technical adviser to the Ministry of Agriculture. Beekeepers are invited to this and to the first of a series of demonstrations at the Grange to-morrow afternoon.

NEP 12 Jul 1943
Leather Cricket Bag, small size, 22/6. Wanted, Beehive.   Phone 43630

NEP 23 July 1943
Two Beehives with Bees, Supers, Smoker Frames, etc.
                                            Telephone evenings Hucknall 247.

NEP 31 Jul 1943
Beehive Wanted. Particulars to "Kasanga", Old Newark-road, Mansfield.

NEP 31 Jul 1943
Thousands of Nottingham people flocked to Wollaton Park today for the three days' show in aid of the Notts. Comforts Fund. To-day's programme included the official opening by the Lord Mayor (Ald. Ernest A Braddock). The Duke of Portland, presiding at the opening ceremony, said those present were supporting a cause of great importance, and by not travelling on the railway they were helping railwaymen to win the war. The NBKA also added to the educational interest. Mr. Icke, the show secretary, stated that there were about 400 entries.

Ald Braddock was Sheriff of Nottingham in 1939/40. Since 1928 the title "Lord Mayor" was given to the leader of Nottingham Council.

William Arthur Henry Cavendish-Bentinck, 7th Duke of Portland (1893-1977). He was elected MP for Newark in 1922, a seat he held until he succeeded his father in the dukedom in April 1943.

NEP 30 Aug 1943
The Horticultural Show of exhibits produced by the personnel of 'D' Division of the National Fire Service, Nottingham has been arranged in the Scouts' Hall, North Church-street, on Thursday, Friday and Saturday. The prize money and expenses are being covered by firemen from their own recreational club funds, leaving all income for the Red Cross Fund. Exhibits will include a beehive built in glass, collections of foliage, plants, loses, fruit and vegetables.

NEP 21 Oct 1943
Pure-Bred R.I.R., laying, 20/- each; Haystack, 1943, approx. 11 tons, offers.
                                        Beehive. Spring-lane, Mapperley Plains

The Rhode Island Red (RIR) type of hen was developed as dual-purpose breed, to provide both meat and eggs.

NEP 23 Oct 1943
Gent's 24in. Royal Enfield Cycle, £5/10/-   Beehive, Spring-lane, Mapperley Plains

In 1907, after serious losses from their newly floated Enfield Autocar business, Eadie Manufacturing and its pedal-cycle component business was absorbed by Birmingham Small Arms Company (BSA), whose chairman was to tell shareholders that the acquisition had "done wonders for the cycle department". The Raleigh company based in Nottingham bought BSA's cycle interests in 1957.

NEP 1 Nov 1943
Merionethshire County BKA alleges that some amateurs got extra sugar for bees but used it for private purposes.

NEP 11 Nov 1943
Lady's-upright, 3-speed, oilbath, gearcase, £6/10/-.
                                          Beehive. Spring-lane. Mapperlev Plains
NEP 14 Apr 1944
Sale, stocks Bees, also Honey Ripener.   Beehive. Spring-lane, Mapperley Plains

Six WBC hives for sale, brood boxes and supers.   M22, Evening Post

NEP 13 June 1944
Mechanics Cinema: "Bees in Paradise" Arthur Askey on the island run on beehive lines where men are secondary considerations. Four airmen crash land on an island populated by beautiful women, but find their idyll has its problems.

NEP 5 July 1944
NBKA in conjunction with the Derbyshire, Leicestershire and Rutland BKAs, the Annual Convention will held the Midland Agricultural College, Sutton Bonington, on July 8th.

    2.45pm.   Opening by Principal Robinson.
    3.15pm.   Demonstrations.
    4.15pm.   Interval for tea.
              Cups of tea at a nominal charge but bring your own food.
    5.15pm.   Brains Trust. Question Master: Mr. Reginald Gamble.
    6.15pm.   Lecture by Miss Ironside, Warwick.   M. Bartle. Hon. Sec.

NEP 19 Jul 1944
Sale, Honey Extractor (6-frame), new Beehive.   Spring-lane, Mapperley Plains

NEP 30 Nov 1944
Auto Cycle, perfect, 12 L. x R.I.R Pullets, large quantity Wire Netting, non sectional fowlhouses, three beehives.   Phone 55031

NEP 23 Mar 1945

BEEKEEPING BETWEEN TWO QUEENS

Notts. War Agricultural Executive Committee. All Beekeepers are invited to Lecture on "Diseases of Bees," to be given in the Nottingham Corporation Gas Show Room Theatre, on 24th March, 1945, at 5.45pm, Mr. A. Worth, instructor in Beekeeping, Dorsetshire. The lecture, which will be illustrated by lantern slides, has been arranged in conjunction with the NBKA.

NEP 10 May 1945
Beehive with bees complete for sale. Call afternoon.   275 Valley Road

NEP 29 Jun 1945
Observation hive as new, takes three standard brood frames, can be seen by appointment. Price £5   Dew. 9 The Green, Lowdham Grange

NEP 7 Jul 1945
NBKA hon. secretary. GR Lucas, Inglenook, Edmonton-road, New Clipstone

NEP 5th Feb 1946
It was revealed at a meeting of the County Council to-day, that the NBKA had offered a trophy to the Secondary Education Sub-committee to encourage interest in the subject in schools by means of competitions. The silver trophy was offered to the association by Mr. SC Lane, of Ruddington, in memory of his wife, who was a keen member for a number of years. The Council acknowledged acceptance of the offer.

NEP 9 Feb 1946
Wanted Beehive, WBC type.   27 Brook-road, Beeston

NEP 24 Aug 1946
The Very Rev. JJF Dockery, has been elected secretary of the East Midland Federation of Beekeepers' Associations, which comprises over 2,000 beekeepers. Fr. Dockery is the superior of the Church of St. Edward's, Nottingham.

NEP 9 Sep 1946
Beehive, complete, WBC, telescopic "Taylors" strainer, ripener, extractor and accessories, £11 or nearest.   33 Eakring-road. Bilsthorpe

NEP 14 Feb 1947
Chairman of the BBKA, Dr. AL Gregg, was the special lecturer on Saturday to members of the NBKA in the Gas Theatre. The lecture was under the auspices of the Nottingham Parks Committee in connection with their scheme of horticultural education.

This was a bit premature – the lecture did not take place until a week later!

NEP 21 Feb 1947
City Of Nottingham. Special Lecture Bee-Keeping. "Mind and Method" by Dr. AL Gregg, chairman of the BBKA will be given in the Gas Showroom Theatre, Parliament Street, on 22nd February. Chairman, Mr. WG Ayres. Admission Free

NEP 3 Mar 1947
Although 1946 was disappointing from a beekeeper's point of view, a record number of new members were enrolled during the year, it is stated in the 62nd Annual Report of the NBKA. The organisation maintains a healthy financial position and the committee is doing its utmost to encourage local activities in other parts of the county. In spite of the bad year, when it was so cold that queens ceased laying when they should have been at their best, a honey gift scheme was carried out whereby honey was sent to children in Mansfield and Nottingham hospitals.

28 Mar 1947
Minute Book - The Retford and District Beekeeping Circle met at the Packet Inn, Grove Street, Retford. Mr HE King, Chairman, and Mr CL Robinson, Secretary. The committee included Colonel James.

NEP 21 Apr 1947
Write to Mr. ER Poole, 7, Sandringham-crescent, Wollaton. He is hon. secretary and treasurer of the NBKA.

It is with great regret we record the death of Mr ER Poole, one of our Life members.

Mr Poole held many offices and was well known to beekeepers throughout the country. In spite of failing health forseveral years, he was always delighted and very willing to give freely of his knowledge and advice. Although Mr and Mrs Poole left Nottingham several years ago, many members and friends will remember with gratitude the work they undertook for the Association and the encouragement they gave to newer members of the craft. To Mrs Poole and her family we offer our sympathy and prayers at this very sad time.

May G Vigrass BM January 1972

NEP 16 May 1947
The first of a series of beekeeping demonstrations will be given in the municipal apiary, Woodthorpe Grange Park, Nottingham, to-morrow by Mr. JH Northage, a local expert on bees. He will be giving further demonstrations there on July 5th and August 16th.

NEP 13 Jun 1947
Beekeepers will find themselves very much home at Woodthorpe Grange to-morrow afternoon, the occasion being a Field Day organised by the Nottingham Parks Committee. Members of the NBKA have been invited, and the committee extend invitations to all bee-keeping enthusiasts. The Vice-chairman of the Parks Committee, Councillor William E Maltby will welcome local apiarists at 3pm. Demonstrations will be given by Miss A. Ironside, bee expert of the National Agricultural Advisory Service and former member of the Warwickshire County Council, and Mr. Reginald Gamble, the BBC's "Bee-Man." A tour of the Municipal Apiary and Model Kitchen Garden will be made and at 5.30pm, Mr. Reginald Gamble will conduct a "Brains Trust". Mr. PH Greaves (Newark), Mrs. Marshall (Mansfield), Mr. ER Poole and Parks' Superintendent Mr. WG Ayres of Nottingham, receiving the apiarists' problems. Written questions for submission to the Trust should be handed in to the steward at Woodthorpe Grange tomorrow afternoon before 5.30pm.

BEEKEEPING BETWEEN TWO QUEENS

William E Maltby was Sheriif of Nottingham in 1955/56

A typical BBC broadcast of 16 November 1940 had this introduction::
            "The handyman among the bees"
With bee-keeping playing such an important part in agriculture, Reginald Gamble is deeply concerned not only about the rise in prices of bee appliances but also about the shortage of timber, metal, and so forth required for their manufacture. It may well be that many a bee-keeper, old or new, can now neither afford nor procure the manufactured article. And that is where the handyman comes in, Necessity being the mother of invention, he uses his wits. He will utilise a length of timber here, a sheet of metal there, and Gamble in his talk will give him a hint or two about cutting them up and putting them together. Bee-keepers, like everybody else, are planning ahead, war or no war.

NEP 23 Jul 1947
Mr. ER Poole, secretary, NBKA 7, Sandringham-crescent, Wollaton, will advise you on forming a beekeeping club.

NEP 4 Aug 1947
Some MPs and many beekeepers, well possibly, Sir Stafford Cripps, must have been more than a little puzzled by a question to the latter, on the House of Commons Order Paper, in the name of Lt Col. Basil Nield. What the honourable and gallant member asked was "If he will sanction the import from the USA of Italian queen bees in order to improve the strain of bees in this country and to combat the disease known as Fowl Brood."

The implication of some esoteric association between bees and chickens arose from a mis-spelling, 'Fowl' should, of course, have been 'Foul.' Our swarms before the war were often replenished with blue-blooded American bees, recruitment formerly practised in other spheres of social activity. Nowadays even queen bees must have import licences.

NEP 8 Aug 1947
For NBKA write to Mr. ER Poole, Ivern, Bramcote-lane, Wollaton. Members have use of literature.

NEP 14 Aug 1947
The last bee-keeping demonstration of the season will be given by Mr. JH Northage at the municipal apiary, Woodthorpe Grange Park, Nottingham on Saturday.

JH Northage was born in Arnold in 1879. In the 1911 Census he was listed as a Brewers Worker.

NEP 27 Jan 1948
For NBKA communicate with Mr. ER Poole, Ivern, Bramcote-lane, Wollaton

NEP 16 Feb 1948
How fruit and seed growers were assisted by migratory bee-keeping shown in the Report of NBKA. Responding to a request from Sir William Starkey, the Newark Beekeeping Circle visited his Bramley Seedling orchard at Southwell, where there

are over 1,000 fruit trees. Twenty-five hives were placed in position and the bees were released, and were soon working and bringing in pollen. When the county association hold a Field Day there, the hum of the bees could be heard in every part of the orchard. Later, about 50 colonies were supplied to a farmer who was growing 46 acres of seed clover and 60 acres of mustard. Colonies were also sent to pollinate other crops. The report, which is to be presented at the Annual Meeting in Nottingham on March 6th, states that the association experienced a satisfactory year.

NEP 18 Feb 1948
City Of Nottingham. Special Lecture on "The History of Bee-Keeping in England." by Dr. H Malcolm Fraser, PhD, BA. (Vice-President of the BBKA) will be given at the Gas Demonstration Theatre on 19th February, at 7.30pm. Admission Free. Chairman, Councillor AW Norwebb.

Arthur William Norwebb was born in Nottingham in 1896 and died there in 1967. In 1940 as President of the Master Butchers' Federation (his father was also a past President), commenting on the shortage of meat, he said, "We shall have to crave the indulgence of the public and ask them to take mutton and lamb in preference to beef." He was Sheriff of Nottingham 1962/3.

Scottish BKA magazine, February 1948
The Very Rev JJF Dockery, MA, of Nottingham, lectured to an audience of 70 members of the Newcastle and District BKA on "Mendelism, the background to Queen Rearing" on 13th December 1947.

NEP 8 Mar 1948
The big-brotherliness of the NBKA was commended on by the Director of Education for the county, Mr. J Edward Mason, when, at the association's Annual Meeting, he presented the EA Lane Memorial Cup to Balderton County School were winners of the cup for the second year. The William Herrod Perpetual Challenge Cup was won Mr. Goward, Mansfield.

Nottinghamshire Schools Rural Science Panel   June 1948
Beekeeping in Schools Part 1
The teaching of beekeeping in schools has much to recommend it. Not only is there a satisfactory blend of theoretical and practical learning in the subject but it suggests a worthwhile occupation to our pupils and will set some of them on the road to acquiring a hobby which will profitably occupy many leisure hours for the rest of their lives. In school, the subject allows ample room for experiment, most of the equipment can be made in the woodwork room, and we have at hand ample material to give our School Science and Nature Study the 'fashionable' biological twist.

And yet it would probably be a fair statement of the present position to say that whereas most children are taught something about the natural history of the honey bee in Nature Study or Science, few of them get their knowledge through a practical approach to the subject. They are seldom given the chance to do any practical beekeeping.

The reason for this state of affairs is undoubtedly the fact that the honey bee, the only one of over 70,000 varieties of insects that has been 'domesticated' by man, possesses a sting. Not until we have accepted the challenge of this weapon does work with bees become a possibility in the school curriculum.

I hope my own case is not typical, for with me ten years elapsed between the time when I first became interested in bees and the time when I acquired my first stock. The bees in this soon proved to me that a bee sting is not a very dreadful thing after all and my constant regret is that I did not start beekeeping when my interest in the subject first awakened.

I have found that boys at school, having been warned that they must be prepared to accept an occasional sting and having been told how to deal with it, are as pleasantly surprised as I was when they get their first. In fact, with my boys it is an occasion for joy rather than sorrow for they have undergone this 'initiation' into the gentle art.

I do not think that beekeeping is a subject which the teacher can learn along with his pupils and I would advise any teacher to get at least a full year's experience with his own bees before beginning to teach the practical side to children.

How to begin
There are three usual ways of commencing:
1. by buying or otherwise obtaining a swarm of bees and having them in suitable hive.
2. by buying a small colony, called a nucleus, consisting of from four to six combs, containing food, brood in all stages, a young fertile queen and plenty of bees, in fact a complete stock in miniature.
3. by buying a full size stock.

The first is a very common way of beginning but unless you know the previous history of the bees in the swarm, his method is not generally advised by beekeeping authorities. The risk of importing disease into an apiary through a stray swarm is a real one, but, in spite of this, few beekeepers seem able to resist the temptation to take a stray swarm if they find one. You would probably buy a swarm by weight, a reasonable price being five shillings per pound of bees, and there are roughly five thousand bees in a pound. A small swarm would be of little use and if you adopt this method of starting, try to get a swarm over four pounds in weight. Try also to get it early for the old rhyme says
"A swarm of bees in May Is worth a load of hay;
A swarm of bees in June Is worth a silver spoon;
A swarm of bees in July Is not worth a fly."

I may have much to say on the subject of swarming and swarm prevention in a subsequent article for it is this natural impulse developed in the bees through the thousands of years of their existence on earth which more than any other thing in the bees' make up, militates against a satisfactory honey harvest.

However, in spite of my warnings above, and other disadvantages which I would attach to this method of beginning beekeeping, many of you will commence your beekeeping career by obtaining a swarm and you will be rewarded by one of the most wonderful sights in nature as you watch the thousands of bees taking possession of their new home, your first hive.

The second method is the best way of beginning. It will yield you no honey in the current season but will build up into a full-sized stock for next year, and your skill in handling the bees, and your knowledge, will grow along with the size of the colony. It is far easier to manipulate a small colony than a large one. The cost of the bees this way would be from £2 to £4.

The third method of starting should give you a crop of honey in the first season but would have the disadvantage referred to above. If you can enlist expert assistance, then you can begin this way and reap a quick reward. But I hope the expert insists that you do the work under his supervision, for in that way you will learn far quicker than by watching him do the necessary work. Outlay for bees under this method would be from £5 to £10.

Whichever way you choose, I would advise that you contact a local reputable beekeeper and place your order with him. If you are buying or making your first hive, I would advise that you get the National type of single-walled hive, and, while you wait impatiently for your bees to arrive, do these three things.

Join the NBKA (Hon. Sec. ER Poole, "Ivern", Bramcote Lane, Wollaton, Nottingham), which will put you in touch with the nearest centre of activity in beekeeping matters, where you will find much willing help. The subscription is five shillings and sixpence.

Get an elementary book on beekeeping, in which you will find listed the equipment you will need and the practical details of bee management.

Stick to your local beekeeper like a brother, for he will allow you to observe and help when he is working with his bees and you will have made a beginning or even undergone your 'initiation' before your own bees arrive. W.Wood, Balderton

Over a long and rewarding life Bill Wood had many interests. He was a perfectionist and would not settle for anything less. His calling was that of Teacher and in this he finished as the head of the Grove Secondary Modern School at Balderton. A feature of this school was the 'Bee House' which Bill insisted be included as part of rural education.

He was deeply involved with cricket in his earlier years and was a strict disciplinarian both of himself and of the boys. He would place an old penny on the pitch at the spot for a good length ball and each time a boy hit the penny whilst bowling, that penny became the property of the boy concerned. He was a very good all round performer himself and played for very many years for the Newark first team.

As a beekeeper he ran his bees in a commercial manner, for very many years being involved with migratory beekeeping both in the orchards and seed crop pollination. He was also active in the Beekeepers' Association and his school teams won the Lane Cup several times. He gave me full support throughout the time I was responsible for the running of the Beekeepers' Marquee at the Newark Show. Ken Torr, BM November 1989

BEEKEEPING BETWEEN TWO QUEENS

NEP 6 Apr 1948
A letter was read from the Beeston Bowling Association requesting permission to include the Council's beehive crest on their buttonhole badges and blazers. This request was acceded to.

NEP 8 May 1948
Sale beehive, good condition.   34 Leighton-street, Nottingham

NEP 4 Jul 1948
Sir Jack Drummond (1891-1952), a director with Boots Pure Drugs Co., will address the North East Midlands Federation of the BBKA who are holding their Annual Meeting on the School of Agriculture. Sutton Bonington.

Sir Jack was a distinguished biochemist, noted for his work on nutrition as applied to the British diet under rationing during the Second World War.

NEP 4 Jul 1948
A Beekeepers' Field Day will be held at the Nottingham Municipal Apiary tomorrow, when a demonstration and talk will be given by Mr. HP Young, recently non-beekeeping instructor to the Hampshire WAEC, In the evening there is to be a Beekeeping Brains Trust, at which 'Twenty Questions' will be put. Members of the NBKA and the Peterborough District BKA will be present. The public are invited.

NEP 27 July 1948
Last Thursday I spent the day cycling in the vale of Belvoir. Near the village of Owthorpe I found swarm of bees hanging from a low hedge, and knowing a beekeeper in the neighbourhood I informed him. Together we returned, and with great interest I saw him safely secure them. But, relating my experience with some friends later in the evening, I was told this swarm was of little value. An elderly man with bee-keeping experience in his younger days said: "A swarm of bees in May is worth a load of hay; A swarm of bees in June is worth a silver spoon; A swarm of bees in July is not worth fly." It seems rather strange to me. Perhaps one of our bee experts could explain.   Chas. E. Jennings, 5, Manvers-road, W.B.

Nottinghamshire Schools Rural Science Panel September 1948
Beekeeping In Schools Part 2

Previously we discussed ways of setting up a stock of bees. We now consider the handling of bees, which is the most important and the most difficult part of beekeeping to teach to children. In other subjects our aim is not primarily to turn out musicians, artists, carpenters and so on, but with beekeeping we hope that our pupils will become beekeepers either before or after leaving school and however good their Natural History, they cannot become beekeepers until they can handle bees.

The children have, as we had, a fear of being stung. Before we approach the apiary let us get our boys and girls in the right frame of mind, for they feel this, their first attempt to handle bees, as probably the biggest adventure they have faced in their school lives. My boys react best to the idea of the challenge of the sting. They realise that they have to face it before they can take the first steps in practical beekeeping: if they go through the performance correctly, doing as they have been told, they probably win, while if, on the other hand, they make a 'bloomer' the bees will certainly tell them about it!

Off we go, then, on a fine warm day in April to open up our first hive. We have a hive tool, a pair of manipulating cloths and a smoker, and we wear veils and have our sleeves rolled up.

First we have to frighten the bees. That is the basic idea behind all manipulation. When bees are frightened, when they think they have to leave their homes in a hurry because of some approaching calamity, they gorge themselves with honey and, having honey-full stomachs, feel, perhaps, in a more docile frame of mind and find it more difficult to bend their bodies into the proper position for stinging. Smoke is the best 'frightener' that we have. Bees have lived through the ages in hollow trees. Forest fires were frequent even under natural conditions. The wind would drive along the fire and smoke would herald the coming disaster to all living things. The bees could not move their homes but they could and did take with them as much as possible of their store of food with which to make fresh cells in another locality. Since prehistoric times man has been aware of this reaction of bees to smoke and has used smoke whenever he wished to look into their homes. So let us light up the smoker and begin.

We blow a few puffs into the entrance of the hive, then give the bees a full two minutes to gorge themselves. After this period we take off roof, lifts and any packing, leaving the quilt exposed. We 'peel' off the quilt, puffing a little smoke across the tops of the frames, and cover with one of the manipulating cloths as soon as possible. With practice this is easily done in one operation by holding one edge of the cloth up to the edge of the quilt where we begin the 'peeling' and drawing the cloth over the frame tops as the quilt is removed.

So far so good. The preliminaries are over and only half a dozen bees have flown out from the top of the brood box, and we noted with relief that they did not seem unduly angry as they flew to the front of the hive. (We are at the back all through the operation). Now we must take the plunge! All we have been taught about handling bees rushes through our minds. We are probably a little excited unfortunately, but that is to be expected on our first attempt to inspect a colony and we do our best to keep the excitement bottled up inside so that it does not upset the deliberate and smooth movements we have planned and practised.

We roll back the manipulating cloth, exposing the end frame. We can see bees on both sides of it. We loosen it with the hive tool, grip the lugs firmly between the first two fingers and thumb of each hand, and gently lift it out of the brood box. We examine the side of the comb facing us and then turn it as we have practised many times marvelling that the bees seem not a bit worried at being turned through a hundred and eighty degrees! All the time holding the comb over the brood box we examine the second side, then, after returning it into its original hanging position, gently place it in a spare box so that we have a vacant space in the brood box to give us the necessary room to examine the other combs.

Next boy, please!

So the lesson goes on. As one manipulating cloth is rolled back, the second is brought into play and unrolled so that only the frame to be examined is exposed.

All too soon we have been right through the brood nest. When bees settled on our hands and arms we kept very still indeed as we were taught — and were relieved when they flew away. We have seen honey and pollen in the combs, brood and even eggs have been pointed out, and, most exciting of all, we spotted the queen bee and actually saw her lay an egg! We should like to go on all afternoon but we have been told that bees do not tolerate lengthy disturbances, so that we see that the combs are properly arranged in their original positions and then we reluctantly replace the quilt and close the hive.

We summarise what we have seen. The bees are using eight combs either for food or brood and very soon we must provide for the growing colony or they will probably begin preparations for swarming.

One boy has been stung. He unfortunately trapped a bee between his finger and the frame lug. The sting was scraped out with a penknife and he told us that it isn't so dreadful after all. He is lucky! He is now a fully fledged member of the Bee Club!
W. Wood, BA, Balderton.

NEP 4 Oct 1948
Glass and oak observation beehive for sale, taking 10 brood frames, 18 supers, beautifully made. 93 Blake-road, West Bridgford.

NEP 7 Oct 1948
Beehive with strong colony bees, all complete, for sale, £7. Tel 66388

NEP 29 Apr 1949
NBKA has an interesting programme for the summer. Their next Field Day is one of the most popular, the orchards of Lt.-Col Sir William Starkey, of Norwood Hall, being thrown open to members on May 7th.

NEP 4 May 1949
Bees for sale in National hives Atkinson Tel. Hucknall 405
Beehives with bees for sale 275, Valley-road. Phone 66388

NEP 11 May 1949
Although the first year of county reorganisation has been described as "probably the worst honey year for many beekeepers," the NBKA was able to arrange a full programme of acvtivities and enjoyed a highly successful period generally. Disclosing this in his Annual Report at a regional meeting held in Nottingham last night, the honorary secretary (Mr JE Cheetham) referred to a number of Field Days which had proved decidedly helpful and instructive, especially for beginners

NEP 12 May 1949
Bees. Two 10-frame stocks (young queens) in WBC telescopic hives with super boxes and excluders - £8 each. 4-frame Nucleus in WBC plinth-type hive. £3
28, Farm-road,Chilwell

NEP 4 Jun 1949
Nottingham and District beekeepers to-day attended an interesting lecture and demonstration, given at the Municipal Apiary, Woodthorpe Grange, by Mr. W Herrod-Hempsall, formerly technical adviser on beekeeping to the Ministry of Agriculture

NEP 14 July 1949
A Practical Beekeeping Demonstration will be given at the Municipal Apiary, Woodthorpe Grange Park on July 16th. Demonstrator: Mr. JH Northage.

NEP 17 Aug 1949
Every minute busy, talk about clocks, criminals or Gilbert and Sullivan, to Nottingham solicitor, Mr. RA Young, and you are heading for an interesting conversation. At 76, Mr. Young has a motto: "Keep working and you'll never get rusty." As if all this were not enough, "I like to scrape on my violin now and again, and still have some honey left from the beehives I once kept."

NEP 19 Sep 1949
At the National Honey Show, organised by the BBKA in London, on Saturday, Mr. H. Vigrass of 19, Waldemar-grove, Beeston, won second prize in the open class for light honey.

NEP 26 Sep 1949
Winners at the Autumn Honey Show of the NBKA at Retford on Saturday were:
Light Honey – 1st. Mrs E. Brooksbank, Beeston; 2nd, R. Neville W. Kay, Sutton-cum-Granby; 3rd, H. Markham. Ranby; commended, HO Hilder, Mattersey, HE King, Torworth
Medium or dark honey – 1st. E. Ingham, Forest Town; 2nd. WB Buttrlck, Retford; 3rd. HE King
Granulated honey – 1st. H. Osborne. Tuxford. 2nd. Charlesworth and Pearson, Mansfield; 3rd. HO Hilder
Heather or Heather Blended honey – 1st. HO Hilder; 2nd. E. Ingham
Shallow Comb for extraction – 1st. S Taylor, Clowne; 2nd. HE King; 3rd. H Markham.
Cake Beeswax – 1st. H. Osborne: 2nd. Neville. W. Kay; 3rd. HE King.
Block of Candy (1-lb) – 2nd. A. Worthington, Retford
One Pint Mead – 1st. HO Hilder; 2nd. AL Westwood, Sherwood; 3rd. Neville W Kay
Section Honey (1-lb) – 1st. H Markham; 2nd. HO Hilder; 3rd. R. Neville W. Kay
Ladies' Class. Honey Cake – 1st. Mrs. C. Taylor, Nottingham; 2nd, Mrs E???, Retford; 3rd. Mrs. B. Neville

Older members will be sorry to learn of the death in his 90th year of Harry Osborne. Harry started beekeeping before the war and was an ardent exhibitor at honey shows winning numerous certificates and awards. He also judged at honey shows, presiding on many occasion at Southwell and Moorgreen.

BEEKEEPING BETWEEN TWO QUEENS

Harry lived in Tuxford with his sister travelling daily by train to his work in Sheffield. He and his sister held many Field Days for NBKA, giving splendid teas! He was a keen gardener, too, and entered hive and garden produce in local shows.

Having served on the NBKA Council for many years as a much respected Show Secretary. He was made an Honorary Life Member for his years of dedication. BM November 1988

NEP 12 Oct 1949
A Honey Show was staged last night by the Ruddington Region of the NBKA. There was a good attendance of members and an excellent display of honey. The various winners were:
    Granulated honey – 1st, Mr. Catling: 2nd, Mr. PD Taylor
    Liquid honey – 1st, RW Marston. 2nd, Mr. Hollingworth 3rd, Mr. Stacey
    Novices class - Mr. P. Stacey and Mr Bell (in order of merit)

Certificates and prizes were presented. Mr. JH Northage, one of Nottinghamshire's most experienced beekeepers, judged the honey.

It is with much regret that we record the death of Mr A Catling of Ravenshead. He had been a member of the Association for a number of years and on behalf of NBKA I would extend deepest sympathy to Mr Catling's family.   *LV Woodhouse, BM November 1971*

"About the year 1950 I was assisting the late Mr Sam Waddingham to pollinate 20 acres of wild white clover for seed grower, RRDK Bradley at Thorpe Lodge Farm near Newark. Mr Bradley had a field growing the new Aberystwyth strain of red clover SL 123 which he had top mown when strongly grown and about to flower. The resulting second crop of flower was much smaller clover knobs. Bees were moved on to it in late August (all double brood boxes) and yielded a bonus of about a super per stock. It was a wonderful September that year!

That same year I had noted, about $3/4$ of a mile from my home apiary where I had two stocks, a field of red clover which I knew had been cut for forage early in the season. I had been reading about the Russian trials of pollination by 'scout training' and decided to have a go. I gathered a bunch of the clover and cut off the knobs with some stalk and leaves and simmered them in syrup - 1pt x 1lb. Next morning early I put more fresh clover into two cardboard boxes and placed them before the hive. I also spread some around the alighting boards. All of it was well sprinkled with the clover syrup. It was fine and sunny and very soon the clover was alive with nectar gatherers. Soon I put the lids on the boxes and took and set them in the clover $3/4$ of a mile away. Two hours later that field was alive and the 2-way traffic ran unabated until the crop was failing. No, I didn't get much surplus from that lot, but the bees did alright. Try it sometime."   CG BM July 1973

I'll let Barrie Ellis describe some facets of his beekeeping career.

"I started teaching at the Robert Thoroton Secondary School, Flintham, when it opened in 1950 having been adapted from the Nissen huts of the RAF Officers' Mess of Syerston Aerodrome. The Headmaster, John Beard, who was a beekeeper, lived in the old Flintham vicarage and invited me in the Spring of 1951 to look at his bees. In fear and trepidation I went along and found the experience fascinating and, after repeating the experience several times, I decided I would like to have my own bees and made my first hive, a National, in the school workshop.

Before the term ended Mr Beard phoned me at home to say he had a swarm in a tree and would I like to collect it? I went straight out to Flintham, climbed a ladder, shook the swarm into a box and, unfortunately, a twig pressed the veil onto my face and a bee stung me on the nose - I was initiated! Later in the day I collected the bees in the box and took them home where they were duly installed in my new hive. On the way home I was conscious of a change in my features - my top lip hung over the bottom lip and my cheeks were swollen and my parents thought I had been in a car accident!

The following day in school assembly my appearance caused considerable amusement amongst the pupils and the headmaster said, "Boys and girls we don't have a new member of staff – that IS Mr Ellis over there!".

John Beard, head of my first school at Flintham in 1950, kept two hives in the vicarage garden there. Invited to assist him one day, I went in fear and trepidation - found it fascinating - made two National hives and started with a nucleus later the same year.

The following year, having read Manley's books, I bought six Modified Dadant hives, took swarms, obtained a couple of WBC hives and by 1953 I was the proud owner of 10 colonies. At first I kept the bees at the bottom of my parents' garden in West Bridgford, but later moved them to an out-apiary in the grounds of Bunny Hall, where I produced 'runny Bunny honey!'

During the fifties I became secretary of the Ruddington and West Bridgford Region of NBKA and a member of the Council. As Chairman of Council, I instigated **BEEMASTER** (NBKA newsletter) in 1958 and chaired the editorial sub-committee. I showed honey, wax and mead at local shows and demonstrated beekeeping at the Royal Show at Wollaton Hall.

In 1954 I formed the Junior (School Children's) Section of NBKA, introducing many young people to the joys of beekeeping. Regular meetings were held each year throughout the county and between 100 and 150 children attended the annual Beekeeping Field Day for Schools at Brackenhurst Farm Institute. Amongst those children were Andrew Barber (NBKA Secretary) and Graham Hardingham (Council member). I had many helpers Geoff Hopkinson, Len Taylor, Frank Kemsley, Bill Leslie, Ken Percival, Bill Wood, Miss MIG Jeans and many others."
   Janet Cousins, BM January 1991

NEP 13 Jan 1950
The early birds under the roof of a shed at Southwell a pair of blackbirds have built their 1950 nest - surely a record for any wild birds of any species in a premature anticipation of spring? This week, too, a local beekeeper reports that her bees were out in full strength as late as half-past four.

NEP 27 Jan 1950
Hon. Secretary of the NBKA is Mr. LC Taylor, 48, Teesdale-road, Nottingham.

It is with regret that we must report the death on 20th April of Mr Leonard Taylor. He was an enthusiastic beekeeper and our Association Secretary at one time. Most members will have seen the *wonderful display panels with the dressed-up bees that he and Mrs. Taylor constructed, and which are still occasionally shown. At one time they both travelled extensively with this display to shows all over the country. It was of course in our tent at theMoorgreen Show last year. Our sincere condolences go to Mrs Taylor.   BM May1980*

Beekeepers will learn with regret of the passing of Leonard C Taylor of Nottingham. He had been in hospital and in a wheelchair since suffering a stroke in 1977. A cheerful and energetic man, he quietly suffered his enforced illness which must have been a sore trial to him, especially as his speech was badly impaired.

He and his good lady will long be remembered for the magnificent 30-foot beekeeping educational stand which they put up at the Scottish National Honey Show from 1967 until his illness, 10 in all, and each one different. In 1976 the Taylors were awarded the John Anderson Memorial Award by the Scottish BKA for their contribution to Scottish beekeeping, something of which they were very proud.

Scottish BKA magazine, July 1980.

Mrs Audrey Taylor who died in August was the wife of Mr Leonard Taylor and both will be remembered for the splendid educational display which they created and took to beekeeping events all over the country. It was a unique exhibition, lovingly made and unsurpassed in its purpose to portray beekeeping to the public. BM September 1986

Mrs Taylor was an excellent needlewoman and made the charming bee dolls which were part of the exhibition. The following poem was written by her and I am sure you would like to read it.

I pray that risen from the dead, I may in glory stand
A crown perhaps upon my head but a needle in my hand.
I never learned to sing and play so let no harp be mine.
From youth until my dying day plain sewing has been my line.
Therefore accustomed to the end in plying useful stitches,
I'll be content if asked to mend the little angels' breeches.

BM November 1986

Minute Book – Minutes of the Council Meeting held at the Shire Hall on 28th January 1950. Present: Mr JP Marshall (Chairman), Rev Father Wilkins, Colonel James, Miss Foster, Messrs Templeman, Goward, Keall, Osborn, Mr Needham (Librarian), Mr Hayes, NEC and the secretary. Apologies from Mr Parkinson attending the BBKA Meeting in London and Mr Marston attending a Leicester Beekeepers' meeting in Leicester.

The minutes of the last meeting were read and confirmed.

The Financial Statement was handed round to the members and passed later in the afternoon.

Arising out of the previous minutes. A letter was read from the Notts. AEC explaining the delay in the Mansfield "Foul Brood" case was owing to a change in the Horticultural Advisory officer, and agreeing that diseased colonies should be quickly disposed of. Mr Needham reported on the Rules sub-committee.

The arrangements for the AGM were completed. The presentation of the Lane Cup to be made by Mr HJ Lane and the William Herrod Perpetual Challenge Cup by Lady Starkey. The secretary reported that if Mr Blakeman found the return journey to Bowden at all difficult, Mr Marston would be pleased to offer hospitality.

The secretary felt that, owing to the clashing of busy times at business, he would not seek re-election at the Annual Meeting, but recommended that the position of treasurer should be separated from that of secretary. The whole question was discussed and a sub-committee was formed to see if the duties of secretary could be lightened. The committee to be Messrs. Needham, Templeman and Taylor.

The question of setting up an Examination Board was raised by Mr Goward. This was supported by Mr Hayes and it was resolved that Mrs Marshall, Mr Vigrass and Mr Wood be asked to serve on the Board.

Sutton-on-Trent Show. The details of the honey section in conjunction with this show were approved. This being one to count for the Herrod Cup.

The name proposed by the Newark region for the position of Advisor was Mr GJ Grevatt, proposed by Mr Keall, seconded by Mr Goward.

Research Liaison Officer. A letter from Mr Parkinson was read stating why he thought he should relinquish this position in favour of Mrs Marshall.

Mr Needham reported that the library funds were exhausted and Mr Keall proposed and Mr Goward seconded, that a further grant of £5 be made. Also, Mr Needham asked for permission to reprint the Library Rules for the library books and asked for the alteration of Rule 3 dealing with the period of loan.

The Chairman welcomed the new secretary of Retford Region, Lt Col James and Mr Hayes, BSc. the senior lecturer in horticulture to the Notts. County Council.

"It is with a feeling of deep personal loss that I record the passing of George Grevatt, one of our stalwart beekeepers. For many years he was a part time Brood Disease Inspector for the Ministry. a past Chairman of the NBKA and also of the Newark Region. He was a Council member for many years, honey judge, friend and advisorto any in need of help. He was also one of the first doing migratory beekeeping to aid fruit and seed growers in Nottinghamshire with the pollination of their crops. BM March 1979

Robert Charles Patrick James was born in Clifton St. Andrew near Bristol in 1898. In 1901 he lived in Chipping Sodbury but by 1911 he was registered in Bristol. From 1926-8 he had moved to Northumberland. He pursued a career in the Royal Engineers (Army Number 15218). He was a Lieutenant in 1919, a Captain in 1928 (when he lived in the North), and a Major in 1937 just before WWII. He was discharged on 4th August 1947 and was granted the honorary rank of Lieutenant Colonel which is the title by which we knew him. He died in 1979 as shown on his gravestone below in Leverton churchyard. He was churchwarden at this church for many years and conveniently lived in the nearby Old Vicarage shown on the right.

BEEKEEPING BETWEEN TWO QUEENS

The following lines appear in the Centenary Booklet:
In addition to wholesale honey production (Col James) specialised in beeswax and mead making. He was a regular attender at conferences and local shows and well-known at the National Honey Show. His other interests included music and the arts and he also enjoyed the odd race meeting.

NEP 6 Mar 1950
Gardening Brains Trust comprising members of the Parks Department staff will be held in the Gas Demonstration Theatre. Lower Parliament-street at 7.50 pm. on March 9th. Question Master: Councillor AW Norwebb (replaces Beekeeping Lecture scheduled to given by Mr. R. Gamble).

Minute Book – The AGM was held at the Milton Restaurant, Milton Street on 11th March 1950. The guest lecturer was Mr AJ Blakeman, President of the BBKA. His Honour Judge Hildyard, our President was present and introduced the speaker.

In order that the largest number of members would be able to hear the whole lecture, the whole of the afternoon session was taken up with Mr Blakeman's very engrossing subject "The Importance of the Drone in the Hive" and how to improve stocks. The attendance was very good and nearly 140 members were present.

After the tea interval came the business meeting. His Honour Judge Hildyard apologised for having to leave and Mr JP Marshall took the Chair.

The Council's Report and Financial Statement having been printed and circulated to all members, the Chairman made a few observations on the Report and, as no questions were raised, the 65th Annual Report and Financial Statement were adopted.

The election of the following office bearers for the ensuing year proceeded.
         President   His Honour Judge Hildyard
         Honorary Vice-President   Sir William Starkey, Bart.

Beekeeping in Nottinghamshire and nationally, suffered a loss in the passing of May Vigrass on 25th November at the age of 72. May who had not been well for about six months made a great effort in October to attend the National Honey Show of which she was Draw Secretary - a position she had held for 13 years. Despite advice to the contrary, she insisted on carrying out this last service to beekeeping. May moved from Leek to Nottingham in 1942 when she married Harry and they started beekeeping together in 1946.

A keen and capable administrator, May developed her interest in beekeeping in the field of service to beekeepers. I remember the vital role she so competently played in helping to organise a very successful residential weekend course at the Nottingham University School of Agriculture in the 1950s. For over 20 years she was Treasurer of the NBKA and also Secretary for a short time besides being the delegate to the BBKA.

Although in later years May's beekeeping activities were mainly administrative, she enjoyed her contact with the craft and was always so friendly and helpful to those who practised it. She gave talks on various aspects of beekeeping and had many beekeeping contacts and friends in Sweden, Austria, France, Spain and Portugal. She was also well-known to many beekeepers in Ireland.

The service that May Vigrass gave to beekeepers and beekeeping was outstanding and she will be missed by her many friends. No one will miss her more than Harry with whom she enjoyed her work for the craft and we offer him our sincere condolences.

Barrie Ellis, BM January 1986

The position of honorary treasurer and secretary was now dealt with and the retiring secretary stated that, owing to time limitations, he wished to relinquish the position but recommended that the post of treasurer should be separated. The Chairman asked for nominations or volunteers and Mrs MG Vigrass offered to take office as honorary treasurer. This was gratefully accepted and the resolution for Mrs MG Vigrass to take office as honorary treasurer of this Association was carried unanimously.

No decision was reached with regard to the position of honorary secretary and the Council proceeded to elect the remaining Office Bearers.

Honorary Auditors	Mr Booth, Mr J Oakesford
Librarian	Mr Needham
BBKA delegate	Mr Parkinson
Research Officer	Mr PH Reeve
Press Secretary	Mr PH Keall

The Chairman again raised the problem of honorary secretary and thought that possibly something could be done to ease the work. The secretary reported that he had made the necessary enquiries to introduce an 'Addressograph' system. This would be efficient and relieve the work of addressing notices, etc. and the cost would be under £20. A proposal was made that this be carried out and the meeting approved unanimously.

Association Field Days. It was suggested that approach should be made to Sir William Starkey, Boots' Experimental Station at Thurgaton, and Prof. HG Robinson at Sutton Bonington Agricultural College for visits to these places, but Mr Waddington strongly opposed the first of these and said that the colonies of bees were mostly his and another members' and he would not allow them to be touched as he thought the previous years' Field Day was abused. The members strongly disagreed and it was resolved that the secretary write to the aforementioned.

A proposal from Mansfield Region that Mr G Hall be made a Life member of this Association was carried without dissent.

BEEKEEPING BETWEEN TWO QUEENS

The Herrod Cup was won this year by Mr Osborne of Tuxford and this was presented to him by Mr Blakeman who also presented the replica to Mr E Goward, last years winner.

The Lane Cup for schools and book prizes were presented to the team from Forest Town County School by Mr JP Marshall.

His Honour Judge Gerard Moresby Thoroton Hildyard inherited Flintham Hall in 1928. His ancestors had the family name Thoroton but one of them adopted the Hildyard part in pursuance of a marriage settlement. He was a Vice-President of the committee which organised the coronation festivities in Newark in 1936. Born in Eastbourne in 1874, he married in Chelsea in 1911 and died in Bingham in 1956.

He retired from the County Court in September 1943 having served there for 15 years. He took silk in 1920. In the same year he was fined 20s at Newark Borough Police court for failing to conform with a halt sign whilst driving!

NEP 20 Mar 1950
Beware of the sweet young thing in the late spring! This is not the advice of a mother to her son, but that of an experienced bee-keeper to anyone who making a start in that fascinating hobby. An article in the Year Book of the NBKA warns that the worst kind of sting is one from a "sweet-seventeen-days old" bee who is taking over guard duties at the entrance of the hive, and who may be a little over-zealous. The colony to avoid is the queenless one at a time when pollen is available in large quantities, says the writer. Least painful is a sting from a quiet foraging bee from a queen-right colony, in the early spring. Unfortunate people who are stung can take comfort in the fact that it takes 430 bees to do as much harm as one rattle-snake, and that experienced beekeepers become more or less immune to bee venom. I wish I could say the same of myself.

Minute Book – Minutes of the Council Meeting held at the Shire Hall on 1st April, 1950, at 2.45pm. Present: Miss Winfield, Miss Foster, Lt Col James, Messrs Templeman, Goward, Reeve, Richardson, Osborne, Keall, Marshall, Andrews. Apologies from Mr Needham (flu), Rev Father Wilkins (Retreat) and Mr Parkinson.

Mr JP Marshall was elected Chairman for the year proposed by Mr Goward and seconded by Mr Osborne and carried unanimously.

The minutes of the last meeting were read and confirmed.

Financial Statement. The secretary reported that he had handed over these duties to the new treasurer, Mrs Vigrass, but unfortunately she had an engagement in London.

Report of BBKA Delegate. Mr Parkinson wished to report to the Council that he is an elected member of the Bee Diseases Committee of the BBKA not as a representative of the county. The meetings take place in between the ordinary BBKA meetings.

Field Days. A letter was read from the Research Department of Boots saying that all bee stocks are now concentrated at Lenton Experimental Station and that the whole of the current season is booked. It was resolved that a date be booked for 1951.

The William Starkey Orchard. A letter was read from Sir William regarding the difficulties regarding a visit this year owing to Mr Waddingtons refusal to co-operate, and it was resolved to leave the event for this season but make some arrangement for another season.

Col James suggested that the Retford Region Field Day be at Miss Huntsman's apiary at West Retford Hall and his own apiary be given over as an Association fixture on 3rd June and the Nottingham Region had an invitation to Woodthorpe Municipal Apiary on this date. It was decided to accept this arrangement. Mr Richardson also suggested that the Newark Field Day to Kelham might also be taken as an Association Field Day.

No reply had been received from Sutton Bonington.

The question of shows to count for the Herrod Cup was discussed and the following were accepted:
 Blythe Show   Mansfield Summer Show   Sutton-on-Trent Show   Autumn Show
 and a possible early Chrysanthemum Show for Nottingham and West Bridgford

Details to be sent to Mr Goward in order to be printed in the Summer Programme.

The Autumn Show and Conference will this year be at Mansfield, on 14th October at the Technical College and this led to a little discussion owing to two shows being at Mansfield were perhaps a handicap to other Regions; but it was decided that this could not be helped and that the shows would attract a good number of entries.

Lane Cup. An examinations sub-committee was appointed – Mrs Marshall, Mr Wood and Mr Vigrass.

The secretary suggested that the honorarium paid to the secretary should be now divided equally between the treasurer and the secretary.

Spraying and Dusting of Fruit Trees. Col James stressed the need for additional warnings respecting poisonous sprays and thought that periodicals like the "Fruit Grower" and "Farmers' Weekly" should have seasonal warnings to safeguard the bees.

The secretary suggested that it might be best if the question of Life Membership be referred to the Council before taking the resolutions direct to the General Meeting.

NEP 14 Apr 1950
All those who are now spraying their fruit blossom should be wary in their choice of spray, remembering that for a bumper crop they need not only freedom from insect pests but also a plentiful supply of pollinating bees. It is heartbreaking to beekeepers to see round the hive entrances daily increasing numbers of dead and dying bees because someone in the neighbourhood has been spraying with a liquid which is

poisonous, not only to the pests which they wish to wipe out, but even more certainly to their benefactors, the bees.   LM Winfield, 48 Aspley Park-drive, Aspley

NEP 10 May 1950
The recent frost severely damaged the fruit blossom in Southwell, I hear. A Southwell bee-keeper states that there will be a loss of about 80% of the anticipated apple harvest in the orchard which is generously lent annually to beekeepers of the county. Although there will be no beekeeping in the orchard this coming season, hives will be placed beneath the eight sheltered acres less affected by the frost in order to assist in pollination. Bee-keepers all over the county will doubt feel sympathy with the market gardeners of Notts. The annual visit to Southwell of the keepers of hives is valued by them as one of the most enjoyable in the honey world.

NEP 17 May 1950
Municipal Apiary Practical demonstration on "Beekeeping for Beginners" will be given at Woodthorpe Grange Park at 5pm on May 20th. Demonstrator: Mr. JH Northage.

NEP 31 May 1950
Municipal Apiary. Beekeeping Field Day will be held at Woodthorpe Grange Park on June 3rd. at 3pm. A practical demonstration will be given by Miss N. Ironside. Admission Free.

NEP 12 Jun 1950
Beekeeping Demonstration for beginners will given by Mr. JH Northage at Woodthorpe Grange Park, 7pm, 15th July. Admission Free.

NEP 6 Jul 1950
The Bee's Hum. A bee has four wings attached to the thorax or centre part of its body, and when in flight or when fanning, as they do to transmit the odour of the colony or to cool the interior of the hive, they beat the air with their wings very rapidly. It is this which causes the humming sound. An experienced beekeeper can tell from the tone of the hum whether the bee is angry or happy; in other words, if it is going to sting or just wants to play.   Stanford H Clarke, 9 Walter-street, Nottingham

NEP 11 Jul 1950
Bees with four hives with honey must sell.   70 Warren-avenue, Sherwood.

NEP 11 Jul 1950
Practical Beekeeping Demonstration for Beginners will be given at the Municipal Apiary, Woodthorpe Grange Park, at 3pm on 15th July. Demonstrator: Mr. JH Northage. Admission Free

NEP 15 Jul 1950
St Anns Rose Show   Bee-keeping equipment was shown by the NBKA.

Minute Book – Minutes of the Council Meeting held at the Shire Hall on 12th August, 1950 at 2.45pm. Present: Mr JP Marshall in the Chair, Messrs GE Templeman, CH Hone, PH Keall, BW Whitehorne, RW Marston, FE Needham, PH Reeve, JE Foley, H Osborne, E Goward, Miss Winfield and the secretary.

The Chairman welcomed Mr Whitehorne, the new secretary of the Radcliffe and Bingham Region.

The minutes of the last meeting were read and confirmed. Arising out of the minutes the secretary reported that the Field Days at Boots Experimental Station and to Sir William Starkey's orchards at Southwell had to be deferred for this season but the date ? June 1951 had now been booked for the visit to Boots Experimental Station at Lenton.

The Chairman stated that it was no surprise that the secretary had asked to be relieved of his duties as it was now many months since the intimation had been given. He understood that Mr Dennis of Hoton had agreed to undertake the duties of secretary should no other be found or willing to take office. A member asked if Col James had been approached and Miss Winfield proposed and Mr Marston seconded a resolution that Col James be written to inviting him to undertake the post of honorary secretary of this Association. It was also resolved that the Chairman should write first to Col James, and to Mr Dennis if the first reply was negative.

Arrangements for the Autumn Show. Mr Pavard of Abergavenny should be asked to lecture on "Rearing of Queens". It was resolved that Mrs Poole be asked to judge the honey competition and ask if she requires a second judge. Mr Goward to be steward of ????? Shows.

It is with regret that we have to announce the death on 14th April, of Mr JE Foley who was for some years Chairman of the Ruddington and West Bridgford Region of this Association. He was very well-known among beekeepers south of the Trent and in recent years many happy Field Days had been spent in his apiary and orchard. Winter meetings in the Region have been held in his home from time to time, and very happy memories will remain with us for many years. We extend our sympathy to Mrs Foley and family.

Ken E Norman, BM May-June 1958

It was in March 1986 that Mrs. IP. Poole, wife of Mr. ER. Poole, a previous NBKA secretary, died in Ludlow where she lived with her son. Both she and her husband were very active in the Association and will be remembered by our more senior members. We owe much to those who have gone before and served so well. May Davis, BM July 86

Registration of Beekeepers. This question was freely discussed and many points put forward, viz.
        the question of location of stocks
        numbers of stocks
        temporary removal of colonies
but one of the most important gains was that of knowing where beekeepers lived from the point of view of "Foul Brood" Inspections.

It was resolved to make a panel of judges and the names of Mr Markham, Mr King, Mr Hone and Mr Goward be placed on this panel.

Rules for Herrod Cup Competition Shows. The matter of having uniform schedules for these shows was discussed, and Miss Winfield was asked to bring the matter before the next Council Meeting.

Editor of the Annual Report and Year Book. It was resolved that Mr Leith as Editor should be invited to attend Council Meetings in order that he may select various matters for inclusion in the Year Book. Mr Keall proposed and Mr Templeman seconded and the proposal was carried unanimously.

Miss Winfield proposed that a "Film Week" be held again as last year. This was agreed and the first week of November was selected.

The Librarian, Mr Needham, asked for a further grant for library funds and a grant of £5 was made unanimously. He also asked permission to purchase filmstrips from the Visual Association should these be suitable and Mr Templeman proposed and Mr Hone seconded the motion for these to be obtained. Mr Needham also suggested that a leaflet be prepared for beekeeping enquiries at shows.

It was recommended that the price of honey should be 3/6d per pound.

Panel of Judges. At the Council meeting held in August 1950, the following judges were appointed – Mr Markham, Mr King, Mr Hone and Mr Goward. It was decided to add Mr Vigrass and Mr W Marshall to the list.

A letter from the secretary of the BBKA Examination Board was considered and discussion postponed to the next meeting.

Bee Research Association. It was decided to consider the advisability of renewing the subscription at the next meeting.

It is with regret I have to report the death of Mr Leith. He was in his 'eighties' and had served the Association for many years. Just when he joined the Association I don't know but it must be around 25 years ago that I first came in contact with him. He attended most of the beekeeping meetings held in the county whether in members' homes, Field Days or county meetings. He often spoke at those meetings and always added something to the occasion. He was very knowledgeable about bees and beekeeping and was for several years a BBKA examiner. For some time he was Editor of **BEEMASTER** *struggling hard during difficult times to keep this news sheet going out regularly to all our members. This Association was served both faithfully and enthusiastically by him. I am sure we are all individually grateful for his service to the NBKA and as members of the Association expressour sympathy with his family who mourn his passing.*
KE Norman (Chairman of Council) BM April 1974

NEP 15 Sep 1950
West Bridgford Chrysanthemum Society Exhibition of Early Flowering Chrysanthemums. Vegetables. Flowers. Fruit and Preserves, Sept. 16th, 1950 at 2.30pm at Rectory-road Schools. Buses Nos. 11, 14a and 21. NBKA will stage Honey Show. Adm 6d.

NEP 16 Sep 1950
The sixth Annual Exhibition of the West Bridgford Chrysanthemum Society, which was to have been held in the Parochial Hall to-day, was transferred to the Rectory-

road schools because of alterations being carried out to the floor at the Hall. There were 230 entries in the show, a slight increase on last year. NBKA held a show in conjunction with the Chrysanthemum Society and had 60 entries. The standard of honey there was exceptionally good, the Post was told by the judge, Mr. C. Hone, of Cropwell Bishop.

NEP 26 Oct 1950
Both successful beekeepers, Mr. and Mrs. H. Vigrass of 19, Waldemar-grove, Beeston, won several awards at the National Honey Show of the BBKA in London. In sections open to the British Isles Mr. Vigrass, with three jars of light honey, won third prize. Mrs. Vigrass was awarded second prize for three jars of medium honey. Mr. Vigrass also won a section open to members of the National Honey Show with three jars of light honey.

Minute Book – Minutes of the Council Meeting held at the Shire Hall, Nottingham on 4th November, 1950. Present: Mr JP Marshall in the Chair, Messrs J Leith, E Goward, GE Templeman, FE Needham, BH Whitehorne, C Hone, H Osborne, RW Marston, Mrs Vigrass, Miss Winfield, Miss Watkin and the secretary.

The minutes of the last meeting were read and confirmed.

Herrod Cup Competition. It was proposed by Mr Marston, seconded by Mr Hone, that only one prize per class be awarded to any one competitor, and after considerable discussion it was decided that points for the cup would be awarded only under the rules of the Association, Rule 2 remaining unaltered.

Subscription to Bee Record Association. The treasurer was authorised to pay the annual subscription of £25. This is the minimum subscription for an Association.

AGM. This is to be held on March 10th, 1951. The Scouts' Hall was suggested as the place for the meeting and Mrs Vigrass said she would make enquiries as to whether suitable arrangements could be made and also enquire about catering. It was decided to write to Mr Armitt and ask him if he would give a lecture.

The next Council Meeting was proposed for January 27th, 1951.

Annual Report and Year Book. Mr Leith kindly consented to undertake the preparation of this work again.

BBKA Report. This was considered and no names were submitted for election to the Council of the BBKA.

Financial Statement. The treasurer read a statement which was approved by the meeting but Regional Secretaries were reminded that they must obtain sanction from the Council for any expenditure not authorised by the Rules of the Association, this sanction to be obtained before the expenditure is incurred.

NEP 22 Dec 1950
Nottingham Region of the NBKA held their Annual Meeting last night. In his Report,

the secretary, Mr PH Keall, said that there had been a slight fall in membership from 240 to 216 during the year. Miss L Winfield was elected secretary and Mr F Vigrass treasurer. The chairman was Mr FE Needham.

Honey (Notts. Year Book, 1950 written by M. Bartle)

To most people sugar is just the sweet substance used in the household for a variety of purposes from jam making to the sweetening of a cup of tea. The chemists call this sugar sucrose and whether the domestic article is brown, primrose, white, granulated, lump, castor, or icing, it owes its sweetness to sucrose as does treacle. Sucrose occurs in many plants and fruits but notably in sugar cane, sugar beet and sugar maple. Practically all the commercial sugar is obtained from the sugar cane and sugar beet. At one time all our sugar was obtained from the sugar cane only and was in consequence called cane sugar but it should be noticed that the sugar obtained from the sugar beet is absolutely identical with cane sugar.

Sucrose is but one of a large family of sugars known as the saccharides (or saccharoses). The differences between the members of this family are of great interest to scientists, but the layman may discern one or two differences as the various sugars are not generally equally sweet or equally soluble in water.

Nectar consists principally of a more or less concentrated solution of sucrose. In some nectars sucrose may be present to the extent of 75% by weight and in others as low as 5%. Analyses of nectar are not readily made and the figures should be received with some caution as the amount of water in the nectar from the same source may vary from hour to hour according to the humidity and temperature of the air. If several nectar-bearing plants are in bloom at the same time it is found that the bees visit the flowers of those plants containing nectar having the highest sugar concentration. This explains why bees may suddenly forsake white clover say, for lime.

Almost as soon as the nectar is collected by the bee the sucrose begins to undergo chemical change. At some stage, probably in the swallowing of the nectar, the bee adds a very tiny amount of a glandular secretion known as invertase. This substance acts on the sucrose causing it to take in the elements of water (hydrogen and oxygen) and at the same time break up into two simpler sugars called dextrose and laevulose. This process can be imitated in the laboratory by heating a solution of sucrose with almost of one of the common acids. It is important to note that dextrose and laevulose are produced in exactly equal weights. We can therefore say that:

Sucrose + Water = Dextrose (glucose or D-glucose) + Laevulose (fructose or fruit-sugar)

This shortened way of expressing the change is often misleading to the non-chemical student. It is not possible by merely adding water to sucrose to convert it into the two simpler sugars. As just mentioned a third substance is necessary and the bees provide it as invertase. It is interesting to observe that whenever we take sugar i.e. sucrose, a precisely similar process takes place before the sugar can be assimilated by our blood. It is perfectly true to state therefore that honey is a pre-digested food.

The mixture of dextrose and laevulose is known as invert sugar and the sucrose is said to have been inverted. This inversion goes on after the nectar has been put into cells by the bees and together with the evaporation of water from the nectar constitutes the process known to beekeepers as 'ripening'. It is generally complete in about 72 hours.

If this were the whole story honey would consist of a strong solution of dextrose and laevulose in equal quantities together with much smaller amounts of other substances which give it aroma, flavour, colour and viscosity. Analyses of honey show that the dextrose and laevulose are seldom if ever present in equal amounts. Compare the following analyses given in round figures only:

	Clover Honey %	Tupelo Honey % (American White Gum)
Dextrose	38	24
Laevulose	40	47
Water	17	18
Sucrose	2	5
Other substances	3	6

Various reasons have been suggested as to why dextrose and laevulose are not found in equal amounts in honey. One reasonable explanation would appear to be that in the original nectar either dextrose or laevulose was present in addition to sucrose. If dextrose was present besides sucrose the resulting honey would contain more dextrose than laevulose and vice versa. Analyses of nectar seem to confirm this idea, but the problem is rather more complicated than is indicated here and cannot be discussed in a brief article.

As dextrose is not as soluble in water as laevulose we should expect a honey rich in dextrose to granulate rapidly and one rich in laevulose to granulate slowly or perhaps not at all, for it is only the dextrose which crystallises out. And such indeed is the case. Charlock honey granulates rapidly because the dextrose in it exceeds the laevulose amount. In the case of tupelo honey the laevulose greatly exceeds the dextrose and this honey has been kept for years without showing signs of granulation. It is generally accepted that the ratio Dextrose/Laevulose is a measure of the rate of granulation. Any values of D/L exceeding .85 imply rapid granulation, and values .75 or less imply slow granulation.

Minute Book – Minutes of the Council Meeting held at the Shire Hall at 2.45pm on January 2nd, 1951. Present: Mr JP Marshall in the Chair, Father Wilkins, Mrs Vigrass, Miss Winfield, Messrs Needham, Leith, Osborne, Hone, Marston, Richardson, WA Booth, Goward, Reeve, Foley and the secretary.

The minutes of the last meeting were read and confirmed.

The treasurer reported that she had found the records of members who had paid subscriptions and the list of members, incomplete and inaccurate. She had written to every member listed in the Year Book and had compiled a new list of members.

BEEKEEPING BETWEEN TWO QUEENS

This new list was agreed to by the Council. The accounts for the year were not ready yet, but would be ready for publication in the Year Book.

Mr Leith said he had made good progress with the Year Book and that it would be printed as soon as he received the statement of accounts.

The secretary said that the Education Authorities were not satisfied with the working of the Advisers Scheme. Many Advisers had not sent in reports. Consequently the Education Authorities found difficulty in making payments to the Advisers owing to lack of evidence in their office of what work had been done. The secretary had, therefore, visited Mr Lyth and Mr Ayres and had modified the rules for Advisers and had prepared a pro-forma for the Quarterly Reports. This would simplify the work of the Advisers. The scheme for the county had been approved by the Director of Education and allowed for seven Advisers. The scheme for the city allowed for two Advisers and was awaiting approval by the City Education Authorities. Both schemes were approved by the Council.

Election of Chairman. This was postponed until the next meeting.

Next Meeting. The Chairman and secretary were authorised to fix a date in April.

Engraving of Cups. The Herrod Cup and the Lane Cup were given to Mr Needham who undertook to have them engraved and brought to the AGM.

Review of Regional programmes. It was decided to put this on the agenda of the AGM, Mr Matson to start the discussion.

Newark Region asked for the authority to spend 25/-. This was approved.

AGM. Mrs Vigrass undertook all arrangements with regard to the YWCA Hall. The secretary would send with the Year Book cards addressed to the treasurer stating the numbers of teas at 2/- required. The hire of the hall would be £4 4s, and the hall would be available from 2.30 – 6.15pm. The following programme was agreed.
      2.30pm   Meet
      2.45pm   Lecture by Mr Armitt - "Matters of Prime Importance in Beekeeping"
      4.00pm   Tea
      4.45pm   Business Meeting

Minute Book – The AGM was held at the YWCA Hall, Shakespeare-street on 10th March 1951. Both the President and Vice-President had written to say they much regretted that they would be unable to be present, and the Chair was taken by Mr Marshall.

A very useful lecture, full of practical information, and containing the most recent results of research, was given by Mr JH Armitt, FRES. 102 members were present.

After the tea interval, Mr Armitt presented the Herrod Cup to Mr W Marshall, the winner of the competition last year, and the replica to Mr Osborne, winner the year before. He then presented the Lane Cup to the winning team of this competition,

Balderton School 'B' team.

The business meeting then took place. The minutes of the last AGM were read and confirmed.

Mr Reeve asked on what authority the Council had voted for the 'Registration of Beekeepers' as this was an important matter which should have been decided at a general meeting. In the course of the discussion that followed, it was clear that this could not have been done as general meetings only took place once a year. The Council had, therefore, discussed the matter at a Council Meeting at which Regional Representatives were present, and the balance of opinion at the meeting being in favour of registration and they had instructed their delegate to the BBKA to vote for it. This discussion was no doubt influenced by the fact that it was understood that the late secretary had circularised Regions about this matter prior to the Council Meeting.

The Financial Statement was then considered. The Chairman pointed out that a proposal for an increased subscription was made two years ago but had been turned down and, in view of the loss incurred this year, asked for proposals to be made to deal with it. Mr Guyler pointed out that there had also been a loss of some £30 the previous year. Mr KE Norman proposed that economies should be made. Mr Oakesford, seconded by Mr Marston, proposed that the subscription for single members be raised to 7/6d and for members of a family to 10/-. This was passed by the meeting, with the proviso that the matter be brought up for further discussion at the Autumn Conference.

It was proposed by Mr Goward, seconded by Mr Norman, that Mr Northage be made an honorary Life Member. This was carried unanimously. It was proposed by Mr Leith, seconded by Mr Dennis, that both Mr J and Mr W Herrod-Hempsall should be made honorary Life Members. This was passed unanimously subject to a check being made to ensure that this had not already been done.

Appointment of office bearers.

		Proposed	**Seconded**
President	His Honour Judge GMT Hildyard	Mr Marston	Mr Pride
Vice-President	Lt Col Sir William Starkey	Mr Foley	Mr Sutherland
Honorary Treasurer	Mrs Vigrasss	Mr Foley	Mr Leith

All the above were carried unanimously

On Mrs Vigrass showing reluctance in continuing in the appointment, a vote of confidence was immediately proposed and carried unanimously.

		Proposed	**Seconded**
Honorary Secretary	Lt Col RCP James	Mr Marshall	Mr Sutherland
Librarian	Mr FE Needham	Mr Leith	Mr Stacey
BBKA Delegate	Lt Col RCP James	Mr Parkinson	Mr Brothwell
Press Delegate	Mr Leith	Mr Norman	Mr Sutherland
Liaison Officer	Mr Reeve	Mr Parkinson	Miss Winfield

	All the above were passed unanimously		
		Proposed	**Seconded**
Honorary Auditors	Mr Oakesfield and Mr Booth	Mr Norman	Mr Leith
	Mr Guyler	Mr Booth	Mr Westwood

On being put to the meeting Mr Norman's proposal was carried by 15 votes to 13

A vote of appreciation of the excellent work he had done in compiling the Year Book was given to Mr Leith.

A discussion was started on the organisation and activities of Regions but as the time had come to terminate the meeting, it was proposed by Mr Reeve, seconded by Mr Oakesford, that this matter should be brought up at the Autumn Conference and should be considered by Regions in the meantime. This was carried.

The meeting welcomed Mr Henson, from the Farm Institute at Brackenhurst. Mr Henson then outlined what he hoped to arrange for the future in conjunction with the NBKA.

It was proposed by Mr Norman, seconded by Mr Grevatt, that the accounts be accepted as they stand but that an accurate Balance Sheet should be produced in future. This was carried.

It was with sadness that I heard of the death of WH. Brothwell. 'Bill' was already a very active member of the NBKA when I joined just after the war. We became very close beekeeping friends, and after the death of Mr Northage, we were appointed jointly as Beekeeping Advisors to the Parks Department. For many years we carried out this work - the main focus being on practical demonstrations of handling bees. We were able to do this at the many meetings held at the Park Dept. Apiary at Woodthorpe Grange. Bill had a very great love for bees and was always more than willing to help anyone who was in trouble. Those of us who were privileged to know him are saddened by the knowledge of his passing and our sympathy goes out to his wife and friends.
Harry Vigrass BM May 1987

Uppermost in our minds must be our sense of loss at the death, at 81, of our President and Honorary Member, Ken Norman. These notes were written on the day of his funeral.
He was honest, generous, kind, and genuine with a firm handshake, a sense of fun and a deep clear voice. His was a musical family and a love of music was enhanced by a good singing voice. He was actively involved with a home for the elderly - all this and more - besides his business as a printer and family commitments.

A keen gardener as well as a beekeeper, his troublesome hip was a double blow but he maintained his interest in his hobbies to the end. Ken Norman belonged to the Association for 56 years. He gave it a lifetime of service. He became Chairman and, in 1983, President. On behalf of NBKA I can only express our gratitude for his life and work and offer heartfelt sympathy to his wife and family. He will be sorely missed and long remembered.   *David Guthrie, BM November 1989*

Minute Book – Minutes of the Council Meeting held at the Shire Hall, Nottingham on April 7th, 1951 at 2.45pm. Present: Mrs Vigrass, Miss Winfield, Lt Col JCP James,

Messrs JP Marshall, WA Richardson, RW Marston, H Osborne, E Goward, J Leith, FE Needham, PH Reeve, EH Hone and BW Whitehouse.

Mr Reeve was elected to the Chair, proposed by Mr Leith, seconded by Mr Goward.

The minutes of the last meeting were read and confirmed.

The purchase of a duplicator for 30/- was approved.

An invitation to a Field Day at Woodthorpe received from Mr Ayres was accepted. Miss Winfield agreed to recommend to Mr Ayres the name of Mr Theaker, Secretary, Lincolnshire BKA, or failing him Mr Worth, as demonstrator. The other Association Field Day would be at Boots Experimental Station at Lenton.

It was proposed by Mr Marston, seconded by Mr Marshall, that Mr Goward should be appointed Show Secretary to include the running of the Herrod Cup. Mr Goward agreed and this was approved. The following shows are to count for points for the Herrod Cup subject to confirmation by Regional Secretaries and Mr Goward respectively:

> Blyth, Sutton-on-Trent, Autumn Conference
> Nottingham and District Summer Flowering Chrysanthemum Society
> Mansfield and West Bridgford

The secretary withdrew the proposal he had submitted with the agenda for the meeting. It was proposed by Mr Marshall, seconded by Mr Goward, that one member is appointed to run the competition and that this should be Mr Reeve. This was approved and Mr Reeve consented to act.

It was decided that in future the secretary should notify the City Director of Education as well as the County Director concerning the Lane Cup competition.

Autumn Conference. This is to be held at Worksop and the date was fixed for October 13th. The show is to include ten classes, prizes as last year.

The payment of Mr Armitt's fee of £2 2s for his lecture and 15s 7d for expenses, was approved.

International Beekeeping Conference. It was decided not to appoint an official delegate.

The Financial Statement submitted by the treasurer was accepted. It was proposed by Mr Marshall, seconded by Mr Marston, that a sub-committee be appointed to consider the financial position and see whether we can carry on with our existing subscription and to make recommendations. This was approved and a sub-committee was appointed consisting of Mr Leith, Mr Marshall and the treasurer. Mr Leith to be in charge of the committee. The treasurer reported that she was having difficulty with Mr Taylor regarding the honoraria and it was decided that the secretary should write to Mr Taylor on receipt of the relevant correspondence from the treasurer.

A letter from the County Horticultural Officer regarding "Foul Brood" was considered and it was decided that Regional Secretaries should write to the secretary and give their views.

A letter from Bedfordshire BKA was considered and it was decided not to support their proposal.

A letter from the Home Produce Council was passed to Mr Whitehouse for him to deal with and letters from the secretary of the Kingston Show were considered.

It was decided that the outstanding amount due to Mr Needham should be paid.

Charlie Hone died on 5th April. Mr Hone lived at Cropwell Bishop and died at the age of 80. Few of us will be able to boast that we kept bees for 65 years. The knowledge of bees and their ways, which he acquired, will not be lost.   *BM May 1988.*

Minute Book – Minutes of the Council Meeting held at the Shire Hall on 28th July 1951. Present: Mrs Vigrass, Messrs PH Reeve, RCP James, Leith, Hone, Goward, Marston, Bealby, Osborn, Booth, JE Foley, Phillips, Needham, JP Marshall, Henson.

The minutes of the last meeting were read and confirmed.

A letter from the Honey Marketing Committee of the BBKA was considered, and a resolution, proposed by Mr Goward, seconded by Mr Phillips, was carried, that a reply should be sent pointing out that it was impossible to give the information required.

The Report of the sub-committee on finance was considered and considerable discussion took place. The recommendation of the sub-committee to raise the single subscription to 7/6d and the family subscription to 10/- was agreed to. As this resolution was passed at the AGM, it will be on the agenda of the Autumn Conference.

Autumn Conference. This was to take place at Worksop at the Technical College. The secretary is to write to the Principal and ask for the use of two rooms of which one must be a lecture room. Mr Goward, assisted by Mr Osborne, agreed to arrange the Honey Show. The secretary is to write to the Principal at Rothamsted and ask for a lecturer – Mr Butler if possible. The above resolution was proposed by Mr Marston and seconded by Mr Phillips.
>Judges for the Honey Show. Proposed by Mr Marston, seconded by Mr Booth, that Mr Hone and Mr Goward should be judges. This was carried.
>Agenda. This is to include the proposed rates of subscription and Regional organisation.
>Arrangements. The lecture is to be at 3pm, and the meeting after tea. Members to make their own arrangements for tea.

During the discussion on finance, it was proposed by Mr Booth, seconded by Mr Bealby, that the treasurer send to Regional Treasurers lists of members who had not paid their subscription, with the request that Regional Treasurers should endeavour to get them paid. This was carried.

It was proposed by Mr Marston, seconded by Mr Phillips, that a statement should be included with the notification of the Autumn Conference, mentioning our acute financial difficulties, and appealing for more members to become Life Members, thus providing funds to tide up over our immediate difficulties until the new rates of subscriptions, if approved, became effective. This was carried.

It was decided to discuss the 'Foul Brood' organisation at the next Council Meeting.

The secretary informed the meeting about the present arrangements regarding the magazine "BeeCraft" and it was agreed that they were unsatisfactory. It was decided to leave it to the secretary to make what arrangements he thought best, provided there was no loss to the Association and that the attempt to make a profit by obtaining a bulk supply and issuing individual members with a copy could be dropped.

Minute Book – The Autumn Conference and Honey Show was held at the County Technical College, Worksop on October 13th, 1951.

A first-class lecture was given by Dr Butler and about 80 members were present.

Tea was provided by Messrs Cheetham of Worksop.

A meeting was held after tea, at which 43 members were present.

The resolution proposed by Mr Oakesford at the AGM that the subscription be raised to 7/6d for single members and 10/- for a family, was carried by 40 votes to 3.

It was proposed by Mr Norman, seconded by Mrs Brooksbank, that a sub-committee be formed to go into the question of the finances of the Association, and to bring forward at the next AGM a resolution to increase the subscription for 1953, if necessary. This was carried.

Regional Organisation was then discussed and Mr Marshall, seconded by Mr Stacey, proposed that Regions remain as they are and that the Council do their utmost to stimulate their activity. This was carried.

Minute Book – Minutes of the Council Meeting held at the Shire Hall on November 10th, 1951, at 2.45pm. Present: Mr Reeves in the Chair, Father Wilkins, Col RCP James, Mrs Vigrass, Miss Watkin, Messrs Stacey, Foley, Osborne, Leith, Needham, Templeman, Marston, Booth, Langford, Henson.

The minutes of the last meeting were read and confirmed.

It was decided to support the Bath and West Show, proposed by Mr Marston, seconded by Mr Booth. Mr Leith was appointed secretary of the Bees and Honey Section. The appointment of stewards was left until the next meeting. Proposed by Mr Booth, seconded by Mr Langford.

The secretary was instructed to write to Mr Goward and thank him (and the judges) for his work in organising the Honey Show at the Autumn Conference.

It was decided that the balance of 10/- left over from the prize money at the Autumn Conference, due to there being insufficient entries in some classes to award a third prize, be credited to the funds of the Association.

The Financial Statement was approved, proposed by Mr Foley, seconded by Mr Leith.

It was decided that the secretary should write to Dr Crane and say that, owing to the state of our finances, we were unable to continue our subscription to the BBKA at present but that we hoped to be able to do so later on.

The resolution passed at the Autumn Conference raising the annual subscription to 7/6d for single members and 10/- for a family will take effect from January 1st, 1952. The proposal passed at the Autumn Conference to appoint a sub-committee to report whether the 1953 subscription should be raised was complied with and a sub-committee was appointed consisting of the treasurer and Mr Needham.

The lack of activity in certain Regions was left to be considered at the next meeting.

The date of the AGM was fixed for March 1st, 1952. Mrs Vigrass was asked and consented to make the arrangements as per last year but with a buffet tea, getting the YWCA Hall if possible and, if necessary, altering the date of the meeting. It was decided to have a Film Show and Mr Needham agreed to make all the arrangements and to select a programme. Failing the Film Show it was proposed that Mr Theaker or, failing him, Miss Ironside, be asked to lecture.

No nominations were made to the BBKA Council.

On the proposal of Mr Marston, seconded by Father Wilkins, it was decided not to publish a Year Book this year but to make out a "Roneo'ed" list of officers and members, statement of accounts with the corresponding figures of the previous year, notice of meeting and explanatory remarks by the Editor, Mr Leith. Mr Leith agreed to undertake this and Mr Stacey said he could get the duplication done free of charge.

It was agreed that samples of labels and their prices be submitted at the AGM for selection of a new label.

Minute Book – Minutes of the Council Meeting held at the Shire Hall on January 19th, 1952. Present: Mrs Vigrass, Messrs RCP James, FE Needham, H Booth, H Osborne, J Leith, P Stacey, CH Hone, WA Richardson, AH Henson.

In the absence of the Chairman, Mr FE Needham was elected to the Chair.

The minutes of the last meeting were read and confirmed.

A letter from Mr Lyth was read, in which he said that the Advisers Scheme had been a failure owing to the difficulty in getting the Advisers to send in their reports.

Instead of this, the Agricultural and Horticultural Committees had decided, in future, to make a grant of £50 a year to the Association, subject to the approval of the County Finance Committee. The secretary was instructed to write to Mr Lyth and thank him for this grant.

A letter concerning Apimondia was considered and it was decided to reply that the Association was in favour of the scheme but could not support it financially at present owing to lack of funds.

A letter from Mr Whitehouse was read in which he said that he had to give up beekeeping on medical grounds and was forced to resign his membership, and it was decided to write and express the Association's sympathy and to thank him for his past services.

It was decided that the Sutton-on-Trent Show on October 1952 and the Nottingham and District Summer Flowering Chrysanthemum Society Show should count for points towards the Herrod Cup.

The treasurer submitted a Financial Statement which was approved.

On the proposal of Col James, seconded by Mr Booth, it was decided to deduct 10% depreciation from the previous years value of equipment for the December 1951 Balance Sheet.

AGM. This is to be on March 15th in the YWCA Hall in Shakespeare-street, Nottingham.

Programme.
    2pm Meet,
    Chairman's remarks
    Business meeting.
        1. Minutes of the last AGM
        2. Council's Report and the Financial Statement
        3. Election of officers
        4. Presentation of Herrod Cup, Lane Cup and Examination Certificates
        5. Any other business.
    4pm Tea
    5pm Films. Mr Hanson agreed to bring a projector and to fit it up, and Mr Needham said that he had arranged to obtain the films and would bring a projectionist.

It was decided to ask the President to present the prizes and if he could not do it, Mr Henson kindly agreed to present them.

Mr Osborne and Mr Hone were unanimously elected stewards for the Bath and West Show and it was left to Mr Leith to arrange for additional stewards as required.

Mr Needham read a letter from Mr Ashley and a discussion took place, after which the meeting terminated.

On 31 January 1952, despite advice from those close to him, the King went to London Airport to see off Princess Elizabeth, who was going on a tour of Australia via Kenya. On the morning of 6th February, George VI was found dead in bed at Sandringham House in Norfolk. He had died from a coronary thrombosis in his sleep at the age of 56.

APPENDICES

Herrod Perpetual Challenge Cup winners
Trophy awarded for most points gained in designated Honey Shows

Year	Winner	Year	Winner
1909	W Lewin Betts, Mansfield Woodhouse	1931	CH Hone, Kinoulton
'10	Geo Marshall, Norwell	'32	CH Hone, Kinoulton
'11	Geo Marshall, Norwell	'33	No record of award
'12	Geo Marshall, Norwell	'34	JJ & A Chambers, Teversal
'13	Geo Marshall, Norwell	'35	Mr & Mrs CH Hone, Cropwell Bishop
'14	James North, Sutton-in-Ashfield	'36	
'15	James North, Sutton-in-Ashfield	'37	
'16	No inscriptions of any awards for these years were on the trophy nor could they be found in any records of NBKA shows being held during this period. This could be due to the devastation of the "Isle of Wight" disease and latterly WW1. NBKA members did show at local village events but this did not qualify for this Cup.	'38	No inscriptions were found on the trophy for this period probably due to the onset of WW2
'17		'39	
'18		'40	
'19		'41	
'20		'42	M Bartle, Mansfield Woodhouse
'21		'43	H Markham, Ranby
'22		'44	HE King, Torworth
'23		'45	Mr & Mrs CH Hone, Cropwell Bishop
'24		'46	PH Reeve, Newark
'25	G Ward, Ollerton	'47	E Goward, Mansfield
'26	Mr & Mrs W Tinder, Edwinstowe	'48	E Goward, Mansfield
'27	J & G Ward, Ollerton	'49	H Osborne, Tuxford
'28	B Watson & W Dodd	'50	W Marshall, Sutton-on-Trent
'29	B Watson, Wellow Green	'51	W Marshall, Sutton-on-Trent
'30	JJ & A Chambers, Teversal	'52	P Bint, Colston Bassett

Information supplied by Maurice Jordan

Lane Cup winners

Year	Winner
1946	Balderton School
1947	Balderton School
1948	Forest Town School
1949	Forest Town School
1950	Balderton School
1951	Newark Mount CofE School
1952	Newark Mount CofE School

Information supplied by Alison Knox

IN MEMORIAM

Remembering those beekeepers who are no longer with us but who were active during the age covered by this book. They contributed to the success of NBKA in their various ways.

Frederick Albert Richards *(1913-2005): Founder of the National Diploma in Beekeeping (NDB) Fred was the son of a Nottinghamshire farmer. He acquired his first colony of bees at the age of 14. The story has it that this first foray into beekeeping was not a great success as when he got the WBC home it was found to contain no bees! However, Fred's interest in beekeeping continued and he took the BBKA examinations, being examined at times by no less than Herrod Hempsall. On leaving school, he joined the Nottinghamshire police and served until 1946, when he moved to Devon as County Beekeeping Adviser.*
In the early 1950s he was, with FS Franklin, instrumental in developing the National Diploma in Beekeeping, seen at the time as an essential qualification for those aspiring to be a Beekeeping Advisor.

Right to the end of his life Fred maintained his interest in bees and beekeeping. He was greatly moved by being made an honorary member of the BBKA in 2004.

Paul Metcalfe 2005 via Yvonne Bullivant

Philip Bint (1929 – 2017)
On 18th May, I attended the funeral of **Philip Bint***, at Colston Bassett church. Phil was a stalwart of the NBKA for as long as I can remember. No matter where in the country myself and my friends were attending a beekeeping event, he would be there. He would invariably travel down in his battered car and we often overtook him as he was chugging home along the motorway. Phil was a dab hand at anything mechanical which was why his car managed to keep on going and going.*

Philip was born 88 years ago in Stoke, but whilst he was still very young his parents took up employment in service at Colston Bassett Hall, his father as a footman and his mother as a maid. His formative years were spent in the Nottinghamshire countryside learning about nature and farming. After his much-enjoyed National Service in the RAF, he worked at Langar airfield for Avro's, working on Lancaster bombers and later at Raleigh Cycles in Nottingham. Phil had great entrepreneurial skills and could turn his hand to most things, such as contract farming, fabrication of trailers and tractor parts, tractor restoration, turkey farming at Christmas and of course beekeeping, selling somewhat cantankerous colonies and honey of course.

He was always cheerful with his flat cap perched on the back of his head. Always a source of laughter and good humour and forever a thorn in the side of his local MP Kenneth Clarke, usually over some farming policy that Philip didn't agree with. Phil always threw himself whole heartedly into whatever he was doing: when contract ploughing for local farmers, Phil would often plough through the night and often woke up to find himself ploughing at right angles to where he was supposed to be going.

Unusually for a funeral service the church was filled with laughter as his eulogy was read, which neatly summed up this unique character. I and many others who knew him will miss his cheery, 'Now then me lad'.

John William Cooke Marshall (1895-1970) *Sutton-on-Trent BM March 1970*
It is with very deep sorrow we have to announce the death of Mr (Bill) Marshall, after a short illness, on January 10th 1970 at the age of 75. Bill had been a stalwart of our Association and beekeeping in general, ever since he could walk. He was one of the real old school of beekeepers and will be missed by many.   *Andrew Barber, Chairman, NBKA*

With the passing of Bill Marshall beekeeping in general, and the NBKA in particular, has suffered a great loss also a link is severed with the late William and Joseph Herrod-Hempsall brothers, as it was Bill's father who introduced the then W Herrod to the art of beekeeping. Incidentally, W. Herrod-Hempsall was Bill Marshall's godfather. As a small boy Billy used to accompany the Herrod brothers on their swarm driving and collecting excursions round about Sutton-on-Trent and would sometimes arrive home in the late evening fast asleep among the skeps of bees in the back of the horse and trap.

About 30 years ago Bill had an accident with a tractor which left him tied down for some time, but not idle. Coming from a basket making family it was quite natural his thoughts should turn to skep making, of which he supplied many of his friends who no doubt will have had years of use out of them.

At one time he used to run up to 40 stocks of bees and his only transport was a bicycle with a tradesman's carrier on the front.

Just after the Second World War a start was made along with the local produce association, to stage honey classes, and largely by Bill's efforts this section grew from a few jars on top of the piano to one of the largest in the county. He became a very successful showman in his own and surrounding counties, receiving great help from his wife in preparing and staging honey, wax and mead. One of his great specialities was casting figures and models of animals, etc. in wax for display and sale at some of the shows he attended.

Up to his retirement at the age of 70 and after, his whole life was bees, the NBKA and his local show. He was a very competent judge at local county shows. At the time of his death he had a score of hives and was busy planning the coming season's work.

Bill Marshall was very well known to me as he was to most of our active members. I always felt that he had forgotten more about beekeeping than many of our certificated brethren ever knew but due to the age in which he was raised he could not put down on paper the practical experience he had gained. This fact was brought home to me when he used to visit me at Crow Park signal box before British Rail closed it last autumn. In fact he caused me to return to print when he informed me that he thought I was one of three members in the Association who had this power in print. So for the sake of all our members I ought not to withhold this capacity of communication but to give of that which I was blessed as he would have done had he been able. Farewell my friend - for with your passing we end an era but our memories of you will never end.
   Ken Torr, BM August 1970

Sadly, we have to report the loss of a veteran, well-respected beekeeper from the Newark Region. **John Thomas (Jack) Rushby** passed away on 27th January aged 76 years. He was a native of Farndon and had lived there all his life. Jack's experience of beekeeping went back some considerable period before the 2nd World War, so he had been active in the craft some 50 years or more. Jack was helpful to his fellow beekeepers, not only with information and advice, but also in practical ways.

BEEKEEPING BETWEEN TWO QUEENS

It was approximately 28 years ago that Jack started me off as a small beekeeper by a practical demonstration of how to take a small cast from a hedgerow and 'run' it into a nucleus hive I had made at evening school woodwork classes. The cast was one from his own apiary but he chose to give it to me, to give me a start, rather than reunite it with his own bees to maximise his own potential honey crop.

Jack had a wealth of stories he could recount on incidents in the village where he had lived and worked over a very long period. For most of his life he was chauffeur gardener to the late Mr Wood, who was also a beekeeper. Both Jack and his employer had their respective hives sited around the gardens and orchard. They cooperated closely in their hobby until Mr Wood's decease. Jack continued as a beekeeper, reducing his number of hives as the years advanced. I believe he still retained one hive to maintain his status as a beekeeper until his final departure from our midst. Those of us who knew him and had dealings with him are the poorer for his passing.

JR Hatton, BM March 1982

It is with sadness that I have to report the deaths of two more of our well-known elderly members. Mr. **JS Selby-Smith** *of Shelford, a beekeeper of very long standing, died after illness in July. We send our sympathy to Mrs Selby-Smith. The other member is Mrs Audrey Taylor who died in August. Mrs Taylor was the wife of Mr Leonard Taylor and both will be remembered for the splendid educational display which they created and took to beekeeping events all over the country. It was a unique exhibition, lovingly made and unsurpassed in its purpose toportray beekeeping to the public*

It was a shock to hear of the sudden death from natural causes of **Bob Boone**, *who was so well known and respected among beekeepers. Last year, in a motor accident, he sustained severe injuries which caused him tohave as much metal inserted into his leg and hip as had been used to enable Barry Sheene to walk again. He was fighting back magnificently and had reached the stage of walking with only one stick when he was tragically struck.*

Bob, who was totally committed to beekeeping, had his first hive of bees when he was twelve years old. He became so absorbed that he eventually made beekeeping and the manufacturing of bee equipment his livelihood. There are many beekeepers who owe much of what they know about beekeeping to Bob and he will be greatly andsadly missed in the beekeeping world.

May Davis, BM May 1987

Some people will remember Mrs. **Marjorie Needham** *who died, aged 97 years, on 20th June 1994 at St. Catherine's Nursing Home, Nottingham. Both she and her husband worked tirelessly for our Association giving legion service and fellowship on committees and in the show tents. Our present is built on the past. May we emulate their excellent example.*

May Davis, BM September 1994

It is with sorrow I inform you of the death of Mr **Frank Harding** *of Toton. He had been a member for many years but unfortunately his work, travelling for a seed company, prevented him attending meetings. He had become very much involved since hisretirement and many local members would have known him. We offer his wife and daughters our deepest sympathy in their loss.*

Alan Lewis, BM May-June 1958

A Vice-President's tribute to a friend
It is with sadness that I have to report the passing of **Dennis Cater**. *Dennis died on 26th July*

1988 and at the funeral on 1st August the NBKA was represented by the Chairman and several members. I have known Dennis for several years since he came to live in Southwell and started keeping bees. We have moved bees for pollination work and worked together on those jobs where two pairs of hands and two heads are better than one.

He served on the Council for several years, and was a valuable helper at Newark and outhwell Shows. With the Southwell Show due so soon after his passing he will be very much missed.
Arthur Spolton, BM September 1988

We have lost one of our long standing and very much liked members of the Association. **Arthur Spolton** was born in Kirkby-in-Ashfield. He spent most of his working life at the Co-op. and ended up manager at Southwell until his retirement. During the war he served his National Service with the RAF. He sadly leaves a wife, son and daughter.

Arthur Spolton - an appreciation
It is nearly 40 years since Arthur Spolton came to give us advice and help in the early days of our venture into beekeeping at the now-closed National School in Southwell.

Whether it was an open day at his apiary, beekeeping meetings at his house, swarm collecting or setting up the beekeeping tent at the Southwell Show, he was there giving unstinting help, with his quiet and friendly manner. He had a wide knowledge and experience in all beekeeping matters and was painstaking and thorough at all times. J. Glendinnin, BM January 1992

Miss **Ivy Jacques** of Misterton who died on the 5th November last, aged 91. She was formerly the Bee Disease Officer for Notts. and Derby and member of the Bee Farmers' Association. She was a great authority on all aspects of beekeeping. In talking to the members who knew her they were all unanimous in their praise for her ability as a beekeeper. She was a very precise and independent person who did not suffer fools gladly.

According to our president, Ken Torr, she dragged Mr Don Peatfield of Olston up to his house after he had collapsed during an inspection of his colonies.

So, although only a slight person when I met her two years ago, she certainly had inner strength, with a mind as clear as crystal. She was one of beekeeping's characters and will be greatly missed.
Andrew Barber, BM January 1993

Bill Wood memory – personal communication from Andrew Barber, 6th June 2018.
"We were regarded as his slave workers extracting honey for him etc. We didn't think that way at the time, but when I became a teacher I met a teacher from my old school who said 'I remember you, you were one of Bill Woods' slave workers'.

A nationally recognised expert and authority on beekeeping for over 60 years, **John James Leverton**, of Skegby, died at his home on 13th December at the age of 83. .Mr. Leverton was born at Kilbourne, Derbyshire in 1872. Well known to beekeepers and gardeners alike for the terraced garden on a hillside was made to bear prolifically, a colourful and fruitful background to the long rows of white-painted beehives which bordered the lower paths.
Coming from a beekeeping family, "Jack" Leverton was keeping bees when he was 12 years old. This year he again wintered his bees although "officially" he had long ago handed them over to the

care of his eldest son, George, also a keen beekeeper. A second of his four sons, Ted, is also a beekeeper.

He was for many years a member of the NBKA and many young - and old - newcomers to beekeeping acquired their enthusiasm and craftsmanship from his patient - and sometimes not so patient tuition. Jack Leverton did not like to see bees roughly handled.

A colliery winding engineer, he began his career in his late 'teens. and did not retire, until he was 70, being asked by the management at Sutton Colliery, where the whole of his working life was spent, to continue owing to the manpower shortage.

During his early years he was Sunday School Superintendent at St. Michael's Church, Sutton, and was for many years a governor of Skegby Church School.

Notts Free Press, 16th December 1955

www.ingramcontent.com/pod-product-compliance
Lightning Source LLC
Chambersburg PA
CBHW080421230426
43662CB00015B/2169